# Lecture Notes in Economics and Mathematical Systems

503

Springer
*Berlin*
*Heidelberg*
*New York*
*Barcelona*
*Hong Kong*
*London*
*Milan*
*Paris*
*Singapore*
*Tokyo*

Alan Kirman
Jean Benoît Zimmermann (Eds.)

# Economics with Heterogeneous Interacting Agents

Springer

Editors

Prof. Dr. Alan Kirman
Prof. Dr. Jean-Benoît Zimmermann
GREQAM/EHESS
2, Rue de la Charité
13002 Marseille, France

Cataloging-in-Publication data applied for

Die Deutsche Bibliothek - CIP-Einheitsaufnahme

Economics with heterogeneous interacting agents / Alan Kirman ;
Jean-Benoit Zimmermann (ed.). - Berlin ; Heidelberg ; New York ;
Barcelona ; Hong Kong ; London ; Milan ; Paris ; Singapore ; Tokyo :
Springer, 2001
(Lecture notes in economics and mathematical systems ; 503)
ISBN 3-540-42209-9

ISSN 0075-8450
ISBN 3-540-42209-9 Springer-Verlag Berlin Heidelberg New York



Springer-Verlag Berlin Heidelberg New York
a member of BertelsmannSpringer Science+Business Media GmbH

http://www.springer.de



Typesetting: Camera ready by author
Printed on acid-free paper SPIN: 10842307 55/3142/du 5 4 3 2 1 0

*This book is dedicated to*
*Mauro Gallegati,*
*the creator of WEHIA*

# Introduction

**Alan Kirman* and Jean-Benoît Zimmermann****

*GREQAM - Université d'Aix-Marseille, Ecole des Hautes Etudes en Sciences Sociales, Marseille

**CNRS - GREQAM, Marseille

This book continues in a tradition which has developed in the annual Workshops on Economics with Heterogeneous Interacting Agents, (WEHIA). The purpose of that workshop was to analyse situations in which individual agents, who might be different from each other, interact and produce behaviour on the aggregate level which does not correspond to that of the average actor. This rupture with the well-established tradition of the "representative individual" is far from having established a central position in economics. That the relaxation of that assumption and the introduction of heterogeneous individuals might change the relationship between micro behaviour and macro phenomena is clearly spelled out by Forni and Lippi (1997). The modelling of the actual interaction between these individuals has been and remains a challenge. Early analytical attempts to introduce direct and local interaction between agents such as in the pioneering work of Follmer (1974) met with little response and the recent wave of interest associated with the names of Aoki (1996), Durlauf (1997) and Blume (1993) amongst others was twenty years in the making. Yet, the influence of other disciplines such as physics and biology in which complex interactive systems and their analysis play a central role is now making itself felt in economics and the papers in this volume reflect that fact.

Paradoxically, the opening up of economics to these ideas has also permitted a reconciliation with other disciplines such as psychology and sociology. The restricted notion of rationality based on no more than introspection and without real empirical content is giving way to other cognitive approaches which can be thought of as representing recent thinking on the notions of "bounded rationality" developed by Simon. Thus there has been a move towards a more limited version of rationality, one which has some of its foundations in the cognitive sciences[1].

As our vision of individual rationality has changed so has our way of looking at the interaction between individuals. Once we abandon the idea that economic agents

[1] For a full and interesting account of this sort of approach see Rubinstein (1998 ).

anticonformism means that the anticonformists become locked out and that their opinions cannot prevail. Homogeneous behaviour will not occur if there is sufficient anticonformism. However the anticonformists can become isolated when their anticonformism is too strong. Here we see clearly how interaction and its nature influences aggregate outcomes.

Martin Zimmermann and his co-authors look at how networks evolve. They provide their players who play a Prisoners' Dilemma game with their neighbours with simple rules. The players imitate their best neighbour but can also change their neighbourhood. This they do according to the strategy played and the level of utility attained. Without perturbations the system reaches a steady state. There are many cooperators in that state but the few predators earn higher profits than their cooperative partners. The network organises itself with leaders who are highly connected appearing. The latter guarantee the persistence of cooperation. However slightly perturbing them can induce cascades and the system swings from almost total cooperation to almost total defection and back again. The inhomogeneous nature of the aggregate state is particularly interesting as is its vulnerability to small shocks to central players.

## From micro to macro behaviour

The contributions of Paul David (1985) and Brian Arthur (1989) to the modelling of technological innovation formalised the notion that "history matters" in determining the state in which an economy or industry will finally wind up. The results obtained there were based on a simple urn model which resulted in a "lock-in" effect. However, although this was one of the early contributions to the idea that the direct interaction between agents may lead to macro situations which are not socially desirable, the assumptions that led to these conclusions have frequently been criticised. De Araujo pursues the line of his earlier research on this subject. By imposing more realistic assumptions on the stochastic process underlying the diffusion of innovation, he casts doubt on the significance of the equilibrium notions usually employed and suggests how one might model, in a more convincing way, the process of technological diffusion.

The famous Kyotaki-Wright model was introduced with a view to providing a justification for the existence of money. The existence of this macro phenomenon is explained by the interaction between traders of different types. However, their analysis is essentially static in nature. Aoki introduces a dynamic structure into the model which allows one to study the behaviour of the system out of equilibrium, and most importantly he allows the number of money traders to vary endogenously.

Thus the heterogeneity of the traders is an emergent feature rather than one imposed ex ante.

One of the advantages of considering the economy as a complex interactive system is that the usual exacting requirements on individual rationality are no longer needed to obtain coherent macro behaviour. Leombruni, Delli Gatti and Gallegati study the "zero intelligence" traders introduced by Sunder as a sort of benchmark. They show how in a contestable market in which the agents follow very simple rules quite rich dynamics may occur. Here the interplay between the simplicity of the rules and the limited heterogeneity of the agents is particularly interesting.

## Finance

One of the ways of characterising aggregates is to look at the distribution of certain microeconomic features. It may well be true that despite a lot of microeconomic movement these distributions remain rather stable over time. Indeed the notion of looking for equilibrium distributions is widely accepted. In their contribution Richmond and Solomon suggest that given certain assumptions the distribution of wealth can take on one of two rather different forms. Which of the two regimes occurs depends on the level of overall wealth. What is interesting is that while overall wealth can evolve in a very complex way, the distribution of wealth is time independent. The power law form of the distribution also characterises stock market returns.

Marsili and De Martino examine a situation in which heterogeneous agent with adaptive behaviour are aware of their market power. They find an abrupt change in collective behaviour as agents give up their price taking attitudes. They pursue this question in a model with trend chasing chartists and find that the resultant behaviour is closely associated with that of the Hopfield model in statistical physics. The behaviour of a system which has rather complicated interaction at the individual level remains well characterised at the aggregate level.

Bubbles have long attracted the attention of economists. The classical position has been to assert that they do not exist and that those phenomena that are described as bubbles can, in fact, be explained by underlying fundamentals. This is the position maintained by Garber (2000) for example. This has been followed by an argument which suggests that if bubbles do occur they can never burst. Recently there has been a tendency to study the interaction of people using different but simple rules to predict in financial markets, (Frankel and Froot (1987) is an early example of this trend). Goldbaum's contribution to this volume is in this spirit. He allows individuals to choose between a fundamentalist view and a technical trading rule. He shows how bubbles can grow and burst. The explanation is that profits are to be made from technical trading but as the technical trading rule gains popularity its efficiency diminishes and there is a collapse in its following leading to losses. This sort of aggregate behaviour would not appear in an economy in which there was no endogenous heterogeneity.

Marco Gallegati also gives a convincing explanation for business fluctuations which arise as a result of the financial heterogeneity of firms. These fluctuations occur as a result of the changes in the financial fragility of the system which result from a shift in the distribution of firm's financial characteristics. Here it is the shifting heterogeneity of the firms that is at the root of the aggregate fluctuations rather than some sort of exogenous shock as in more standard analysis.

## Coalitions

One important feature of economic activity is the tendency for agents to form coalitions. These coalitions themselves become actors and the resultant aggregate behaviour is the result of the interaction at various levels between individuals and coalitions. This interaction may be purposeful, strategic and sophisticated as in many game theoretic analyses or may be governed by simple rules which may be modified by learning.

A good example of this sort of approach is the analysis proposed by Haeringer who examines cases in which local jurisdictions have decision rules to decide upon the level of production of a local public good and in which individuals are free to choose where to settle. He shows that a stable coalition structure exists for a wide class of decision rules and he shows that the core of this coalition formation game is not empty.

Fiaschi and Pacini analyse a world in which there are increasing returns to scale in the technology which pushes people to coalesce for producing and dividing an output. They show there exist at least a Strong Nash Equilibrium, where the richest coalesce among themselves and the poorest do the same. There is a high probability that agents play SNE, but this decreases with the size of the population, due to increasing coordination difficulties. Indeed, the more unequal is the initial resources distribution, the less efficient is the SNE. More generally inequality in the initial distribution of endowments amplifies inequality the inequality in the individual payoff utilities.

In the concluding chapter of this section, Durieu, Solal and Thoron discuss a situation in which individuals choose where to locate themselves geographically. There are benefits to belonging to a large city and contributing to the public facilities it produces. However, there is a counterbalancing effect which is that individuals have an incentive to free-ride, for example by living outside a city, thereby avoiding taxes but profiting from the facilities of that city nevertheless. The equilibrium allocations of individuals to towns can be very different from each other. For example, as the relative weight of the public effect of the facilities produced by towns increases, individuals have a tendency either to group themselves into large towns or to live in one of a large number of small towns. However, unless the public effect is very strong indeed, coalition or town structures with a number of intermediate sized towns are not viable. Here we see, in a non-cooperative setting, quite different aggregate outcomes can arise from an initial situation in which all the individuals are symmetric. The extreme types of town formation show that the "average behaviour or arrangement" is, in general, not viable and that an average view of aggregate outcomes would be very misleading.

## Interactions and Knowledge

The knowledge necessary to make optimal decisions is often very demanding. In fact. individuals frequently make decisions when there is a lack of information and a great deal of uncertainty. In such case we have not only to abandon optimisation but many would argue that the maximisation of expected utility does not describe behaviour. What individuals often do is to profit from the experience of others to make their choices. Thus they simplify their problem by imitating others. the first paper of this section by Charbit and Fernandez studies this situation. They claim that only learning provides a suitable basis for decision making in situations of great complexity and uncertainty. They make a detailed analysis of mimetism which is, of course, one way of learning. They go on to characterise rational mimetism and show which social influences can bear on the way in which individuals learn.

Once one allows for interaction such as imitative behaviour one has to specify who can be imitated. This brings us back to the network problem. The specific aspect of interest in the article by Grossetti and Bès is the way in which information is exchanged in collaborative arrangements. They look at empirical evidence between CNRS laboratories and the private sector in the field of engineering. They develop

# Part I.

# Social Networks

# A Simple Model of Fads and Cascading Failures on Sparse Switching Networks

**Duncan J. Watts**

Santa Fe Institute, 1399 Hyde Park Road, Santa Fe, NM, 87501 and Operations Research Center, E40-149, MIT. 1 Amherst St. Cambridge, MA, 02132 - duncan@santafe.edu

**Abstract:** The origin of large but rare cascades that are triggered by small initial shocks is a problem that manifests itself in social and natural phenomena as diverse as cultural fads and business innovations (1-5), social movements and revolutions (6-8), and even cascading failures in large infrastructure networks (9-11). Here we present a possible explanation of such cascades in terms of a network of interacting agents whose decisions are determined by the actions of their neighbors. We identify conditions under which the network is susceptible to very rare, but very large cascades and explain why such cascades may be difficult to anticipate in practice.

How is it that small initial shocks can cascade to affect or disrupt large systems that have proven stable with respect to similar disturbances in the past? Why did a single, relatively inconspicuous, power line in Oregon trigger a massive cascading failure throughout the western US and Canada on 10 August 1996 (9), when similar failures in similar circumstances did not do so in the past? Why do some books, movies and albums emerge out of obscurity, and with small marketing budgets, to become popular hits (3), when many apparently indistinguishable efforts fail to rise above the noise? In this paper, we propose a possible explanation for this general phenomenon in terms of binary-state switching networks.

Our model is motivated by considering a population of individuals each of whom must decide between two alternative actions, but whose decisions depend on those of other individuals. Decision makers often pay attention to each other either because they have limited knowledge of the problem, such as when deciding which movie to see or restaurant to visit (2), or else because the nature of the problem itself provides incentives for coordinated action, as is the case with social dilemmas (6,7) or competing technologies like personal computers or VCR's (1). Although the detailed mechanisms involved in binary decision problems can vary

$$G_1(x) = \frac{\sum_k k\rho_k p_k x^{k-1}}{\sum_k kp_k} = \frac{G_0'(x)}{z}.$$

In addition, let $H_0(x)$ be the generating function for the vulnerable cluster size distribution; that is, $H_0(x)$ generates the probability that a randomly chosen vertex will belong to a vulnerable cluster of size $n$. Finally , let $H_1(x)$ generate the cluster size distribution for vertices that we arrive at by following a random edge. It follows that $H_1(x)$ satisfies the self-consistency equation $H_1(x) = 1 - G_1(1) + xG_1(H_1(x)$ , from which $H_0(x)$ can be computed according to $H_0(x) = 1 - P + xG_0(H_1(x))$.

The quantity of primary importance to the problem of cascades is the average vulnerable cluster size $\langle n \rangle = H_0'(1)$ as this is the quantity that diverges at percolation. Substituting the expressions for $H_0(x)$ and $H_1(x)$ above, we find that $\langle n \rangle = P + (G_0'(1))^2 / (z - G_0''(1))$, which diverges at the critical point defined by:

$$G_0''(1) = \sum_k k(k-1)\rho_k p_k = z \qquad (1).$$

When $G_0''(1) < z$, all vulnerable clusters in the network are small, hence we do not expect to see any cascades. But when Equation 1 is satisfied, many vulnerable clusters become connected to form a percolating vulnerable cluster, from which it follows according to our claim above that cascades should be possible. Although similar in form to the condition for standard percolation on a random graph (27), the vulnerability condition $\rho_k$ (analogous to the occupation probability in percolation) is a function of both the degree and threshold distributions; a feature that has important consequences for cascades.

We illustrate the cascade condition for the special case of a uniform random graph (19), in which any pair of vertices is connected with probability $p = z/N$. Further, we assume initially that all vertices have the same threshold $\phi_c$; that is, $f(\phi) = \delta(\phi - \phi_c$ . A characteristic of uniform random graphs is that $p_k = e^{-z}z^k/k!$, the Poisson distribution, in which case our cascade condition (Equation 1) reduces to $zQ(K_c - 1, z) = 1$, where $K_c = \lfloor 1/\phi_c \rfloor$ and $Q(a,x)$ is the incomplete gamma function. Figure 1 expresses the cascade condition graphically as a boundary in the $(\phi_c, z$ phase diagram and compares it to the region (outlined by solid circles) in which cascades are observed in the full dynamical model. One distinctive feature of both the analytical and dynamical models is that the resulting cascade window has both a lower and an upper boundary. In the language of critical phenomena, for each value of $\phi_c$, the system has two critical points, as a function of the average degree $z$, at which the characteristic time scale of the dynamics diverges (see Figure 2a). The lower critical point is similar to the standard percolation transition for random graphs that occurs when the average connectivity exceeds $z = 1$ (19,26). But the upper critical point is different: here the system is highly connected; so highly connected, in fact, that most vertices are stable, and the percolating vulnerable cluster is ultimately lost.

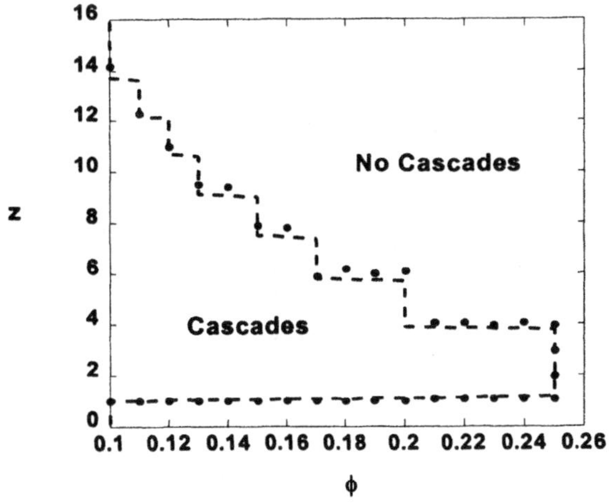


**Figure 1.** Predicted and actual cascade window for a uniform random graph with a homogenous threshold distribution $f(\phi) = \delta(\phi - \phi_c$. The dashed line encloses the region of the $(\phi_c, z$ plane in which the cascade condition (Equation 1 in text) is satisfied. The solid circles outline the region in which global cascades occur in the full dynamical model for $N = 10{,}000$ (averaged over $100$ random single node perturbations).

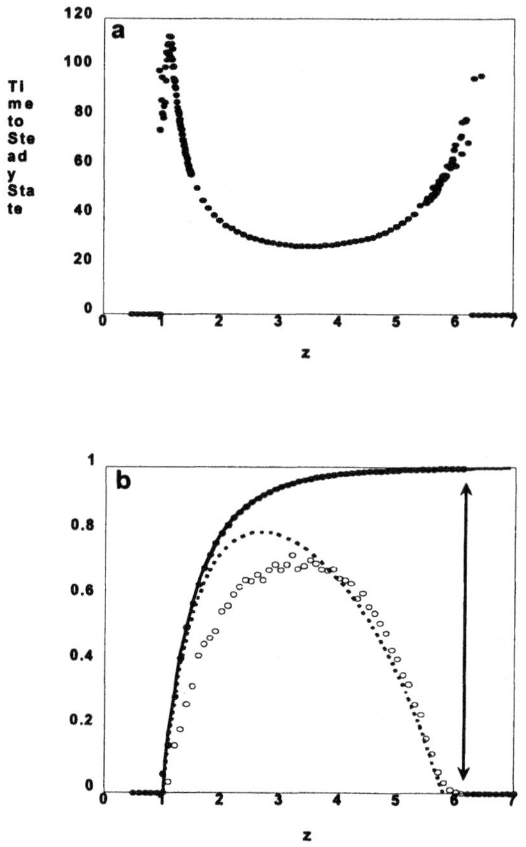


**Figure 2.** Two cross sections of the cascade window from Figure 1, for $\phi_c = 0.2$. (a) The time scale of the dynamics (that is, the average time required for a cascade to terminate) diverges at both the lower and upper boundaries of the cascade window, indicating that the dynamics has two critical points. (b) Rare but large cascades occur at the upper boundary. The dashed line represents the fractional size of the largest vulnerable component, and the open circles represent the frequency of cascades that result from a single site being switched on at $t = 0$ in the full dynamical model (averaged over 1000

random perturbations for N = 10,000). The solid (upper) line is the average fractional size of the entire connected component of the random graph, and the solid circles correspond to the average size of global cascades in the dynamical model, when they are triggered. The appearance and disappearance of the vulnerable component correspond to the lower and upper boundaries respectively of the cascade window. At the upper boundary, the vulnerable component is small prior to vanishing (i.e. the transition is continuous), corresponding to rare cascades, but the graph is highly connected; hence the expected size of any cascades that do occur remains large right up to the boundary and then drops to zero discontinuously (arrows).

To understand the nature of the upper transition, we solve exactly for the fractional size $S_v$ of the vulnerable cluster inside the cascade window. From the definition of the generating functions, $S_v = 1 - H_0(1) = P - G_0(s)$, where $s$ satisfies $s = 1 - G_1(1) + G_1(s)$. In the special case of a uniform random graph, with homogeneous thresholds, we obtain $S_v = Q(K_c + 1, z) - e^{z(s-1)} Q(K_c + 1, zs)$, in which $s$ satisfies $s = 1 - Q(K_c, z) + e^{z(s-1)} Q(K_c, zs)$. We contrast this expression with that for size of the entire connected component of the graph, $S = 1 - e^{-zS}$ (26), which is equivalent to allowing $K_c \to \infty$. In Figure 2b we show the exact solutions for both $S_v$ (dashed line) and $S$ (solid line) for the case of $\phi_c = 0.2$ $(K_c = 5)$, along with the relevant order parameters of the dynamics: the probability and expected size of cascades. It is clear that while the probability of cascades (open circles) is approximated by $S_v$, the average size of the cascades that are actually triggered (solid circles) is given instead by $S$. The reason is that once a cascade has commenced, vertices can have multiple neighbors in the on state, and so even those vertices that were deemed stable with respect to the initial shock can be toppled, allowing the cascade to occupy the entire connected component of the graph.

The phase transition at the upper boundary of the cascade window thus exhibits a dual nature, depending on which order parameter one observes. The probability of a cascade increases continuously, in a second-order phase transition (28), as the critical point is approached from above, but the expected cascade size jumps discontinuously from zero to one in the manner of a first-order transition, once cascades become possible. In this region cascades, like fads (3), will occur only rarely and thus unpredictably, but when they do occur they will be extremely large. The unpredictability of cascades has another aspect: Instead of many potential innovations, only a few of which actually succeed, consider a single product or innovation that is introduced repeatedly until it is successful. Now there are two time scales to the problem: the introduction period $T_i$, given by the inverse of increasing threshold heterogeneity increases the network's vulnerability to cascades. (b) Heterogeneity of degree. The solid line is again the cascade window from Figure 1. The dashed line is the equivalent window for a random graph with a degree distribution that is a power law with exponent $\tau$ and exponential cut-off $\kappa_0$, as described in the text. For this case, varying $\tau$ is equivalent to the varying the mean connectivity $z$, but the width of the corresponding distribution is in general much greater than for an equivalent Poisson distribution that is characteristic of a uniform random graph. Hence increased heterogeneity of degree shrinks the cascade window, making the network less vulnerable.

These qualitative results are quite general within the class of random networks, applying to arbitrary distributions both of thresholds $f(\phi)$ and degree $p_k$. Variations in either distribution can affect the vulnerability of the system considerably, as demonstrated by Figure 4. Figure 4a compares the cascade window from Figure 1 with corresponding windows for $f(\phi)$ that is normally distributed with mean $\phi_c$ and increasing standard deviation $\sigma$. Figure 4b again draws a comparison with the cascade window from Figure 1, but now the thresholds remain homogeneous and $p_k$ is varied according to $p_k = Ck^{-\tau}e^{-k/\kappa}$ $(k > 0)$, where $C, \tau, \kappa$ are constants. This class of graphs has attracted much recent interest (15-18,26,27) as a model of many real networks, ranging from social networks to the Internet and the World Wide Web. The power law distribution has the effect that the mean connectivity $z$ is dominated by the presence of a few highly connected nodes, while many nodes have only one edge. Because Equation 1 has the effect of excluding both these extremes from the vulnerable cluster, random graphs with power law degree distributions tend to be much less vulnerable to random shocks than uniform random graphs with the same $z$, a point observed elsewhere (17,18,27) with respect to the random deletion of nodes. Considered together, Figures 4a and 4b make the additional point that the introduction of heterogeneity to a system is a mixed blessing: On the one hand, an increasingly heterogeneous distribution of thresholds renders the system more vulnerable to cascades; but on the other hand, an increasingly broad distribution of degree can make it more stable.

A significant theoretical challenge is to extend the results of this paper to include networks that exhibit local structure, such as clustering (14), which in general violates the assumption that vertices initially can have at most one neighbor in the on state. One possible extension is to assume that individuals are assigned to small groups, within which interactions are dense, and that the groups in turn interact randomly (29). This approximation has the effect of replacing $z$ in the above analysis, with the density $z_g$ of group interactions, leaving the model qualitatively unchanged, but making cascades generally more likely. A more socially realistic approach represents individuals and the groups to which they belong as different kinds of vertices in what is called a bipartite graph (19). Random bipartite graphs have been used recently (26) to model a number of real social networks, from boards of directors of Fortune 1000 companies (4) to collaboration networks of movie actors (14) and scientists. In this case, the analog of Equation 1 must include two threshold distributions, one each for individuals and for groups, but otherwise the analysis remains unchanged.

Quantitative data on cascades in social networks is rare and is generally concerned only with a few successful innovations (4-6), rendering direct empirical validation of the model another challenging exercise. At a qualitative level, however, the results presented here are highly suggestive of phenomena observed in real-world examples of cultural fads and technological innovations: not only are the successes hard to separate a-priori, from the failures, but some cascades happen almost instantaneously while some, like cellular pagers or the 1989 Leipzig parades (6), exhibit latency periods that are decades long. We hope that the introduction of this simple framework will stimulate theoretical and empirical efforts to analyze more realistic network models, and obtain comprehensive data on the frequency, size and time scales of cascades in real networked systems.

## References


1. Arthur W. B. and Lane D. A. Information contagion. *Structural Change and Economic Dynamics* **4(1)**, 81-103 (1993).
2. S. Bikhchandani, D. Hirshleifer, I. Welch. A theory of fads, fashion, custom and cultural change as informational cascades. *J. Pol. Econ.* **100(5)**, 992-1026 (1992).
3. M. Gladwell. *The Tipping Point: How little things make can make a big difference.* (Little Brown, New York, 2000).
4. G. F. Davis, and H. R. Greve. Corporate elite networks and governance changes in the 1980s. *American J. Soc.* 103(1), 1-37 (1997).
5. T.W. Valente *Network Models of the Diffusion of Innovations* (Hampton Press, Cresskill, N.J., 1995).
6. S. Lohmann. The dynamics of informational cascades: the Monday demonstrations in Leipzig, East Germany, 1989-91. *World Politics* **47**, 42-101 (1994).
7. N. S. Glance, B. A. Huberman. The outbreak of cooperation. *J. Math. Soc.* **17(4)**, 281-302 (1993).

8. M. Granovetter. Threshold models of collective behavior. *American J. Soc.* **83(6)**, 1420-1443 (1978).

9. D. N. Kosterev, C. W. Taylor, W. A. Mittelstadt. Model validation for the August 10, 1996 WSCC System Outage. *IEEE Trans. on Power Systems.* **14(3)**, 967-979 (1999).

10. M. L. Sachtjen, B. A. Carreras, V. E. Lynch. Disturbances in a power transmission system. *Phys. Rev. E.* **61(5)**, 4877-4882 (2000).

11. J. M. Carlson, J. Doyle. Highly optimized tolerance: a mechanism for power laws in designed systems. *Phys. Rev. E.* **60(2)**, 1412-1427 (1999).

12. The analysis here assumes synchronous updating of agents, but the results are qualitatively unchanged if an asynchronous, random updating procedure is used instead.

13. Multimodal distributions $f(\phi)$ can exhibit stable equilibria that are intermediate between zero and one, but in this case, small shocks will always trigger cascades. In either case, the dynamics of the system is predictable.

14. D. J. Watts, S. H. Strogatz. Collective dynamics of 'small-world' networks. *Nature* **393,** 440-442 (1998).

15. A. L. Barabasi, R. Albert. Emergence of scaling in random networks. *Science* **286,** 509-512 (1999).

16. L. A. N. Amaral, A. Scala, M. Barthelemy, H. E. Stanley. Classes of behavior of small-world networks. http://xxx.lanl.gov/abs/cond-mat/0001458 (2000).

17. R. Albert, H. Jeong, A. L. Barabasi. Error and attack tolerance of complex networks. *Nature* 406, 378-382 (2000).

18. R. Cohen, K. Diaz, D. ben-Avraham, S. Havlin. Resilience of the internet to random breakdowns. http://xxx.lanl.gov/abs/cond-mat/0007048 (2000).

19. D. B. West. *Introduction to Graph Theory* (Prentice Hall, Upper Saddle River, NJ, 1996).

20. J. P. Sethna et al. Hysteresis and hierarchies: dynamics of disorder-driven first-order phase transformations. *Phys. Rev. Lett.* 70(21), 3347-3350 (1993).

21. J. Shrager, T. Hogg, B. A. Huberman. Observation of phase transitions in spreading activation networks. *Science* 236, 1092-1094 (1987).

22. P. Bak, C. Tang, K. Wiesenfeld. Self-organized criticality: an explanation of $1/f$ noise. *Phys. Rev. Lett.* 59, 381-384 (1987).

23. J. Adler. Bootstrap percolation. *Physica A* 171, 453-470 (1991).

24. S. Solomon, G. Weisbuch, L. de Arcangelis, N. Jan, D. Stauffer. Social percolation models. *Physica A* 277, 239-247 (2000).

25. C.Moore. Majority-vote cellular automata, Ising dynamics, and p-completeness. *J. Stat. Phys.* **88(3/4),** 795-805 (1997).

26. M. E. J. Newman, S. H. Strogatz, D. J. Watts. On random graphs with arbitrary degree distributions and their applications. http://xxx.lanl.gov/abs/cond-mat/0007235 (2000).

27. D.S. Callaway, M. E. J. Newman, S. H. Strogatz, D. J. Watts. Percolation on random graphs with arbitrary degree distribution. *Phys. Rev. Lett.* (submitted) Available at http://xxx.lanl.gov/abs/cond-mat/0007300 (2000).

28. H. E. Stanley. *Introduction to Phase Transitions and Critical Phenomena* (Oxford University Press, Oxford, 1971).

29. L. Sattenspiel, C. P. Simon, The spread and persistence of infectious diseases in structured populations. *Math. Biosci.* 90, 341-366 (1988).

30. The author is grateful for the support of A. Lo, and acknowledges D. Callaway, M. Newman, and S. Strogatz for illuminating conversations.

the agent's neighbourhood doesn't proceed from a metric, but is grounded within the graph of social relations.

We consider here that a social network is the product of individual relationships, agents are keeping up among themselves, with regard to a given purpose and not the fact of any infrastructure (institution, communication ...), that could only supply a support for those relations. As far as innovation or technological standard diffusion is concerned the network has to be formed from the relevant relationships given the considered problem. For example when it deals with a communication product, links will correspond to the inter-individual communication potential, likely to be satisfied by that considered product. In other contexts, more subjective factors could be taken in account : family links, friendship, same group belonging ...

In former papers (Steyer and Zimmermann, 1996 and 1998-b) we have studied the role of **relational proximity structures** in the dynamics and issue of innovation diffusion. We emphasised the existence of a **transition frontier**, of non zero thickness when the network is not uniform, that determines failure, incertitude and success zones for innovation diffusion (Steyer and Zimmermann, 1998-a). We have shown that the network anisotropy, if it generally tends to slow down the diffusion pace, brings in return a better long term efficiency, that can be identified for example by the lower level of initial adoption required for the successful introduction of an innovation on the market.

Our approach aim is here to explore the role of learning processes into the propagation dynamics within a network structure. In a recent model, formally represented by a neural network, we have introduced a **relational learning** that constitutes a way to set up an endogenous network evolution (Plouraboué, Steyer and Zimmermann, 1998). We proved the existence of a **self organised criticality** phenomenon, where some agents acquire key-positions within the network that bring them a strong structural capacity of influence over the whole population of potential adopters.

In this paper, we study the way how network auto-organisation can lead, under given conditions, to a critical state characterised by macroscopic effects generated from microscopic impulses at the level of the individual agent. It is the peculiar structure of those critical networks that allow macroscopic "avalanches" (Steyer, 1993) to take place, on which the diffusion process is likely to lean. We analyse the way learning process leads endogenously to such a critical state and how it strikes against the finite size of the network.

In a first section of the paper we introduce the notion of self-organised criticality (SOC) and we explain the reasons why it applies very hopefully to our question of innovation diffusion. In the second section we present the foundations of our model of innovation diffusion supported by a social network Then the third section is devoted to the question of social learning and the principles of social network endogenous evolution. In the fourth section we present the results of a series of simulations and we show that the system is able to reach a critical state characterised by a response to micro-impulsions that follows a 1/f law. Then in the fifth section we draw the lessons of such a learning for innovation diffusion and we shall conclude on research perspectives.

# 1 Self organised criticality and innovation diffusion

The transition frontier emphasised in Steyer and Zimmermann (1998-a) offers a way to better explain the different types of structural effects from which the success or failure of the diffusion process is dependent. There appears a double threshold effect related to the initial adoption rate let's say $\tau_0$. Under a certain threshold noted $\tau^-$ , it is not possible to obtain a convergent process (in the meaning of the final generalised adoption by a limit structure). Beyond a second threshold noted $\tau^+ > \tau^-$ , the convergence is always successful. Between those two values the convergence is uncertain and depends from the distribution of the initial adopters among the network structure.

This can be interpreted as stating a double structural effect. The first effect, related to the structure of the network itself is at the origin of the level and thickness of that transition frontier, while the second effect, resulting from the distribution of innovators (or initial adopters) among the network, is the source of the success or failure of the diffusion process on the frontier, i.e. when $\tau_0 \in ]\tau^-,\tau^+[$ . We found that the level and thickness of that frontier depend both of links' strength and network anisotropy.

Firstly, in terms of links' strength, we found that the frontier level and thickness is a decreasing function of average strength. But this effect is not the one that matters for us here and it shall be possible to obliterate it by forcing the average strength to be constant through an appropriate procedure of normalisation. Secondly, translating anisotropy in terms of entropy we found that the more the network is irregular (i.e. entropy is low), the more the level of the frontier is declining and the thicker it is. This implies that when entropy becomes very small, it is likely to diffuse successfully from a very small number of initial adopters (with $\tau_0 > \tau^-$ ). In return, it is likely to fail to diffuse even for a relatively high number of innovators, while $\tau_0 < \tau^+$.

**Figure 1: Phase diagram for f*=0,05**

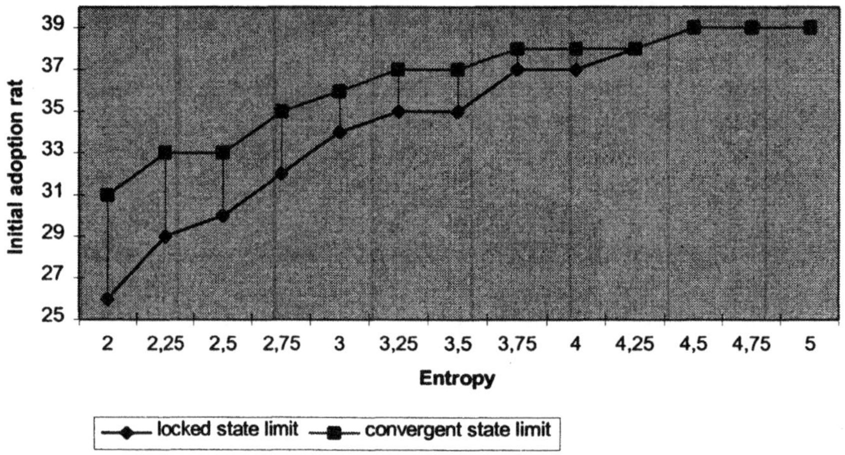


Source: Steyer and Zimmermann (1998-a)

Finally it means that a small number of agents are able to influence a sufficient set of other individuals and give rise to avalanches likely to diffuse on the whole network. In other terms a micro-impulse induced on each of those leaders will produce a macro-observable effect at the level of the global population. Such a phenomenon is very closed to what is described in physics by the notion of "scale invariance", meaning that there is a positive probability to observe a result of any size, in the limits of the population size itself. This has been described, in other terms, as the characteristic of a critical state towards which complex systems composed by a large number of interacting agents tend to evolve. Those chain reactions released from a micro-perturbation at the level of one single constituent are not predictable from local observation; they are not the result of individual features of certain elements but the consequence of the system's global history. It is called **Self-Organised Criticality** (SOC) (Bak, Tang and Wiesenfeld, 1987; Bak and Chen, 1991).

Few applications of Self-Organised Criticality theory have been yet developed in economy, in spite of the numerous observations of the fluctuations of aggregate indicators that conform to a flicker noise (1/f distribution[1]), like the variations of cotton price or Dow Jones index observed by Benoît Mandelbrot. In relation with Per Bak, Scheinkman and Woodford (1994) have developed a model of a large economy where sectors are interacting through input-output relations linking productive units, that are located on a cylindrical lattice. Exogenous shocks are generated in the final-goods demand by the way of orders received by the units. Authors show that the aggregate production of the system converges to a Pareto-Levy stable distribution with exponent 1/3.

As far as our work on innovation diffusion is concerned, our approach is not distant from the latter, in so far as the question is to examine the result on the system of an adoption decision that could be taken, for exogenous reasons, by a sub-group of agents called *innovators club*. More concretely it could be useful to consider the extent of avalanches caused by individual adoption decisions, the question being then to let an endogenous process making the network evolve towards a self-organised criticality state.

Two main conditions are generally required to observe S.O.C.:

- firstly the system has to be stimulated randomly by exogenous events;

- secondly the system must be "open" or dissipative, in order to let the random stimulations leave out.

Then the system will evolve spontaneously towards a critical state which becomes stable over long period. Such a system has been studied by Bak, Scheinkman and Woodford (1994). In our case, we exhibit self-organised criticality in a social network, without requiring the openness condition. A social network has not to be considered as open; influence has not to leave out the system. The working incentive for SOC is the learning process.

In a former paper (Plouraboué and al., 1998) we have shown that SOC state can be identified trough a distribution of avalanches sizes that following a Pareto-Levy law. In the present paper, we use an other mean to characterise a self-organised critical state. We measure the temporal signal of the system under random stimulations; the mark of SOC is the 1/f distribution of the signal power spectrum.

[1] More widely it is the matter of an $1/f^{\alpha}$ distribution, with $0< \alpha < 2$ .

# 2 Social networks, influence and individual behaviours

Within a social network, a given agent receives a cumulated influence from the other agents he is related with, that can be valuated as a balanced sum of those agents "activities" by the intensity of the links carrying their influence. On the base of that total received influence, that given agent will revise its proper state, its proper "activity", through what is generally called an "activation function" or a "transfer function" in neural networks into which category our model can be ranged.

Given I a population of agents, the state of each of them, also called 'activation level', is described by a normalised feature level noted $f_i$ $\forall i \in I$, taking its value in $\{-1, +1\}$. This level indicates if a given agent adopts (+1) or doesn't adopt (-1) the considered innovation or the technological standard.

We draw the way agents are interconnected by describing a matrix $W = (w_{ij})_{(i,j)\in I^2}$ whose general term $w_{ij}$ measures the strength of the connection from j to i. $w_{ij}$ is a valuation of the way i catches the weight of j in its decision taking, and corresponds to the idea that i makes for itself from j, according to its own system of values. Such an influence matrix has consequently no reason to be considered as symmetrical. It can include positive as well as negative links according as that influence corresponds to an incentive to or a dissuasion (aversion) from adoption. By convention, we consider that links absolute values are not greater than 1. In that version of the model we restrict our analysis to an influence matrix with diagonal elements $w_{ii} = 0$ $\forall i \in I$, so as to exclude auto-influence effects. Furthermore, we impose a normalisation condition that express the limited and equal for all capacity of an agent to perceive influence from the other agents (receptivity). Other versions of the model could let that receptivity more open and introduce a certain level of agents singularity in this respect. In this paper, we prefer to impose a uniform receptivity so as to concentrate our attention on the sole structural effects as carrying differentiation among agents. Then we impose a formal condition of vector $w_i = (w_{ij})_{j\in I}$ normalisation that can be written by the general rule:

$$\sqrt[\rho]{\sum_{j\in I} |w_{ij}|^{\rho}} = 1 .$$

In the present version of the model, we choose to fix $\rho = 2$

We consider that agents relational capacity is limited, meaning that one agent cannot be linked with everybody, the actual object of the model being to study agents embedded in social networks. Such an approach implies a limited network connectivity that grounds the agents' bounded rationality. We note C the network connectivity meaning that, if C=3, an individual agent is able to receive a non zero influence from only 3 other agents.

Then, at each period, a given agent i receives a cumulated influence equal to

$$\sum_{j \in I} w_{ij} f_j$$

Hence it is led to revise its state or activity through the relation

$$f_i = F(\sum_{j \in I} w_{ij} f_j)$$

More precisely, at each period of time, the network transition can be described trough the markovian system:

$$f_i^{t+1} = F(\sum_{j \in I} w_{ij} f_j^t)$$

The F function, called activation function or transfer function, describes the way a given neurone or agent is led to change its proper activity level under the influence received from the other agents it is connected with. Such a function can take different shapes, but for simplicity reasons we restrain F to a sign function:

F(x) = + 1 when x > 0

F(x) = - 1 when x $\leq 0$

# 3 Networks and social learning

Generally speaking, learning corresponds to the ability of an individual to modify its behaviour, or performance, taking count of its passed proper experience. Social learning corresponds to a situation where agents or individuals are able to modify their behaviour, or state, or opinion ..., on the base of information derived from the observation of their neighbours (Bala and Goyal, 1998) or more generally from the observation of those agents behaviour and performances. As noted by Vriend (2000), "with individual learning, an agent learn exclusively on the basis of his own experience, whereas the population or social learners base themselves on the experience of other players as well. The difference between these two approaches to modelling learning is often neglected, but (...) for a general class of games or social interactions this difference is essential".

In our approach, we propose to give to social learning a further meaning, by introducing the ability, for an individual, to revise the existence and strengths of their neighbourhood links as a consequence of the evolving degree of affinity or credibility it can feed for its different neighbours. This imply that such networks are basically non-symmetric on the contrary of Young (1999) where links are valuated but symmetrical. In that sense, social learning brings to social network its endogenous character.

By resuming the foundations of the Hebb (1949) approach, we introduce a learning process that express the principle that two agents are led to strengthen or weaken their links, in relation to their opinions' similarity regarding the adoption of a concerned innovation. Through such a reallocation of influence distribution, at each period of time, the social network evolves endogenously towards a greater structural coherence. Then our main hypothesis will be that such a restructuration process will drastically alter the innovation diffusion dynamics, within the concerned population. The main result that could be reasonably expected is that this learning process constitutes a way to build what sociologists call "social cohesion" (Burt, 1987), that is the population ability to carry out an innovation collective assessment. The corollary, in terms of innovation diffusion dynamics, would be that communication supported by the social network, earlier than the actual innovation introduction, can lead to deeply transform the diffusion process.

Learning formally consists in an incremental alteration of the network structure, described through the matrix $W = (w_{ij})_{(i,j)\in I^2}$ , at each period of time. The idea of a trace is introduced here, a kind of remanence of the past that induces a relative increase of $w_{ij}$ connection strength when agents i and j share the same opinion ($f_i \times f_j > 0$). This means that when two consumers share the same opinion they tend symmetrically to reinforce their mutual reliance, hence reciprocal influence. Then at each period, a reallocation of the connection structure strengths is introduced, according to the so-called *Hebb learning rule* : $\Delta w_{ij} = \lambda(f_i \times f_j)$ where $\lambda$ , called Hebb parameter or learning parameter, express the relative weight of learning in comparison with the one of the relational memory.

This is effectively a reallocation, because the total linkage of an agent is constant, following the normalisation principle $\sum_{j \in I} w_{ij} = 1$. , that must be applied at each time step. From a consumer behaviour point of view, this learning rule means that the more two consumers share the same attitude towards an innovation, the more they will consider each other as experts for the future, that is, the more their interaction $w_{ij}$ shall increase.

Actually, such a point of view needs to be balanced at two complementary levels. Firstly, that learning shall apply for any existing link around each agent in the network. This means that, at the end of the final reallocation, a lower relative growth will be transformed in a weakening of the concerned link. Secondly the sequence activation revision - learning, imposes that at each period of time, a given agent i begins with recomputing its proper state on the base of activation levels announced by its relational neighbours, balanced by the strength of influence links that connect them to it. It is then possible for the agent i to modify its relation, that is the trust degree it puts in an other agent j, on the base of the activation level announced by j and its proper recomputed state. Anyway, the influence level $w_{ij}$, that is a private information for i, can then be written:

$$w_{ij}^{t+1} = w_{ij}^{t} + \lambda(f_i^{t+1} \times f_j^{t})$$

# 4 Simulations and results

A series of simulations has been conducted for the described model. The initial network is built of 1000 individuals, each of them being connected to three peers following a random choice. Each individual state is randomly chosen in global equal proportions. All the agents, one by one or following a random order, have then to reconsider their proper state according to their neighbours' state, in an asynchronous process. During the state revision process, a learning procedure is achieved following the Hebb rule and the normalisation of $w_{ij}$ coefficients. Learning is applied following 15 000 individual decisions.

Then the evolution of the social network is stopped and the network is frozen in its issuing structure. But individuals are still randomly stimulated and continue to revise their choice without any further social learning. Every 1000 periods the global state of the network is measured by the sum S(t) of all individual states.

Figure 3 shows the graph of the global signal from the network structure issued from learning, while Figure 2 represents the graph of the global state obtained from the initial network without any learning. The latter appears to be not critical and the signal is equivalent to a white noise evolving under random perturbations.

**Fig. 2: S(t) without learning**

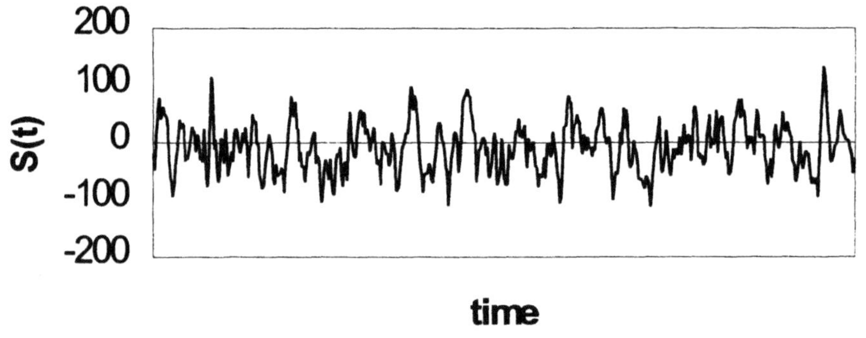


**Fig. 3: S(t) after learning**

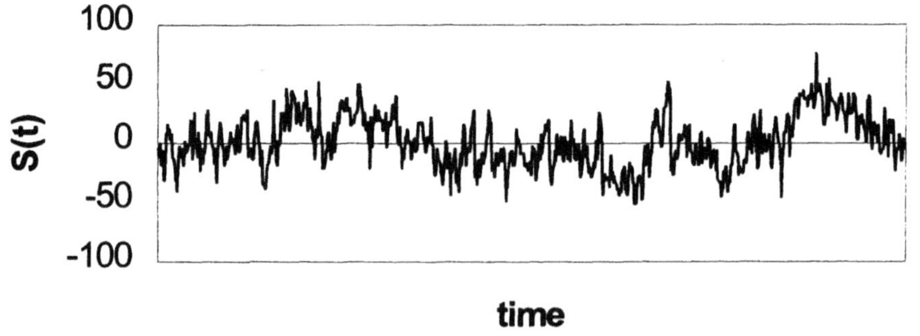

The power spectrum of such a noise conforms to the Bak and al. (1987) expression: $|\text{Fourier}|^2 = 1/[1 + f^2]$ where f is the frequency in an ad'hoc unit. Figure 4 shows that the spectrum of the S(t) network activity without learning is actually well approximated with such a function.

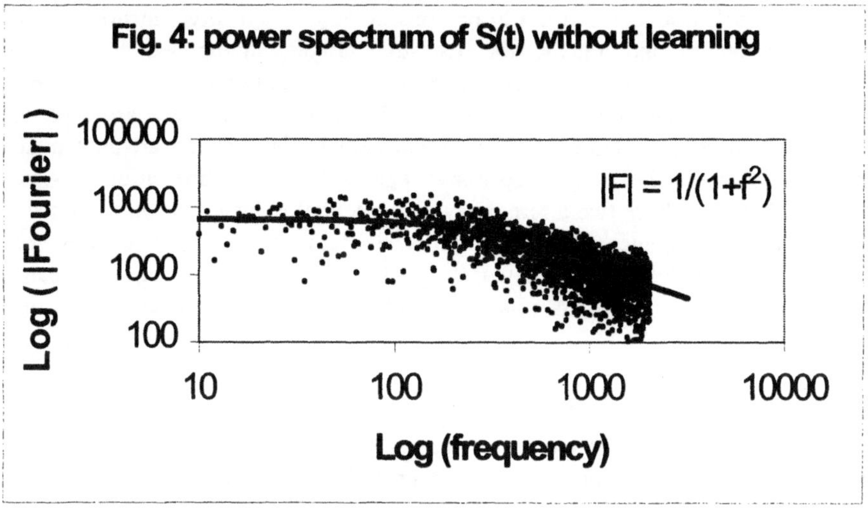


On the contrary, when learning has been achieved, the power spectrum, that is shown in Figure 5, appears definitely changed: it corresponds to a power law. Econometric analysis shows that the exponent of that power law is close to -1, that is the characteristic signature of a self-organised critical state as defined by Bak and al.

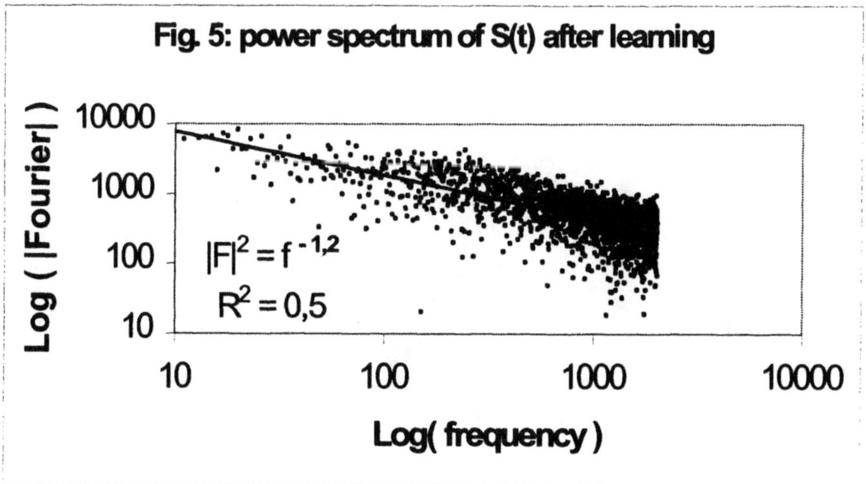

We have emphasised the emergence of criticality through the interpretation of Fourrier's power spectrum, but its origin remains to be elucidated. To main tracks can be invoked, following the nature of opinion leaders' influence power. Those leaders' global influence can firstly be constituted by their direct influence over a large number of individuals, the learning effects having led to the links polarisation around a small set of agents. It can secondly and alternatively result from a structural position of leaders within the network. Then they are capable to indirectly influence a large number of individuals through the advancement of large scale avalanches, e.g. progressive agents to agent stimulation. So as to arbitrate between those two alternatives it is essential to bring our mind to bear on the possible emergence of direct opinion leaders as a consequence of social learning. One individual's charisma, e.g. its direct influence capacity, can be measured as the sum of its relative weights within the other agents' decision process: $\gamma_j = \sum_{i \in I} |w_{ij}|$ . The emergence of direct opinion leaders would show itself in the evolution of the $\gamma_j$ histogram that would reveal high values after learning. On the contrary the histograms obtained before and after learning, as shown in Figure 6, highlight very close distributions enabling us to conclude to the non-emergence of direct opinion leaders. The origin of the emphasised self organised criticality shall subsequently be searched into the network topological structure rather than into local individual occurrences.

**Figure 6: Histogram of Individuals' Charisma before and after learning**

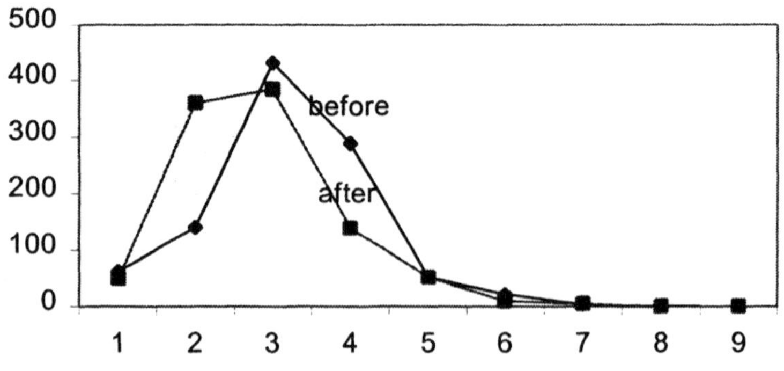

# 5 SOC and network management

The critical nature of the social network issuing from social learning involves substantial economic consequences. It means that small perturbations can entail macroscopic effects as well in terms of the size of the concerned avalanches as of their duration. Hence several individuals are capable to decide, whether knowingly or not, of the future evolution of any diffusion process within the network. In the field of innovation diffusion, if an agent, in place to generate a macroscopic avalanche, decides to adopt the new product or technology, he will act drastically in favour of the diffusion success. Individuals that do not hold the power to trigger a macroscopic avalanche are not convenient for the diffusion success. The reverse argument is also relevant: if an agent in place to generate a macroscopic avalanche induces a negative word of mouth on the new product, the diffusion failure is certain. Individuals that are not in a position to move a macroscopic avalanche do not hold, by their own the power to stop the innovation diffusion.

Nevertheless, it remains important to notice that those critical phenomenon are of very peculiar nature due to the fact that we can observe them in the context of a finite size and not open system. Macroscopic fluctuation, the system appears capable to generate from the achievement of a self-organised critical state, cannot be discharged out of the system, unlike all SOC former models in physics as well as in economics. From that stems a two-way behaviour that makes difficult to manage such a network. In the field of innovation diffusion this can contribute to give an explanation of the success or failure of new products or technology launching that could not be explained in terms of epidemic or threshold models.

# 6 Conclusion

In that paper we have shown that a self-organised critical state can be obtained in a close system like an influence social network. Like for all the identified SOC phenomenon, it is possible to emphasise both a distribution of avalanches following a Pareto-Levy law and a 1/f noise. Those two effects are due to the social learning that involves a reallocation of the diverse weights that any individuals allows to his peers. The precise mechanism of that effect building is not yet explained and further simulations and an analytical reasoning are still required. Classical arguments that are generally used to explain the appearance of SOC are not always relevant for our construction, because of its closed nature. It reveals interesting research perspectives, concerning the endogenous building of reticular topological structures that could spontaneously evolve towards a critical state. The consequences of such a result are but already imaginable in terms of innovation diffusion. We have emphasised the emergence of certain individual positions that are critical in the success or failure of an innovation diffusion. The question becomes that of a cross-checking between those positions and the innovative agents distribution among the population; in other terms, where are innovators situated within the social network?

# References

Bak P., Tang C. and Wiesenfeld K. (1987), "Self Organised Criticality", Phys. Rev. Lett. 59, 381.

Bak P. and Chen K. (1991), "Self-Organized Criticality", Scientific American, January, 264(1), pp.46-53, in French "Les systèmes critiques auto-oragnisés", Pour la Science, Mars, pp. 52-60.

Bala V. and Goyal S. (1998), "Learning from neighbours", Review of Economic Studies, 65, 595-621.

Burt S.R. (1987), "Social Contagion and Innovation: Coherence versus Structural Equivalence", American Journal of Sociology, 92, 1287-1335.

Cohendet P. (1995), "Networks, diffusion of technologies and irreversibilities", Annales des Télécommunications, 50, n°2, pp.234-245.

Cohendet P., Llerena P. et Stahn H. (Eds.) (1998) "Local interactions and global effects", Springer Verlag.

Deroïan F. (1998), "Interactions locales et bifurcations endogènes au sein d'un réseau de type neuronal", Document de Travail GREQAM n°98A41, Octobre.

Hebb D.O. (1949), "The organisation of behaviour", Wiley, New-York.

Plouraboué F., Steyer A. and Zimmemann J.B. (1998), « Learning induced Criticality in Consumers' Adoption Pattern : A Neural Network Approach », Economics of Innovation and New Technology, Vol. 6 pp.73-90.

Scheinkman J.A. and Woodford M. (1994), "Self-Organized Criticality and Economic Fluctuations", American Economic Review, Vol. 84 N°2 (May), pp. 417-421.

Steyer A. (1993) " La Théorie des avalanches; physique sociale des phénomènes de propagation", Thèse de doctorat, Ecole des Hautes Etudes Commerciales, Jouy-en-Josas, Novembre.

Steyer A. et Zimmermann J.B. (1996) "Externalités de réseau et adoption d'un standard dans une structure résiliaire", Revue d'Economie Industrielle, N°76.

Steyer et Zimmermann (1998-a) "On the frontier: structural effects in a diffusion model based on influence matrixes", in Cohendet P., Llerena P. et Stahn H (Eds.) (1998).

A.Steyer et J.B.Zimmermann (1998-b), " Etude empirique de l'influence sociale dans les phénomènes de diffusion - l'exemple du câble et du fax en France", *Economie et Prévision*, 1998-4 N°135

Vriend, N.J. (2000). An Illustration of the Essential Difference between Individual and Social Learning, and its Consequences for Computational Analyses. Journal of Economic Dynamics and Control, 24, 1-19.

# Social Percolators and Self Organized Criticality

Gérard Weisbuch[1], Sorin Solomon[2], and Dietrich Stauffer[3]

[1] Laboratoire de Physique Statistique† de l'Ecole Normale Supérieure, 24 rue Lhomond, F 75231 Paris Cedex 5, France
[2] Theoretical Physics Department, Racah Institute of Physics Hebrew University of Jerusalem
[3] Institute for Theoretical Physics, Cologne University, D-50923 Köln, Germany

**Abstract.** We discuss the influence of information contagion on the dynamics of choices in social networks of heterogeneous buyers. In the case of non-adaptive agents, the dynamics results in either the contagion process being stuck and very few agents actually buying (flops) or in a 'hit' where most agents a priori interested in getting the product actually buy it. We also show that when buyers and sellers try to adjust bids and asks the tatonement process does not converge to equilibrium at some intermediate market share and that large amplitude swings are actually observed across the percolation threshold.

## 1 Introduction

One of the simplest mode of interaction among agents used in economic models is imitation. A number of economists characterized the dynamic outcome of such processes for identical agents: under a wide range of hypothesis, one observes homogeneity of agent choices. These dynamics are often described as herding behavior and empirical phenomena such as bubbles in financial markets are interpreted along these lines. On the other side, we might a priori expect that heterogeneous agents would display a variety of behaviours. The purpose of this contribution is to shed some light on empirically observed phenomena with the help of two concepts proposed in natural sciences, percolation (Broadbent and Hammersley 1957, Stauffer and Aharony 1994) and self organized criticality (Bak and Tang 1989).

A stylised fact observed in a number of economic and social phenomena could be denoted "hits or flops". In situation such as:

- markets such as toys and gadgets (Farrell 1998), and of course stock markets;
- the movie industries;

† Laboratoire associé au CNRS (URA 1306), à l'ENS et aux Universités Paris 6/7, e-mail: weisbuch@lps.ens.fr

- the adoption of technological changes, or political and economical measures,
- the political arena where voters choose parties and political options.

one observes either distributions of market shares (or fraction of adopters or voters) which are biased towards the extreme: either close to 0 (flops) or 1 (hits), rather than a continuous distribution with intermediate values corresponding to the variety of individual situations and preferences. One possible interpretation is the occurrence of percolation across a social network which we discuss in this contribution.

The next section concerns passive actors with fixed preferences for buyers (or adopters or voters) facing passive sellers (or decision makers or parties) with fixed offers: we then consider a "one shot" possible purchase or adoption situation.

In the case of a series of purchases on different occasions, one would expect changes in the hits and flops situation in the presence of adaptive sellers which would want to avoid flops (!) and would be happier if they could obtain sales hits at a lower production cost. Buyers could also adjust their preference according to whether they were able to get what they wanted in the past. One of the question we raise in the later sections is whether a "tatonement" processes would restore equilibrium and balance between supply and demand. In fact, we will show by computer simulation that the resulting dynamics is characteristic of self organized criticality.

## 2 Social percolation

### 2.1 Simple models

Although the present discussion could apply to a number of equivalent situations, let use the case of a market with buyers and sellers to be more specific. We first start from the idea that agents lack full information about a product before buying it. But agents are not isolated. They are members of social networks: any of them is related to a number of other agents, called in our framework neighbours. An agent can get lacking information about a product from those of her neighbours who bought the product since they were able to use it. A simple hypothesis is that information is reducible to a scalar quantity that we call the quality $q$ of the product, transferable from any agent who purchased to all her neighbours. On receiving the information, an agent can then take a decision about buying herself. In fact all agents might not want to buy a product of quality $q$: information is only one necessary condition to take a decision; if it happens that the information from their neighbours is "the quality of the product is less than what you demand", the agents take a negative decision. A simple way to implement such a decision process is to suppose that agents $i$ each have individual preferences $p_i$. Heterogeneous agents of course have different preferences. The two conditions for purchase can be written as:

An agent will buy a product (or adopt a new technology, or vote for a party) when:

- she obtains information about the quality $q$ from one of her neighbours who made a positive choice;
- if $q > p_i$.

As presently posed the problem becomes equivalent to what physicists call site percolation (Stauffer and Aharony 1994). Let us consider a network,

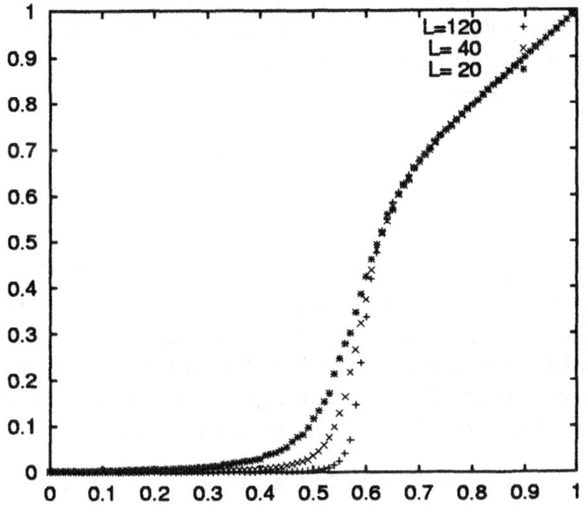


**Fig. 1.** Fraction of purchasers as a function of the percentage of agents with preferences smaller than the quality of the product for square lattices of linear dimension 20, 40 and 120.

lattice or with random connections, with a random uniform distribution of $p_i$'s. If one starts from initial conditions with a large majority of undecided agents (by lack of information) and a small fraction of "early birds" who have already taken a positive decision, contagion proceeds across the net from newly decided agents to their neighbours. The early birds could be individuals who got the product for free or who happened to get direct relevant information in the case of markets, promoters of a new technology in case of the adoption of new technologies or party members in politics. Waves of purchase (or adoption) propagate across the network, new purchasers carrying the information at each time step to their neighbours. We can predict a priori that none of the agents with preferences larger than the quality of the product will participate to the above described contagion process across the social network; they can then be removed as totally inactive. In fact for random distribution of $p_i$'s the outcome of the process is:

- surprisingly predictable: the fraction of actual buyers (or adopters) only depends upon the fraction $f$ of agents with

$$p_i < q$$

  and not upon the initial distribution of early birds[1];
- surprisingly non-linear: the fraction of actual buyers is nearly a step function of fraction $f$ with a threshold $p_c$.

Figure 1 summarizes simulations on square lattices of linear size 20, 40 and 120. $f$ is varied from 0 to 1 by steps of 0.01. Each point correspond to an average over 1000 random samplings. Periodic boundary conditions were used. The initial condition is one early bird in the center of the lattice. The steepness of the percolation transition depends upon $L$: the slope would be vertical for infinite $L$.

The comparison of $f$ to the percolation threshold $p_c$ on the social network allows to predict the success or failure of the product with a very high probability:

- When $f > p_c$ most sites with preferences smaller than $q$ are invaded, which results in a large success for the seller (or the promoters of the new technology, or the political party presenting the platform). The fraction of actual buyers is nearly equal to $f$ in this regime as seen in figure 1.
- By contrast when $f < p_c$, contagion soon stops and the product is a commercial failure with very small market share.

Let us recall some well known figures (Stauffer and Aharony 1994) in the case of uniform random distributions of $p_i$'s: for a square lattice with four neighbours per site $p_c = 0.593$, with eight neighbours $p_c = 0.407$, and for a random network with large connectivity $k$, $p_c = 1/(k-1)$ (the connectivity of a network is the number of neighbours per site). The above contrasted dynamic behaviours resulting from the necessity of contagion for adoption of a new product are very different from the prediction of "perfectly rational economics" which would yield a market share exactly equal to $f$.

This contrasted behaviour between the two dynamical regimes is a generic property independent of the details of the network connectivity. Furthermore it is not restricted to the hypothesis of unique criterion for agent's decision. In management science, a standard approach to decision making is multi-criteria analysis. Several criteria can influence agent decisions: they might

[1] Still prediction is difficult in the transition region, with outcome fluctuating largely for different realizations of the sampling of preferences. Furthermore what we mean by independence of initial distribution, refer to genericity: the property is true with a probability converging to one for neworks of infinite size and for random samplings.

take into account not only immediate or delayed profits but also time spent, risks, or even non-economic considerations such as moral, aesthetic or social values. The qualitative behavior that we above described, hits or flops, is also observed for multi-criteria analysis whatever procedure is used to process the criteria, as long as we suppose that all the information necessary for decision is transmitted at the same time by the neighborhood.

## 3 Adjustment meta-dynamics

Among the situations that we try to model some are occurring rarely such as major technological changes, while some others are recurrent as in the movie industry: people often visit movie theaters (or restaurants) in their home town, and producers are producing new movies for the market at a frequency high enough for both parties to adjust supply and demand according to their previous experience. In fact, one should distinguish among two regimes according to whether the percolation process is faster or slower than the tatonnement process.

### 3.1 Slow adjustment

In the case of the movie industry, a director is producing movies at a slow rate with respect to the time it takes for the movies to eventually percolate across the social network of movie-goers. Readjustment is slower than percolation. The phenomenon can be formalised (Salomon etal 2000) as:

- after opportunities during which they went to the movie, the agents will be more demanding and typically increase their expectations (here the preferences $p_i$); on the opposite, those who did not go, lower their preferences.
- after hits (resp. flops) the movie producers will decrease (resp. increase) the quality $q$ of the produced movie(s), in their effort to remain above the threshold while minimizing expenses.

We have made computer simulations based on a fast contagion process leading eventually to percolation, embedded in a slower adjustment process as described above: we iterated a series of steps each one composed of a contagion process which was left evolving until percolation or its absence was checked, and of the resulting adjustment process. The early birds were agents with $p_i < q$ aligned along one side of the square lattice. Percolation (resp. its absence) was considered to be achieved when one site of the purchasing cluster reached the other side (resp. when the contagion process stopped earlier). All details concerning the algorithms that were used, including the Leath algorithm, are described in Solomon et al (2000).

After a transient adjustment period, we observed an alternation of hits and flops. Such a dynamics is often described as self organised criticality: the adjustment process brings and maintains the system parameters $p_i$'s and $q$ in the neighborhood of the percolation threshold.

The simulations were done on square lattices. We restricted ourselves to the simplest dynamics: the quality of the movie increased by $\delta q$ if no cluster spanned from top to bottom, while it decreased by $\delta q$ otherwise. The viewer's preference $p_i$, initially distributed randomly between 0 and 1, changed by $\pm\delta p$ depending on whether $i$ went to the movie or not.

The results can be summarized as :

- Adaptive movie quality: for fixed agent preferences, $\delta p = 0$, and for adaptive quality $\delta q > 0$, we observed that the quality $q$ moves to the usual percolation threshold; in this limit, the dynamics of our system is reminiscent of self-organization mechanisms arising in thermal critical phenomena [11], where a suitable feedback mechanism may push the temperature towards the critical temperature.
- Adaptive customer preferences: $\delta p > 0$ but fixed quality $\delta q = 0$, the $p_i$ distribution drifts towards a single peak centered on the fixed $q$ value, taken equal to 0.5 (no percolation) or 0.593 (some percolating clusters).
- When both $\delta p$ and $\delta q$ are positive, $p_i$ and $q$ drift towards $p_c = 0.593$, even if the initial $q$ was 0.5. Our dynamic percolator thus drifts towards the standard percolation threshold of 0.593, exhibiting self-organized criticality (for longer times the $p_i$ and $q$ may drift away together).

We thus observe large-amplitude variations of the fraction of movie goers which is reminiscent of the fat tails observed in the distribution of return in financial markets. The above analysis is consistent with the interpretation of fat tails as due to strong cooperative effects, (here the constraints imposed by information contagion on the buy/sell decisions of the agents), in the neighbourhood of a transition[2]. Variants of this model are discussed in Ahmed and Abdusalam (2000), Goldenberg etal (2000), Huang (2000), Das Gupta (2000) and Weisbuch and Stauffer (2000).

### 3.2 Fast adjustment

In the present case, we suppose that the time necessary for the information to propagate across the social net is larger than the time between successive purchases by the same buyer. Although purchases are repetitive, we further suppose that after purchasing a good, a buyer is not active during some kind of a refractory period $m$ in analogy with the phenomena observed in nerve tissues or chemical reactions. The rationale for this hypothesis in economics is the existence of some consumption time for a commodity or some decay time

[2] We do not imply here that the General Equilibrium theory is wrong, but simply that some of its assumptions about convexity don't apply to situations where the decision process gives such large variations as observed around the percolation transition

for an investment between successive purchases. The regime of fast adjustment is observed when the refractory period is smaller than the time it takes for the purchase front to sweep across the lattice. In the iterated percolation model three conditions are then necessary for purchase by an agent:

- The agent has not taken any decision (to purchase or not) for a period at least equal to the refractory period $m$;
- he gets some private information from one of his neighbours;
- the quality $q$ of the product is higher than his preference/ expectations $p_i$.

We further add preferences[3] adjustment dynamics to the above decision process. We suppose that agents which purchased before the refractory period increase their preference coefficient for the next period by choosing $p_i$ through a random sampling between $p_i$ and 1. Alternatively, we suppose that agents who did not purchase before the refractory period decrease their preference coefficient for the next period by choosing $p_i$ through a random sampling between 0 and $q$. Those agents which did not receive the information because they might have been screened by refusers don't readjust their preference.

In the case of repetitive purchases, waiting until one of your neighbours purchases still makes sense for the interpretation of contagion based on positive externalities. For the interpretation based on information propagation, waiting for private information from a neighbour after the first purchase might seem unnecessary, since the agent should already know about $q$. The idea here is that $q$ might change through some adjustment process of the producer, hence the necessity for a prospective buyer to wait for up-to-date private information before any decision. Since only the sign of the quantity $q - p_i$ matters for the decision process, readjusting randomly $p_i$ should be equivalent to readjusting randomly both $q$ and $p_i$. We ran most simulations with fixed $q$ and tested the hypothesis on equivalence by a few runs with adaptive $q$.

**Simulations set-up** The model defined above was run on square lattices with four neighbours and periodic boundary conditions. We ran many numerical simulations (see Weisbuch and Solomon 2000). We used the Leath algorithm to follow the time evolution of purchases. In the description of the results, one time step corresponds to the propagation of information and of the purchasing process from all purchasers at the current time step to all the immediate neighbours (at the next time step). The following quantities were monitored:

- Time evolution of purchases at each time step (figure 2).
- Patterns of purchase in the network at given time steps (figure 3).

[3] quality adjustment discussed later does not change the results

- Average fraction of actual purchases at any given time step. (These average fractions were taken after an initial growth process of 100 time steps).
- The purchase process eventually comes to an end if no purchase is made during a time step. We have measured the average final time whenever the purchase process actually stopped, and the frequency of such occurrences. Averages were most often made over 5 000 runs. (The averages for the fractions of purchase and of agents willing to purchase are only taken for the set of non-stopping runs).
- For the set of non-stopping runs we averaged the power spectrum of purchases (figure 4).

The simulations were made on square lattices of size $L$ equal to 40, 60, 80 and 160. We used periodic boundary conditions. The refractory period $m$ was varied from 4 to $L$. For $m$ larger than $L$ the purchase process stops after one sweep of the network by the purchase front (see further) by lack of potential purchasers. Preferences $p_i$ were initially randomly distributed (uniform distribution). One single run normally goes on for 4096 time steps: this is more than necessary to take fast Fourier transform since coherence is lost after roughly $2L$ time steps.

**Typical results for a representative set of parameters** Let us analyse a first experimental condition with $L = 40, m = 10$ , runtime of 4096 steps and averages taken on 500 runs. The figure 2 displays a typical time plot of sales. One can notice large fluctuations with short (around 25 steps) time scale correlations, very different from white noise.

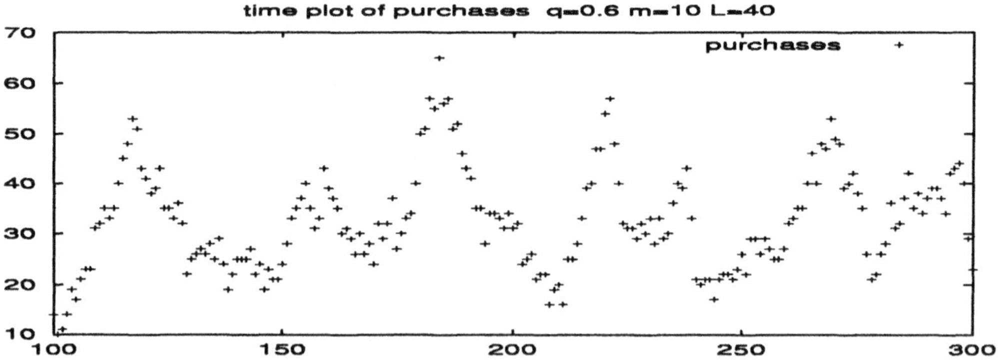


**Fig. 2.** Time plot of purchases after 100 time steps for a 40 × 40 square lattice, with a refractory period of 10 and quality 0.6 slightly above the percolation threshold.

A typical pattern displaying the states of the agents is shown in figure 3. White cells correspond to "awake" agents that are ready to make a choice when triggered by eventual active neighbours. Their previous choices occurred more than $m$ time steps ago. Other cells made their choice more recently: black cells correspond to agents who purchased and grey cells to agents who refused to purchase. On average (500 runs and 4096 time steps) 2 percent of the agents (O(1/L)) purchase at any time step. They are located on the black positions at the edges of the dark clusters, forming a disconnected purchasing front.

On-line monitoring of the patterns dynamics shows the displacement of the purchasing fronts across the lattice. The typical time scale, around 25, observed in figure 2 for the low frequency of the oscillation is similar to the time it takes to a front to move across the lattice and it scales with $L$.

**Fig. 3.** Pattern of purchases on a 40 × 40 square lattice. White cells correspond to "awake" agents that are ready to make a choice when triggered by their active neighbours. Black cells correspond to agents during their refractory period who purchased and grey cells to agents who refused to purchase.

We observed an average fraction of 0.593 potential purchasers, very close to the percolation threshold, which explains why quite often (19 percent of observed runs) the runs stop at an average time of 358, i.e. after a few sweeps across the lattice.

Finally, figure 4 displays the power spectrum of purchases averaged on those 81 percent of the 5000 runs which carried across 4096 time steps without stop. We observe a power law with a -2 exponent between periods 4 and 40. The flat spectrum at larger periods corresponds to the loss of coherence for times larger than the sweep time of fronts across the lattice.

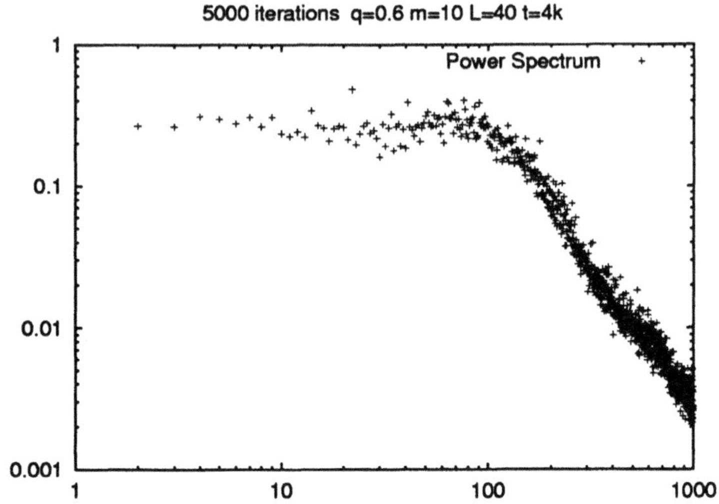


**Fig. 4.** Log-log plot of the power spectrum of purchases computed on a time span of 4096 time steps and averaged over 3300 runs. 1000 on the x axis correspond to 4 periods, 100 to 40 periods. The high frequency (above 100) spectrum obeys a power law with a -2 exponent.

**Influence of simulation parameters and model variants** Systematic tests of simulation parameters and model variants were made.

**The quality parameter $q$**

When $q$ is less than the percolation threshold, purchases stop very early since there is no percolation across the lattice. When $q$ is higher than the percolation threshold, the dynamics soon re-shuffle the distribution of individual preferences $p_i$ such that the fraction of agents with preferences below $q$ gets closer to the percolation threshold.

**The refractory period $m$**

At low values of $m$, the sweeping process of the purchase front across the lattice is less dominant and a lot of re-purchasing can occur locally. Purchase rate is doubled at $m = 4$ with respect to $m = 10$ and nearly all runs persist for long time, except 9 percent which stop very early after 7 time steps on average. The power spectrum does not display a well characterized power law.

**Variants of the initial model**

As mentioned in the introduction, we also tested the equivalence between the model with fixed $q$ and a model with adaptive $q$ to test the equivalence of dynamic properties of both model. We supposed that after each time step, a producer changes the quality of the proposed product by a "tatonnement" process: when a sale is made (resp. refused), she decreases $q$ by $\delta q$ (resp.

increases) through equation:

$$q(t+1) = q(t) \pm \delta q. \tag{1}$$

We tried $\delta q = 0.0001$ and $\delta q = 0.00001$ corresponding to adjustments of roughly 0.1 and 0.01 per lattice sweep for $q$. The obtained Fourier spectra display the same scaling behaviour as for fixed $q$.

In the small world (Watts and Strogatz 1998) variant a small fraction (typically a few per thousand) of the connections were randomly assigned across the whole lattice without taking into account any notion of neighbourhood. The results were qualitatively analogous to those obtained when all connections are periodic.

We also tried to change the readjustment process of those agents which refused to purchase. Rather than randomly redistribute them in the segment $[0, q]$, we redistributed them in the segment $[0, p_i]$. The same dynamics were observed.

Finally another change was a slower adjustment of preferences: purchasers preferences being redistributed in $[p_i, p_i + \alpha(1 - p_i)]$ and non purchasers' preferences in $[p_i, (1 - \alpha)p_i]$ with $0 < \alpha < 1$. Once more we did not observe big changes.

## 4 Conclusions

In conclusion, we have seen that information contagion processes do not average but rather amplify at the "macro" level the consequences of the heterogeneity of individual agents.

In the absence of any readjustment dynamics of prices and preferences, contrasted behavior of either hits or flops are observed on either side of the percolation threshold. The imitation or polling processes, such as Polya urns, often described in the economics literature, also result in contrasted behavior depending upon initial conditions. On the opposite, for the contagion dynamics described here, no sensitivity to initial conditions is observed except in the neighborhood of the threshold.

Rather than driving the economic systems towards equilibrium with small eventual fluctuations, readjustment dynamics brings it close to the percolation transition resulting in large non-Gaussian fluctuations. In the case of slow readjustment, the system alternates from a flop regime with no contagion to the hit regime where the purchasing cluster percolate across the network.

Let us stress that although most simulations reported here were done on square lattices, we have no reason to doubt from the validity of the qualitative results for all sort of networks such as random nets or small world. The percolation transition is a generic property observable on any kind of networks.

One might also ask what are the assumptions that are critical for the observed dynamics. Randomness of the initial distribution of preferences is important. We also suppose that early birds are a minimal fraction of all possible buyers, and that their distribution is random, with no correlation with the distribution of preferences.

An alternative mechanism proposed by Plouraboué etal (1998) and Steyer and Zimmermann (2000) in this volume to explain long term correlations is also based on information diffusion on a social network but implies a Hebbian learning mechanism: connections among neighbours are of varying intensity, and are reinforced when agents take similar decisions. The decision process of individual agents is based on polling their neighbours rather than simply being triggered by any active neighbour. Steyer and Zimmermann report in the presence of Hebbian learning an even more correlated behaviour with a $1/f$ power spectrum instead of our $1/f^2$ result.

Our conjecture based on the two sets of simulations ( Steyer and Zimmermann and ours) and on reasonable arguments is that long term correlations and non-exponential distribution of fluctuation sizes should be expected whenever decision processes are based on information contagion and customer adaptation. Lux and Ausloos (2001) express the same idea by referring more generally to interactions among economic agents as the generating mechanism. Reciprocally, when long term correlations are observed, an information contagion process is a possible explanation for the effect.

These observations about dynamics can be used not only to describe empirical phenomena, but also to build "selling" strategies for sellers or policy implementers.

Acknowledgments: We thank ZhiFeng Huang, Abhijit Kar Gupta, Jean-Pierre Nadal, Antonio Turiel and Jean Vannimenus for collaborations and helpul discussions.

Parts of this work were supported by CNRS, the Arc-en-Ciel program of French embassy in Tel-Aviv, SFB 341 and by Cray-T3E time of the Jülich supercomputer center. Part of it was achieved during visits of D.S and S.S to Ecole Normale which we thank for its hospitality.

## References

Ahmed E. and H.A. Abdusalam, (2000) Eur. Phys. J. B, 16, 569.

Bak, P. and Tang, C. Earthquakes as an SOC phenomenon, J. Geophys. Res. 94, 15635-15637 (1989).

Broadbent, S.K. and J.M. Hammersley, Percolation processes I. Crystals and mazes, Proc. Camb. Phil. Soc., 53, 629-641, (1957).

Farrell W. (1998), "How hits happen", HarperCollins, New York.

Goldenberg J., B. Libai, S. Solomon, N. Jan, D. Stauffer, 2000, Marketing percolation Physica A, 284, 335-347.

Huang, Z.F. (2000) *Int. J. Mod. Phys.* C 11, 287, and *Eur. J. Phys. B*, 16, 379.

Kar Gupta, A. and Stauffer, D. (2000), Int. J. Mod. Phys. C 11, 695.

Levy H., Levy M., and Solomon S., (2000) *Microscopic Simulation of Financial Markets*, Academic Press, New York.

Lux T. and Ausloos M., (2001), "Market Fluctuations I: Scaling, Multi-Scaling and Their Possible Origins" in A. Bunde and H.-J. Schellnhuber (Hg.): *Facets of Universality in Complex Systems: Climate, Biodynamics and Stock Markets*, Berlin.

Plouraboué F., Steyer A. and Zimmermann J.B., *Economics of Innovation and New Technology*, 6, 73, (1998).

Solomon S., Weisbuch G., de Arcangelis L., Jan N., and Stauffer D. (2000) Physica A 277, 239.

Stauffer D. and Aharony A., (1994) "Introduction to Percolation Theory", Taylor and Francis, London.

Steyer A. and Zimmermann J.B., "Self Organised Criticality in Economic and Social Networks: The case of innovation diffusion" proceedings of the Workshop on Economics and Heterogeneous Interacting (2000).

Watts D. J. and S. H. Strogatz , *Nature*, 393, 440, (1998).

G. Weisbuch and D. Stauffer "Hits and Flops Dynamics" Physica A, 287, 3-4, 563-576, (2000).

G. Weisbuch and S. Solomon "Self Organized Percolation and Critical Sales Fluctuations" Int. Jour. Mod. Phys. C, Vol 11, No. 6, 1263-1272, (2000).

# Lock-out in Social Networks*

Frédéric Deroian

G.R.E.Q.A.M., 2 rue de la Charité 13002 Marseille, France. E-mail: deroian@ehess.cnrs-mrs.fr - fax number: 04 91 90 02 27 - phone number: 04 91 14 07 23.

**Abstract:** We introduce negative relationships in a network of social influence whose agents cope with a binary opinion concerning two competing standards, and we explore the possible collective dynamics. A weak proportion of anticonformism leads to well known phenomenon of lock-in (collective agreement uppon one given opinion). Too much anticonformism generates persistent instability in individual opinions about technological standards. Now, we point out that an optimal dose of anticonformism can originate lock-outs: a minimal amount is necessary to depart from homogenous behaviors, but too much would prevent the rest of the society from reaching the opinions of anticonformists, and achieve the lock-out. Hence, our interactionist analysis points out specific conditions of demand as a possible reason for changing consumer behaviors in a context of standardization..

**Keywords:** Innovation, Social Network, Lock-out
**J.E.L. classification:** A13, D11, O30

## 1 Introduction

The interactionist literature clearly explains how increasing returns to adoption can lead potential adopters of a technology to a situation of lock-in (Arthur [1989]). This result is confirmed by the literature on social networks (Bala and Goyal [1998]). The sources of lock-in are well established now, and are principally learning by using, network externalities, scale economies in production, informational increasing returns, technological interrelatedness and evaluation norms. Yet, this phenomenon must be qualified. When direct interactions are not too much uniform, structural properties of social networks can prevail (Steyer and Zimmermann [1998], Young [1999]), originating the possible coexistence of competing technologies. The literature on social learning (Arthur and Lane [1993], Ellison and Fudenberg [1995]), or Gibbs measures models (Durlauf [1993]) also give possible sources of diversity.

Now, opposite to this important volume of interactionist studies about lock-ins, the literature on lock-outs is not dense. Foray and Grübler (1991), in an evolutionist perspective, show how a technological substitution is conditionned by market perequisites, to the extent that innovation keeps on

* I am indebted to Dominique Henriet for very helpful comments about the technical support of the article.

changing after its introduction. Witt (1997) puts ahead conditions on potential demand to explain lock-outs. Models of lock-ins are based on critical masses, whom an innovation must acquire to supplant the incumbent through notably a better intrinsic quality. Finally, Kirman (1993) shows that direct interactions in a large population can originate stable and differentiated but transient attractors, at the expense of a self-conversion component in the individual behaviors. This last explanation is the closest to ours. The price of our endogenous explanation of lock-outs is the introduction of negative links, based on a very simple microeconomic foundation. We assume that social norms, partly responsible for the perpetuation of mature technologies, die under the pressure of internal weariness and anticonformism, and can refer to peer effects (see Pesendorfer [1995] or Cowan *et al.* [1997] for models with both distinction and peer effects).

It is clear that a lack of anticonformism leads to lock-ins. It is also clear that too much anticonformism would generate persistent instability in individual opinions about technological standards (Deroïan [2001]). Now, we will point out that an optimal dose of anticonformism can originate lock-outs, in the sense that a minimal amount is necessary to depart from homogenous behaviors, but too much would prevent the rest of the society from reaching the opinions of anticonformists, and achieve the lock-out.

We set up in section 2 a model of social influence in the spirit of Steyer and Zimmermann (1998). Namely, social interactions are represented with a valued graph-like structure, whose part of links are negative. On this structure of interaction, agents are submitted to a "popularity weight" (Ellison and Fudenberg [1995]), which corresponds to a very simple mechanism of influence. Indeed, regardless to their own experience, agents follow the evolution of the opinion of their neighbourhood. This specification enables us to isolate a pure interactionist mechanism responsible for lock-outs. Precisely, individual receptiveness to network influence, integrating the notion of cumulation of influence, gives rise to the feature of "bottleneck". Broadly speaking, a bottleneck effect can be conceived as a transient attractor that momentarily prevents a given opinion from reaching a fixed point. On this basis, we give in section 3 some results that make the lock-out analogous to a track problem, and conditioning its effectiveness to a very limited range of parameters. An economic interpretation of bottleneck effect is presented in section 4, which exhibits the relation between incremental innovations on incumbent standards and technological disruptions by new entrants. Section 5 concludes.

## 2 The model

There is a finite population of $N$ agents. The agents recurrently actualize their level of opinion concerning the adoption of two competing innovations $A$ and $B$. The expected utility of agent $i$, noticed $u_i$, is formalized by a continuous variable belonging to $[-1,1]$ that expresses the agent's disposal to pay for about his possible adoption of one of the competing technologies. The higher (resp. the less) $u_i$, the higher the agent's propensity to pay for technology $A$ (resp. $B$).

Agents are linked to each other on a valued graph-like structure, representing the social network. We design by the scalar $\omega_{ij}$ the valued link which resumes $j$' influence on agent $i$. In so far, the influence network is not necessarily symmetric, and the values of ties can be either positive or negative, as we integrate aversion relations as possible negative influence. We denote by $\Omega = [\omega_{ij}]$ the matrix associated with the graph of relationships.

On this structure of interaction, agents periodically revise their current opinion according to the opinions of their neighbours in the preceding period. Time is discrete, and the dynamic is synchronous. Agents, submitted to cumulation of influence, deal with a sum of weighted opinions. The expected utility $u_i$ depends on the agents with whom $i$ is linked according to the following rule

$$u_i^{t+1} = \sigma\left(\sum_{j\neq i} \omega_{ij}.u_j^t\right), \textit{ for every } i \in \{1, ..., N\} \tag{1}$$

where $\sigma$ is called "activation function", and is endowed with the following properties: $\sigma$ is increasing, $\lim_{x\to-\infty} \sigma(x) = -1$, $\lim_{x\to+\infty} \sigma(x) = 1$. In this model, we examine a continuous level of individual opinion. In order to derive analytical results, we even need a $C^1$ function. The results are given for all function satisfying this shape. Now, as an illustration, we use the most commonly activation function

$$\sigma(x) = \tanh(\beta.x),\ \beta > 0 \tag{2}$$

Parameter $\beta$ represents the individual receptiveness to external influence. It is the same for all agents. When $\beta \leq 1$, the entering influence is dimmed, since for every $x \neq 0$, $|\sigma(x)| < |x|$. But, as $\beta > 1$, for $|x|$ smaller than $x^*$ solution of $x^* = \tanh(\beta.x^*)$, the activation function amplifies the perceived influence. Let us remark that when $\beta$ tends to infinity, the function approaches a ternary step function, such that individual opinions belong to $\{-1, 0, 1\}$. In some sense, $\beta$ represents the individual autonomous behavior regarding to the issue of adoption, in the spirit of Percolation models such as David and Foray (1993) concerning the standardization of EDI.

Moreover, the sensitivity of the impact of social influence upon individuals is a decreasing factor of the level of social influence in the neighbourhood, and this is of particular importance for the result. This means that agents have a finite ability to deal with external influence, and agents use the content of informations when they are quite indifferent. For instance, when agents receive a great level of influence, they must cope with and balance the new and the former informations, which qualifies the real impact of new informations.

Put differently, the amount of local information is perceived by individuals as legitimating the current level of opinion.

Let us now remark that individual utility is influenced by the utility of others, not by their choice. This is fundamentaly different from most of interactionist models. The reason lies in influence. Agents have direct relationships with each other, and confront with their mutual opinions. On the contrary, in traditional models of direct interaction, agents observe others' decisions, without knowing the reasons for their choices (limited rationality). As mentioned above, this distinction has consequences on agents' rationality. In our approach, agents exert influence on the basis of opinions which are by no means decisions. In that sense, we intend to approach the notion of normative mimetism, consisting in a social pressure on individual preferences.

## 3 A problem of track

We will show in dimension 2 the likeliness of transient lock-ins. Exit of lock-in is modelized as an endogenous phenomenon. We want to show, independently from the ability of an economy to produce innovations, and from the degree of competition, that the demand can be a driving force of exit of lock-in. Particularly, the very agents that entail the lock-out are those who give rise to herd behaviors.

We first define what we call bottleneck effect, and determine conditions on the two dimensional model ensuring its existence.

*Definition of a bottleneck effect*

Let us consider the case $N = 2$. The dynamical system is

$$\begin{cases} u_1^{t+1} = \sigma\left(a.u_1^t + b.u_2^t\right) \\ u_2^{t+1} = \sigma\left(c.u_1^t + d.u_2^t\right) \end{cases} \tag{3}$$

The opinions of agents 1 and 2 are thus linked through the values $a$, $b$, $c$ and $d$, and we impose $b < 0$. Agent 1 wants in some extent to disrupt from agent 2's opinion, whereas agent 2 follows partially agent 1's preferences. It is a track problem. Agent 1 establishes his opinion, and agent 2 suits it, so that a tension is generated in the choice of agent 1. There might be a lot of situations where two types of agents exert on each other antisymmetric externalitites. They can be synthetized in a distinction group / aspiration group problem. For instance, let us consider a polluting firm on intermediary market, which generates a negative externality upon the environment of a second one, and let us imagine that the second one sells on final market. Social influence can also provide good examples, like restaurants or fashion (upper and lower class ; the first distinguishes, the second imitates), the use of a public goods with heterogenous agents (a sport club with both competitors and neophytes, the latters are attracted by the presence of good players, and the former are helped from good entertainment if there are too many neophytes).

How many time will he bear the proximity of agent 2 ? Actually, he can tolerate in a arbitrarily long time agent 2 before disrupting. The symmetric situation can also occur, *i.e.* agent 2 can itself take a delay before following agent 1's opinion. We call such a phenomenon a bottleneck effect on agent 1 or agent 2, and we characterize it dynamically.

**Definition** *A bottleneck effect is defined as the temporary attraction, during an arbitrarily long period, of one's opinion toward a transient fixed point before reaching a real fixed point.*

The general mechanism is very simple. Let us assume a sequence $u_{n+1} = f(u_n)$, with $f$ a $C^1$ function defined as depicted in the following picture:

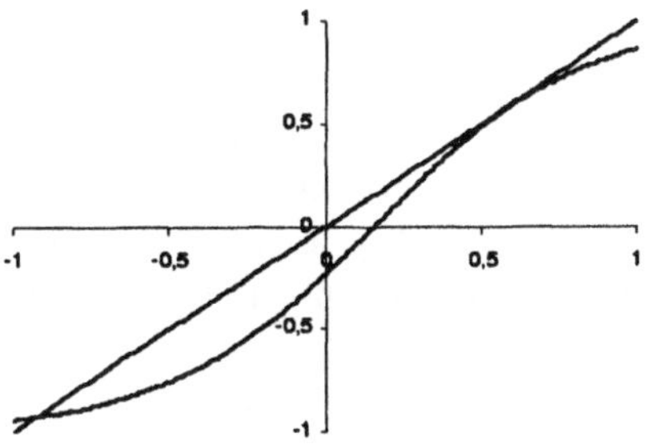


**Fig. 1.**

The function has the following properties. First, there exists a unique positive value, noticed $a$, such that $f'(a) = 1$. Second, the function $f(x) - x$ has a unique and negative root, noticed $b$. Let us now consider $u_0 > a$. In the example depicted, the limit of the sequence, as $n$ tends to infinity, is $b$. Nevertheless, we can adjust the function $f$, keeping its properties true, in such a way that, whatever integer $N$, there exists an arbitrarily small value $\varepsilon = a - f(a)$ ensuring $u_N > a$. Broadly speaking, it is possible to generate a bottleneck, *i.e.* a virtual attractor of the sequence, before reaching its real fixed point $b$.

In the light of this scheme, two characteristics of the mechanism are put ahead. Firstly, it momentarily prevents a variable to reach its fixed point. We argue that the demand can be attracted by a new and better technology without any visible modification of the structure of preferences during a long period. Lots of factors contribute to keep opinions far from the innovation.

For instance, the incumbent can stay attractive by producing incremental innovations, the new technology can be associated with migration costs, its unit costs can be higher, ... We do not forget another type of explanation, consisting in normative mimetism. Namely, the innovation, disturbing individual preferences, finds some obstacles to its diffusion in the habits of the population. Routines of consumption and social pressure are undeniably psychological factors that matter in the process by which an agent becomes persuaded of the quality of an innovation. Secondly, the change of mind is not gradual, but sharp. Once integrated the idea that the innovation can be part of one's own reality, and once repelled the factors freezing individual preferences, agents are submitted to a burst to adoption, which is a well-known phenomenon of diffusion curves. Thus, it cannot be denied that the demand can be a driving force to both perpetuate the position of an incumbent and open opportunities to new entrants.

We have now to find conditions to ensure the existence of bottlenecks[1].

*Bottleneck on agent 1*

We determine conditions on the influence matrix to ensure a bottleneck effect on agent 1. The theorem gives the conditions for the existence of recurrently disrupting opinions of agent 1 and establishes that the length of periods can be arbitrarily large, which represents the bottleneck effect on agent 1. The corollary puts notably ahead that reducing antisymmetry favors the disruptive behavior. Let us first give a numerical example attesting for such dynamics.

Let us consider the following example:

$$\begin{cases} u_1^{t+1} = \tanh\left(\beta.\omega_{11}.u_1^t + \beta.\omega_{12}.u_2^t\right) \\ u_2^{t+1} = \tanh\left(\beta.\omega_{21}.u_1^t + \beta.\omega_{22}.u_2^t\right) \end{cases} \tag{4}$$

For convenience, we notice $a = \beta.\omega_{11}$, $b = \beta.\omega_{12}$, $c = \beta.\omega_{21}$ and $d = \beta.\omega_{22}$. Let us also introduce the following influence matrix:

$$\begin{pmatrix} 1.5191 & -0.234316 \\ 1.34088 & 0.976355 \end{pmatrix} \tag{5}$$

We examine in the figures 2 and 3 (Figure 3 depicts the same dynamics as figure 2 around a disruption) the dynamics of expected utilities ($\beta = 1$). In this computation, agent 1 recurrently changes his mind. Once agent 2 moves, the delay of wait before disruption is close to 750 periods. Let us give conditions on matrician elements that allow such a phenomenon.

[1] The problem is fully explained when the leader takes time before disrupting. The two other cases, namely the bottleneck on the follower (who takes time before joining) and the mixed case, since reproducing the same type of explanations, are not fully developed here.

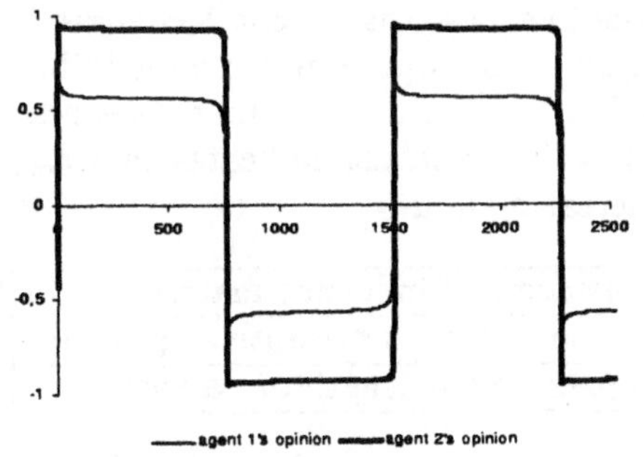


**Fig. 2.**

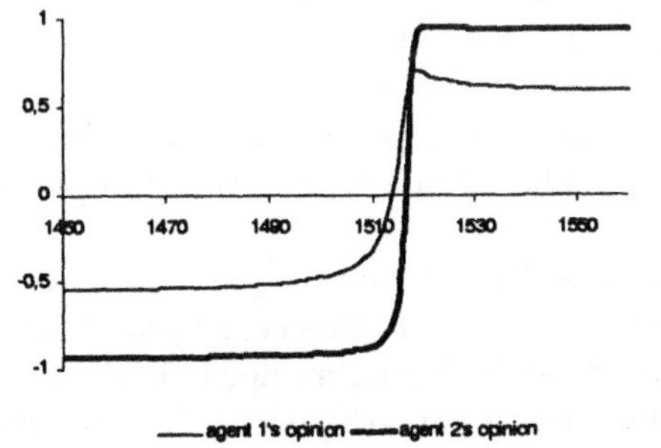


**Fig. 3.**

**Theorem** *If agent 1 is sufficiently self-influent* $(a > 1)$*, if agent 2 is not too much self -influent* $(0 < d < 1)$ *and if the relative dependence of agent 2 vis-à-vis agent 1 is sufficiently high regarding to the intensity of agent 1's irritability and the two self-inertia* $(c > c_{\min}(a,b,d) > 0)$*, then agent 1 is periodically submitted to a bottleneck effect before disrupting from agent 2's opinion, and agent 2 follows agent 1' opinion at once.*

*When considering the case of* tanh*, the technical condition is:*

$$c_{\min}(a,b,d) = \sqrt{\frac{a}{a-1}}\left[\tanh^{-1}\left(-\frac{a}{b}H(a)\right) + \frac{ad}{b}H(a)\right],$$

*with* $H(a) = \sqrt{\frac{a-1}{a}} - \frac{1}{a}\tanh^{-1}\left(\sqrt{\frac{a-1}{a}}\right).$

The proof of the theorem is given in the Appendix. Now, a general property characterizing the impact of the influence matrix on the length of periods, directly derived from the theorem, is given in the following corollary.

**Corollary** *Less antisymmetry impacts positively on the length of periods.*

The sensitivity of periods' length to matrician elements is opposite to the sensitivity of the likeliness that agent 1 changes his mind. Indeed, once established that agent 1 has a periodical palinode behavior, the closest the parameters to the limit, the strugglest the bottleneck. Studying the derivatives of $c_{\min}(a, b, d)$, a little analysis indicates that $c_{\min}(a, b, d)$ is decreasing in $d$ and $|b|$, and increasing in $a$.

| elements of influence matrix | $a$ | $\|b\|$ | $c$ | $d$ |
|---|---|---|---|---|
| impact on agent 1's disruptive opinion | $-$ | $+$ | $+$ | $+$ |
| impact on the length of periods | $+$ | $-$ | $-$ | $-$ |

(6)

Let us assume that agent 2's opinion is in the vicinity of agent 1's opinion. The signs reveal the tension of the track.

When $a$ increases, agent 1's opinion increases, and the value of $c$ which allows the disrupting opinion of agent 1 is greater. Hence, the likeliness of agent 1's palinode is decreased. When $|b|$ grows up, agent 1 is more irritable, so its behavior disrupts more quickly (a weaker value of $c$ was compatible with agent 1's palinode). When $c$ increases, agent 2 is more influenced by agent 1. The changing opinion of agent 1 is thus amplified, which contributes to its next disruption. When $d$ increases, agent 2's opinion is reinforced, what decreases $c_{\min}$.

Conversely, the time during which agent 1 is waiting before disrupting increases when $a$ grows up, since in this case agent 1 is relatively less affected by agent 2's opinion. When $|b|$ increases, agent 1 is more irritable and changes his mind faster. When $c$ or $d$ increase, agent 2 is attracted toward a greatest value, which amplifies agent 1's reaction. The less antisymmetry, the longer the periods.

Therefore, we can predict the existence of periodical dynamics, whose period lenght is arbitrarily large. The reason lies in the fact that agent 1's opinion is attracted toward a transient value before reaching its real fixed point. When agent 1 joins the fixed point, agent 2 follows by changing his opinion, what generates a new situation: the fixed point that agent 1 was reaching disappears and a new (opposite) fixed point is created, but also a new bottleneck (also symmetric).

Let us finally notice that numerical tests exhibit an extreme sensitivity of bottleneck effect to variations of matrician elements. In dimension 2, originating long periods requires to adjust parameters until at least the fourth decimal[2].

*Two other types of track*

[2] We have numerically tested if the sensitivity of period lenght was proportional to a power law of parameter $c$, but the decay appears to obey an exponential law, confirming the extreme instability of the phenomenon.

Actually, three possible cases are exhibited and illsutrated with the use of tanh: the bottleneck can affect either agent 1, or agent 2, or both agent 1 and agent 2. Let us consider the following influence matrix $\begin{pmatrix} 0.8 & -0.4 \\ 1.41256 & 2.6 \end{pmatrix}$. In this situation, agent 2 needs a delay before following agent 1's opinion as depicted in figure 4.

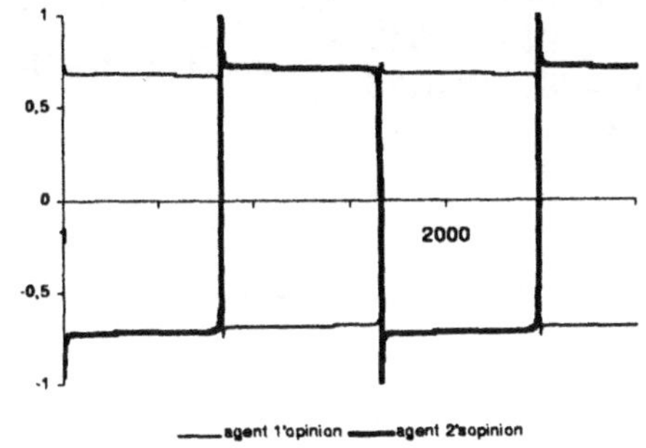


**Fig. 4.**

Agent 2 must cross over a bottleneck before joining agent 1's opinion. At this time, its function, initially generating a fixed point, crosses below $y = x$, what makes agent 2 disrupt. In picture 5, we understand the conditions on agent 2's dynamic:

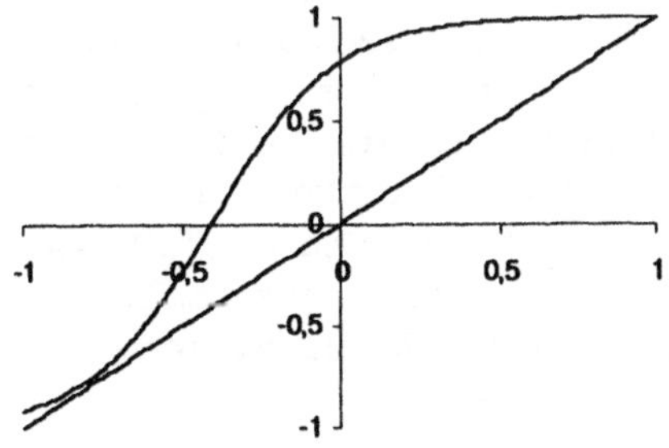


**Fig. 5.**

We observe $y = x$ and the function $\tanh(1.41256 \cdot 0.75 + 2.6 \cdot y)$. In this case, agent 2 takes time before following agent 1. When agent 1 is caught up, he changes his opinion at once, what changes agent 2's dynamic curve.

It is now the symmetric curve (of the depicted one) in relation to $y = x$ axis. A bottleneck, positive this time, and a negative fixed point are generated, perpetuating the phenomenon. A second theorem, similar to the previous one, exists (not presented here), that gives sufficient conditions (on the same basis of the previous theorem) to ensure that agent 2 is recurrently submitted to a bottleneck effect.

To finish the analysis, tracks can occur with bottleneck effects on agent 1 and agent 2. Agian, let us consider the case of tanh and let us consider the influence matrix $\begin{pmatrix} 1.54191 & -0.234316 \\ 1.34088 & 2.9136 \end{pmatrix}$. It originates the following dynamics -figure 6- on agents' activities:

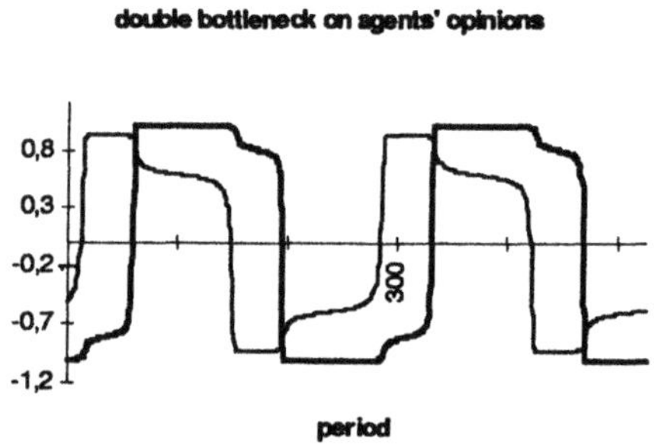


**Fig. 6.**

The dynamics is very simple. In fact, when agent 1 disrupts, agent 2 submits a bottleneck effect before reaching agent 1's opinion, and then agent 1 submits a bottleneck effect before disrupting from agent 2's opinion. Hence, bottlenecks are alternatively originated, and their creation is caused by the disruption of the opinion of one of the two agents. When the two opinions have the same sign, agent 1 submits a bottleneck effect before getting an opposite opinion, which generates a bottleneck on agent 2. We do not develop in this short article analytical conditions for its occurence.

## 4 Economic interpretation

The ability of demand to drive innovation activity is well known (Gerosky [2000]). In this respect, bottleneck effects are interesting for at least two main reasons. A transient diversion of one trajectory by an attractor can be originated, and the existence of such a phenomenon is extremely sensible to initial conditions of the social network.

The feature of bottleneck corresponds to the creation of a virtual attractor of an opinion. The long term behavior is established, but its trajectory is rerouted during an arbitrarily long time. Hence, a lock-in can be a fictitious attractor whose dynamics ineluctably goes away in the long run. The ability of a social system to stay in a lock-in or to exit from it depends in our optic on the delay during which the mature technology keeps attractive. But it also depends on the new technology appeal, and therefore is an incentive to technological competitiveness. The incumbent technology is constantly pushed to optimizate, which contributes to reinforce a lasting comparative advantage. Facing to this situation, a new entrant is more likely to find other types of competition, what can lead to the emergence of new technological paradigms. Market capture shall create a close link between incremental innovations and technological breaks.

Moreover, our approach puts ahead the importance of social norms. Even if the innovation has a higher quality, normative mimetism must be transcended. Actually, the network of interpersonal influence must collectively accept the innovation, and this often requires time (Valente [1994]).

The second remark concerns the extreme sensitivity of bottleneck effects to variations of the elements of the influence matrix. A limited shock on it entails the modification of the long run trajectory in two possible directions: either the bottleneck slackens and the the periods are shorter, or the attractor changes *i.e.* the transient attractor becomes definitive. In other words, a social network in formation or being disrupted can lead to freeze the system in the lock-in or to push it away. What can be such a pertubation ? Influence relationships can change because of the introduction of the new technology, or because of incremental innovations. Let us give some example illustrating simple microeconomic bases to it. The irruption of a technology can increase the irritability of an agent with regard to conformists. Conversely, it can reduce the herd behavior of agents who are submitted to normative mimetism. In the first case, the bottleneck slackens and the new technology can find the opportunity of a demand on the market, in the second the transient attractor can become definitive and the diffusion of the innovation may abort. Hence, the evolution of innovations in the course of their even diffusion can contribute to mould a social network.

## 5 Conclusion

In the light of these analyses, discussions about individual motives to change and the abilities of a social system to regenerate are pertinent concerning issues of diffusions of innovations. The main point is that dynamics leading to lock-ins are not irreversible. The precise correct proportioning of aversion relations in the network originates the renewal of the conditions of demand. It on the leaf mould of interactions that lock-ins are predicted, it is on this even mould that the system evolves. In other words, socials norms live and die, and it is particularly interesting when thinking demand as a driving force of innovative activity.

## References

Arthur, W. B., 1989, Competing technologies, increasing returns and lock-in by historical events, Economic Journal, 99, 116-131.

Arthur, W. B. and Lane, D., 1993, Information contagion, Structural Change and Economic Dynamics, 4, 81-104.

Bala, V. and Goyal, S., 1998, Learning from neighbours, Review of Economic Studies, 65, 595-621.

Cowan, R., Cowan, W. and Swann, P., 1997, A model of demand with interaction among consumers, International Journal of Industrial Organisation, 15, 711-732.

Deroïan, F., 2001, Morphogenesis of social networks and coexistence of technologies, Economics of Innovation and New Technology, 10, forthcoming.

Durlauf, S. N., 1993, Nonergodic economic growth, Review of Economic Studies, 60, 349-366.

Ellison, G. and Fudenberg, D., 1995, Word-of-mouth communication and social learning, Quaterly Journal of Economics, 110, 93-125.

Foray, D. and Grübler, A., 1991, Morphological analysis, diffusion, and the pattern of technological evolution: ferrous casting in France and the FRG, in Nakicenovic, N. , Grübler, A. (ed.), Diffusion of Technologies and Social Behavior (Berlin, New York, London and Tokyo), 407-431.

Geroski, P. A., 2000, Models of technology diffusion, Research Policy, 29, 603-625.

Kirman, A., 1993, Ants, rationality and recruitment, Quaterly Journal of Economics, 108, 137-156.

Pesendorfer, W., 1995, Design innovation anf fashion cycles, American Economic Review, 85, 771-792.

Steyer, A. and Zimmermann, J.-B., 1998, On the frontier: structural effects in a diffusion model based on influence matrices, in Cohendet, P. P. Llerena and Stahn, U., The Economics of Networks (Springer-Verlag).

Valente, T. W., 1994, Network Models of the Diffusion of Innovations (Hampton Press).

Witt, U., 1997, "Lock-in" vs. "critical masses": industrial change under network externalities, International Journal of Industrial Organization, 15, 753-773.

Young, H. P., 1999, Diffusion in social networks (Working Paper, Center on Social and Economic Dynamics).

## Appendix

**Theorem** *If agent 1 is sufficiently self-influent ($a > 1$), if agent 2 is not too much self -influent ($0 < d < 1$) and if the relative dependence of agent 2 vis-à-vis agent 1 is sufficiently high regarding to the intensity of agent 1's irritability and the two self-inertia ($c > c_{\min}(a,b,d) > 0$), then agent 1 is periodically submitted to a bottleneck effect before disrupting from agent 2's opinion, and agent 2 follows agent 1' opinion at once.*

*When considering the case of* tanh, *the technical condition is:*
$c_{\min}(a,b,d) = \sqrt{\frac{a}{a-1}}\left[\tanh^{-1}\left(-\frac{a}{b}H(a)\right) + \frac{ad}{b}H(a)\right]$,
*with* $H(a) = \sqrt{\frac{a-1}{a}} - \frac{1}{a}\tanh^{-1}\left(\sqrt{\frac{a-1}{a}}\right)$.

Proof of the theorem:
The proof organizes as follows. We show in the lemma 1 that when

$$\begin{cases} c > c_{\min}(a,b,d) = \sqrt{\frac{a}{a-1}}\left[\tanh^{-1}\left(-\frac{a}{b}H(a)\right) + \frac{ad}{b}H(a)\right], \\ with H(a) = \sqrt{\frac{a-1}{a}} - \frac{1}{a}\tanh^{-1}\left(\sqrt{\frac{a-1}{a}}\right), \end{cases} \tag{7}$$

the agents recurrently change their mind. Particularly agent 1 disrupts from his own opinion, agent 2 tracks him, so agent 1 changes again his mind, and the system is periodic. We show next in the lemma 2 that the delay during which agent 1 bears the vicinity of agent 2 before departing can be arbitrarily large.

**Lemma 1** *Assuming* $a > 1$, $b < 0$ *and* $0 < d < 1$,
*when* $c > c_{\inf}(a,b,d) = \sqrt{\frac{a}{a-1}}\left[\tanh^{-1}\left(-\frac{a}{b}H(a)\right) + \frac{ad}{b}H(a)\right]$,
*with* $H(a) = \sqrt{\frac{a-1}{a}} - \frac{1}{a}\tanh^{-1}\left(\sqrt{\frac{a-1}{a}}\right)$,
*then agent 1 recurrently changes of mind that is his opinion is attracted toward an opinion of opposite sign.*

Proof of the lemma 1:

**Claim** *Let us notice* $\Psi(x) = x - u(x)$. *Whatever* $\varepsilon > 0$ *and* $b < 0$, *there exists* $y > 0$ *such that for every* $x > 0$, $\Psi(x) > 0$.

Proof of the claim:

1. For every $a > 1$, and for every $y > 0$, there exists a unique $\widetilde{x}(y) > 0$ such that $u\prime(\widetilde{x}(y)) = 1$, since $u\prime(x) = a.(1 - \tanh^2(ax + by))$, and a little calculus gives

$$\widetilde{x}(y) = \frac{1}{a}\tanh^{-1}\left(\sqrt{\frac{a-1}{a}}\right) - \frac{b}{a}y \tag{8}$$

2.Using the fact that $\Psi(\widetilde{x}(y)) < \Psi(x)$ for every $x > 0$, $\Psi(\widetilde{x}(y)) = \varepsilon > 0$ induces

$$y = -\frac{a}{b} \cdot \left[ \left[ \sqrt{\frac{a-1}{a}} - \frac{1}{a} \tanh^{-1} \left( \sqrt{\frac{a-1}{a}} \right) \right] + \varepsilon \right] \tag{9}$$

Now, $H(a) = \sqrt{\frac{a-1}{a}} - \frac{1}{a} \tanh^{-1} \left( \sqrt{\frac{a-1}{a}} \right)$. $H$ is defined, continuous, strictly increasing and positive for $a > 1$. Then, there necessarily exists a positive value of $y$ such that, for every $\varepsilon > 0$, $\Psi(\widetilde{x}) = \varepsilon$.

For every $a > 1$, the value $y_{\min}$ such that $\Psi(\widetilde{x}(y_{\min})) = 0$ is

$$y_{\min}(a) = -\frac{a}{b} H(a) \tag{10}$$

Let us come back to the dynamics. Using the fact that $y$ converges very quickly toward its fixed point (since when $0 < d < 1$, no bottleneck effect can occur on agent 2), there exists $x^* > 0$ such that $y = y_{\min}$, satisfying

$$y_{\min} = \tanh\left(cx^* + dy_{\min}\right) \tag{11}$$

*i.e.*, using (10) and (11)

$$x^* = \frac{1}{c} \left[ \frac{ad}{b} H(a) - \tanh^{-1} \left( \frac{a}{b} H(a) \right) \right] \tag{12}$$

The issue becomes: is $x^* < \widetilde{x}(y_{\min})$? Using (8) and (10),

$$\widetilde{x}(y_{\min}) = \sqrt{\frac{a-1}{a}} \tag{13}$$

Now,

$$\widetilde{x}(y_{\min}) - x^* = \sqrt{\frac{a-1}{a}} - \frac{1}{c} \left[ \frac{ad}{b} H(a) - \tanh^{-1} \left( \frac{a}{b} H(a) \right) \right] \tag{14}$$

and $\widetilde{x}(y_{\min}) > x^*$ is satisfied if

$$c > c_{\min}(a,b,d) = \sqrt{\frac{a}{a-1}} \left[ \frac{ad}{b} H(a) - \tanh^{-1} \left( \frac{a}{b} H(a) \right) \right] \tag{15}$$

But, $\tanh^{-1}(Z) - dZ$ is positive when $Z > 0$ if $0 < d < 1$ (see figure below). Since $b < 0$, $\frac{a}{b} H(a) < 0$ and there exists a positive value of $c$, below which agent 1 does not change his opinion in spite of agent 2's proximity.

The three curves depicted above represent $\tanh^{-1}(Z) - dZ$, for $d = 0.6$, $d = 1$ and $d = 1.5$.

We have thus established that agent 1 disrupt from his own opinion when agent 2's opinion is of the same sign. Now, the system is periodiec, namely agent 1 recurrently changes his opinion under the pressure of agent 2's gregarist behavior, since the function tanh is uneven, what implies the same analysis for $x < 0$ and $y < 0$.

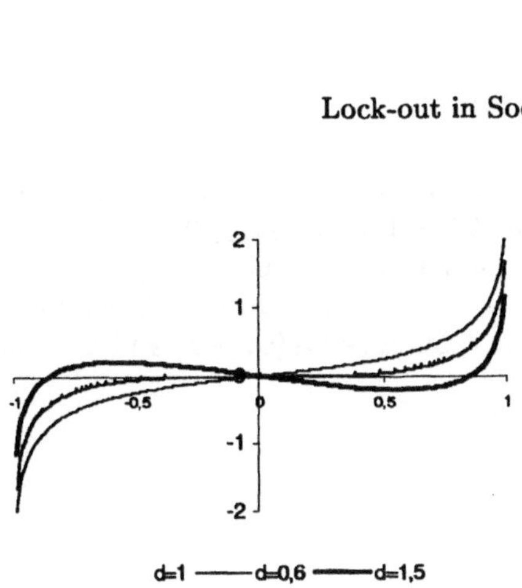


**Fig. 7.**

To finish, the lemma 2 characterizes the length of periods.

**Lemma 2** *Whatever N positive integer, there exists* $b_N < 0$ *such that, for every* $x_0 > \widetilde{x}(y_{\min})$, $x_N > \widetilde{x}(y_{\min})$, *if* $x_N$ *is the position of x after N periods.*

Proof of the lemma 2[3]:

First, we must see if the modulus of the negative fixed point toward which agent 1's opinion is attracted is greater than $\sqrt{\frac{a-1}{a}}$, which is the modulus of the transient fixed point that attracts all $x$ such that $|x| > \sqrt{\frac{a-1}{a}}$. Actually, the value of $y$ such that $-\sqrt{\frac{a-1}{a}} = \tanh\left(-a\sqrt{\frac{a-1}{a}} + by\right)$ is $y = -\frac{a}{b}H(a)$. For any $y > -\frac{a}{b}H(a)$, the fixed point is smaller than $-\sqrt{\frac{a-1}{a}}$. The issue is thus translated and consists in comparing the modulus of the fixed point which attracts agent 2's opinion with $-\frac{a}{b}H(a)$. The condition is basically $c > c_{\min}(a, b, d)$.

Being assured that the fixed point toward which agent 1's opinion is greater (considering the positive case) than the transient fixed point, we can show it easily. It is actually a general assertion, characterizing all $C^1$ functions locally approached by a tangent, beyond the specificity of tanh . Basically, when the system evolves, for all $x > \sqrt{\frac{a-1}{a}}$, $u(x) < x$. Since conditions of the lemma 1 are satisfied, when $x$ decreases, $y$ decreases, translating $\tanh(ax + by)$ to the left, without intercepting the axis $y = x$. Consequently, it struggles the bottleneck. Hence, if the lemma is prooved with the restriction that $y = y_0$, fixing the function $\tanh(ax + by_0)$ as the function that entails the less struggled bettleneck, and determining an appropriate value

[3] A rather long proof can be given by recurrence, which seems to us non necessary to be presented, given the simplicity of the argument.

$b_{N,y_0} < b_{\max}(y_0)$, then continuity and monotonicity of this function in $b$ assure a suitable value of $b \in ]b_{N,y_0}, b_{\max}(y_N)[$.

Now, for a fixed value of $y_0 > 0$, $u_b(x) = \tanh(ax + by_0)$ is continuously translated to the left and to the top when $b$ increases, so that we can arbitrarily control the difference $|u_b(x) - x|$ and thus $\left|u_b^N(x) - x\right|$.

# Cooperation, Adaptation and the Emergence of Leadership

Martín G. Zimmermann[1,2,*], Víctor M. Eguíluz[1,3,**], and Maxi San Miguel[1,***]

[1] Instituto Mediterráneo de Estudios Avanzados IMEDEA (CSIC-UIB), Carretera de Valldemossa 7.5km, E-07071 Palma de Mallorca, Spain
[2] Depto. de Fisica–FCEN, Universidad de Buenos Aires, Pabellón I Ciudad Universitaria, 1428 Buenos Aires, Argentina
[3] Center for Chaos and Turbulence Studies, The Niels Bohr Institute, Blegdamsvej 17, DK2100 Copenhagen Ø, Denmark

**Abstract.** A generic property of biological, social and economical networks is their ability to evolve in time, creating and suppressing interactions. We approach this issue within the framework of an adaptive network of agents playing a Prisoner's Dilemma game, where each agent plays with its local neighbors, collects an aggregate payoff and imitates the strategy of its best neighbor. We allow the agents to adapt their local neighborhood according to their satisfaction level and the strategy played. We show that a steady state is reached, where the strategy and network configurations remain stationary. While the fraction of cooperative agents is high in these states, their average payoff is lower than the one attained by the defectors. The system self-organizes in such a way that the structure of links in the network is quite inhomogeneous, revealing the occurrence of cooperator "leaders" with a very high connectivity, which guarantee that global cooperation can be sustained in the whole network. Perturbing the leaders produces drastic changes of the network, leading to *global dynamical cascades*. These cascades induce a transient oscillation in the population of agents between the nearly all-defectors state and the all-cooperators outcome, before setting again in a state of high global cooperation.

## 1 Introduction

In the agent-based models used in Social Sciences, Economy and Political Economy, agents interact directly with one another, and a social macrostructure emerges from these interactions. The implications of these models are easily studied with computer simulations. In this computational approach it is very natural to implement a network of interactions among the agents [9]. In addition, such computer simulations permit to study the dynamical evolution of the social structures.

In this context, an important question that is being addressed in a number of ways is how the aggregate or global behavior *emerges* from the individual characteristics of the agents. A particular aspect of this question is to understand if

* zeta@df.uba.ar and http://www.nld.df.uba.ar
** martinez@nbi.dk and http://www.nbi.dk/CATS
*** maxi@imedea.uib.es and http://www.imedea.uib.es/PhysDept

the global behavior is determined by average commonly found agents or if a few individual distinct agents can have a strong influence in the emerging macrostructure. In the latter case such special agents play the role of social leaders. Generally speaking, the influence of each agent depends on the network of interactions with other agents, an interaction being represented by a link between two agents. These interactions can be restricted to a set of agents placed in neighboring sites of a regular spatial lattice, can reach arbitrary agents as in a random network or can occur through intermediate "small-world" networks [17]. In most cases, this network of interactions is fixed and given from the outset. However, it is natural to consider situations in which the network of interactions evolves dynamically adapting itself to the emerging global structure.

In this paper we tackle the problem of how cooperation arises in a dynamically evolving network of agents. The network adaptation allows the emergence of an asymptotic state dominated by those special agents which in the course of the dynamical evolution are able to establish a much larger number of links than the average agent.

The paradigm to study the emergence of cooperation has been the Prisoner's Dilemma (PD) game. Using evolutionary game theory [18], it was shown [3,2] that cooperation may be sustained by a population of agents meeting repeatedly through global random interactions. Two agents interact playing the game and, according to their outcome, their strategies are allowed to evolve. A second route to cooperative behavior, pioneered by Nowak and May [1], comes from the consideration of *"spatial games"*. In these games every individual interacts only with a given subset from the whole population (e.g. the neighbors). The neighbors are defined by a fixed network of interactions. The spatial network can promote the emergence of cooperation in situations in which global non-cooperative behavior results if the interactions were random and homogeneous. Here we also consider a spatial Prisoner's Dilemma game. The novelty is that the group of agents with which a given one interacts adapts *endogenously* during the dynamical evolution. The adaptation of the network builds up cooperation.

In our computer simulation of the game we implement two dynamical rules. The first one is an action update: At each time step, each agent plays the same strategy cooperate (C) or defect (D) with all its local neighbors. Then the agents revise their individual strategies and imitate the neighbors strategy with highest aggregate payoff. Only a few agents will be found to be satisfied and will keep their strategy. The second rule is the network update: Unsatisfied agents are allowed to change their local neighborhood. Specifically, we let defectors break with a certain probability any link with other defectors, and replace them with new neighbors selected randomly from the whole network. The motivation behind this rule is that two defectors playing a PD game would certainly prefer to change its neighbor in order to find a cooperator from whom to 'exploit' a better payoff.

The results of our simulations show that the network of agents reaches a steady state with a high degree of cooperative behavior. The fraction of agents that cooperate depends only slightly on the incentive to defect in the individual game. This

[1] See for *e.g.* [14,11,10,6,15]. A cellular automata representation with several learning rules is presented in Ref. [8]. For an extensive study of the dynamics of a PD game with different strategies, evolution of the strategies and different networks see Ref. [4].

behavior contrasts with previous studies on "spatial" PD games where partial cooperation was reached, but it was observed that the fraction of agents which are cooperators strongly decreases as the incentive to defect is increased. This feature results from the adaptation of the network and it is reflected in the non-homogeneous structure that it reaches during the dynamical evolution. We will show that the process of "searching" cooperative neighbors performed by defectors, results in the emergence of a leader agent, defined as the cooperator with the largest number of links with other agents in the network. When the leader is *not* the wealthiest (i.e., the one with largest payoff), the network is in an unstable situation and, depending on the parameter $p$ measuring network adaptation, recurrent *global cascades* may be observed. These cascades induce large oscillations in the fraction of agents which are cooperators, together with a large reorganization of the network. In most cases, a final state with a high degree of cooperation is reached. We have also tested the robustness of such cooperative state. We find that a perturbation (spontaneous change of strategy) on a non-leader usually results in a short transient dynamics returning to the steady state. However, when a leader is perturbed, *global cascades* may be observed in the system before a state with a high degree of cooperation is recovered. This identifies the importance of the highly-connected agents which play the role of social leaders in the collective dynamics of the system.

The paper is organized as follows. The next Section defines the spatial version of the PD game in an adaptive network. Section 3 describes our numerical results on the emergence of cooperation. In Section 4 we discuss the evolution of the network and the emergence of the "leader" agents. Finally in Section 5, we summarize our results.

## 2 Spatial Prisoners Dilemma in an adaptive network

We consider the following framework: $N$ agents sit in the *nodes* of an adaptive network $\Gamma$ where the *links* define their neighborhoods. Each agent plays a PD game only with those other agents directly connected by one *link*. In this paper, we will restrict ourselves to the case of bidirectional or undirected links, and interactions to first neighbors. Indirect interactions with neighbors' neighbors have been studied for example in [5,16,7]. Thus, two agents are *neighbors* if they are directly connected by *one* link. We define the *neighborhood* of agent $i$ as the subset of $\Gamma$ which are neighbors of $i$, and we represent it as neigh$(i)$; its cardinal is $K_i$. The *coordination number*, $K$, is defined as the average number of links per node

$$K = \frac{\sum_{i=1}^{N} K_i}{N} . \tag{1}$$

In this paper we consider random networks $\Gamma$ with coordination number $K$ formed by distributing $KN/2$ links between pairs of nodes $(i,j)$, with the constraint that $(i,j) = (j,i)$ (bidirectional links). The resulting distribution of the number of links in the network is Poissonian with the maximum located at the coordination number $K$.

We denote by $s_i(t) = \{0,1\}$ the strategy of agent $i$ at time step $t$, where $s_i = 1$ corresponds to play cooperation (C), and $s_i = 0$ corresponds to defection (D), and will be referred to as C-agents or D-agents, respectively. The payoff matrix for a

2-agent PD game is shown in Table 1, where it is standard to take $b > \sigma > \delta > 0$ and $b/2 < \sigma$.

**Table 1.** Prisoners Dilemma payoff matrix

| | C | D |
|---|---|---|
| C | $\sigma, \sigma$ | $0, b$ |
| D | $b, 0$ | $\delta, \delta$ |

We consider the situation in which agents seek the largest possible benefit from their local interactions in the network $\Gamma$. We assume each agent plays the same strategy with all its neighbors neigh($i$), and the strategy is updated by all the agents at the same time; *synchronous* update. The time evolution is as follows:

1. Each agent $i$ plays the PD game with each neighbor using the *same* strategy $s_i$ and collecting a total individual payoff $\Pi_i$,

$$\Pi_i = s_i \mu_i \sigma + (1 - s_i)\,(\mu_i b + (K_i - \mu_i)\delta) \ , \qquad (2)$$

   where $K_i$ is the number of links of agent $i$ and $\mu_i$ is the number of neighbors of agent $i$ that are C-agents.
2. Agent $i$ revises its current strategy at each iteration of the game (*i.e.* at every time step), and updates it by *imitating* the strategy of its neighbor with a highest pay-off. Agent $i$ is said to be *satisfied* if it has the largest pay-off in his neighborhood. Otherwise it will be *unsatisfied* and it will revise its strategy.
3. Network Rule: each agent may adapt its local neighborhood:
   *if agent $i$ is an unsatisfied D-agent then with probability $p$ breaks a link with a D-neighbor $\alpha \in$neigh($i$), and replace it with a new agent $\beta$ uniformly from $\Gamma$.*

This scheme leads to a time evolution of the structure of the whole network, but the coordination number $K$ remains constant: for each unsatisfied D-agent $i$, it will replace (on average) $(K_i - \mu_i)p$ D-neighbors by new neighbors randomly chosen for the whole set, and thus its local coordination number, $K_i$, will not change; however, the replaced D-agents will lose one link and the new selected ones will gain one link.

The network rule justification is based on the assumption that given two D-agents playing a PD game, if they are unsatisfied, then they would prefer to exchange D-neighbors with the hope of finding a new C-neighbor from whom to exploit a much better individual payoff. This effectively amounts to ascribe a "searching" capability to D-agents. In our proposed setting the searching is not optimized, in the sense that the searching is random, so D-agents taking its chance to improve its payoff may end up with a new C-neighbor with a larger payoff, forcing it to replicate this new strategy in the next time step. In the same spirit, one could think that cooperators would also have a preference to break links with other D-agents. We prefer to keep this asymmetry in roles so that D-agents may be described as being competitive in nature, while C-agents remain conservative[2].

[2] In [1] agents may refuse to play with other agents irrespective of the strategy played.

The probability $p$ is a measure of the adaptability of the network. Dynamically, this parameter acts effectively as a time scale for the evolution of the network. For $p = 0$ the network does not evolve and remains fixed, while for $0 \neq p \ll 1$ the evolution is very slow as the adaptation takes over only after several trials. For $p \sim 1$ the adaptation is done at the same speed as the game is being played. In other terms, $p$ represents a transaction cost composed of two parts: first, the cost of breaking an agreement and second, the cost of finding a new partner and that this new partner accepts the agreement. One could separate these two costs, and would have a process of breaking links (with a given probability $q$) and another process of generation of links (with a probability $r$). However for the sake of simplicity we consider these two processes as a single one.

We have also investigated other variations of the network adaptation rule. For example, instead of breaking links with any of the unsatisfied D-neighbors, we also tested a rule which allows an unsatisfied D-agent break with probability $p$ solely the link with its D-neighbor with largest pay-off. The qualitative results obtained with this adaptation rule are rather similar [19].

It is worth making the following remarks:

- Links between satisfied agents do not change, which it seems to be a reasonable assumption.
- It is clear from the network adaptation rule, that not only D-agents may *actively* change their neighborhood, but also C-agents will *passively* evolve their own set of players by receiving new links from "searching" D-agents.
- In the present model, we do not take into account spontaneous creation or destruction of links, therefore *the total number of links in the network $\Gamma$ is conserved.*
- In a standard 2-agent Prisoner's Dilemma game there exists a unique Nash equilibrium (D,D). In a previous work [19], we considered a variation of the PD payoff matrix with $\delta = 0$, for comparison with [14,11]. In this situation the 2-agent PD game has as pure Nash equilibria either (C,D), (D,C) or (D,D). However the results in [14,11] indicate that, at least for fixed regular network, no qualitatively difference is found in the spatial games when using $1 \gg \delta > 0$.

In the next sections we present the results of computer simulations of the model described above. We take as a free parameter the incentive to defect $b$. We consider random initial networks with coordination number $K = 4$ and $K = 8$ and several values of the adaptability parameter $p$. We investigate, among others, the following statistical measures: (i) the fraction of cooperators, that it, the fraction of agents which are C-agents, denoted by $f_C = (\sum_{i=1}^{N} s_i)/N$, (ii) the average payoff per agent $\Pi = (\sum_{i=1}^{N} \Pi_i)/N$ of the whole network and the distribution of payoff, (iii) the probability of having a link between two C-agents, $p_{CC}$, between a C-agent and a D-agent, $p_{CD}$, and between two D-agents, $p_{DD}$. These probabilities satisfy:

$$1 = p_{CC} + 2p_{CD} + p_{DD} \ . \tag{3}$$

Throughout this work, the parameter $b$ which controls the incentive to defect was varied in the range $1 < b < 2$, while the other PD payoff matrix elements were fixed to $\delta = 0.1$ and $\sigma = 1$. Finally, if not otherwise stated, the network $\Gamma$ consists of $N = 10000$ agents, with an initial fraction of $0.6N$ C-agents randomly distributed in the network.

**Table 2.** Average fraction of C-agents, $f_C$, for different initial random networks, with and without network adaptation. The results are averaged over 10 different initial conditions after $T = 300$ time steps of evolution.

| $b$ | $K=4$ $p=0$ | $K=4$ $p=1$ | $K=8$ $p=0$ | $K=8$ $p=1$ |
|---|---|---|---|---|
| 1.05 | 0.89 | 0.942 | 0.95 | 0.994 |
| 1.15 | 0.87 | 0.947 | 0.90 | 0.989 |
| 1.35 | 0.59 | 0.920 | 0.58 | 0.988 |
| 1.55 | 0.31 | 0.900 | 0.38 | 0.988 |
| 1.75 | 0.09 | 0.885 | 0.03 | 0.983 |
| 1.95 | 0.04 | 0.889 | 0.01 | 0.962 |

## 3 Cooperation enhancement in an adaptive network

The PD game with local interactions in non-adaptive ($p = 0$) regular lattices, has been previously studied in Refs. [14,15,13]. These studies showed how partial cooperation can be sustained due to the local interactions, in spite of memory-less strategies. Several extensions to this *spatial* model have been studied in the literature. For example, introducing asynchronous updates [6] or introducing errors in the imitation process [12], the basic results persist [13].

For comparison with the adaptive network considered below, we give in Table 2 some numerical results for a fixed ($p = 0$) random network. Such network is our initial condition for the adaptive network. We see that the average fraction of C-agents decreases with an increasing value of the incentive to defect $b$. Notice that depending on the characteristic coordination number $K$, there is a critical $b = b^*$ such that for larger values of $b$ the system reaches a state of all D-agents. That is, for $b < b^*$ *partial cooperation* is supported in these networks. We obtain for $K = 8$ a critical $b^* \approx 1.75$. The numerical results also show that increasing the average size of the neighborhood (average number of links per agent, $K$) the average number of C-agents $f_C$ decreases faster with $b$.

When the network is able to adapt with the outcome of the individual games, either the dynamics settles onto a *steady state* after some transient time, or the system reaches a full defect state where all the agents are playing D and the network is continuously evolving. To compare the asymptotic dynamics of the fraction of cooperators between the adaptive and non-adaptive case, we illustrate in Fig. 1 a time series of $f_C$ evolving for $0 < t < 150$ in a fixed network ($p = 0$), and then for $150 < t < 300$ with a fully adaptive network ($p = 1$). We observe that the fraction of cooperators increases as the adaptation is turned on. Table 2 illustrates this behavior for other values of $b$. We conclude that the network adaptation enhances a highly cooperative network. Also notice that in the non-adaptive case, the fraction of cooperators fluctuates slightly, while in the adaptive case a steady state is reached.

It is important to notice that in order to obtain the above results, the *initial* fraction of cooperators in the network has to be sufficiently large. In all our numeric computations we took an initial fraction of $0.6N$ of C-agents, which proved to be a good number for the coordination numbers studied. This was also noticed in previ-

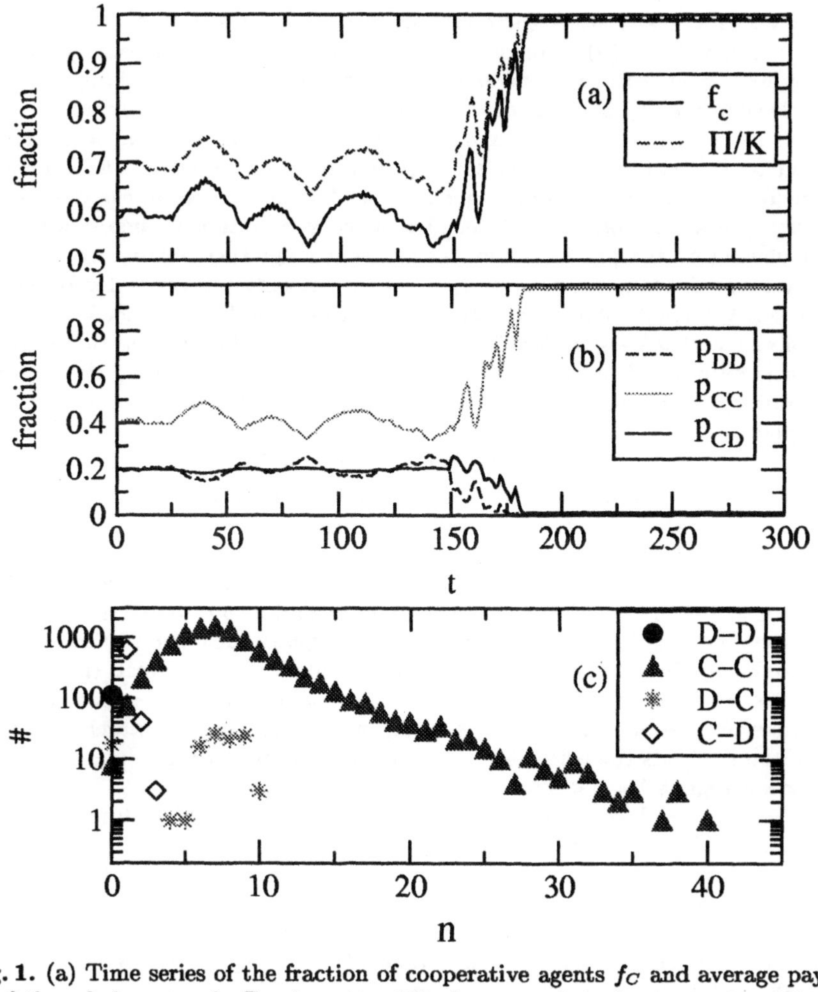


**Fig. 1.** (a) Time series of the fraction of cooperative agents $f_C$ and average payoff $\Pi$ of the whole network. For $0 < t < 150$ the network is fixed $p = 0$, while for $150 < t < 300$ the network is fully adapting $p = 1$. (b) Corresponding time series of the different links probabilities: two C-agents ($p_{CC}$), a C and a D-agent ($p_{CD}$), and two D-agents ($p_{DD}$) having a link. (c) Distribution of D-D-links, C-C-links, D-C-links and C-D-links for the steady state. ($b = 1.35$, $K = 8$).

ous spatial games and reflects the fact that the cooperative strategy will replicate throughout the network, only if (the approximate) average payoff of C-agents is larger than the one of D-agents. We will show below that in some circumstances, perturbations of the network may destroy the cooperative outcome and lead to a full defective network. Such asymptotic state is a dynamical state, since the network is continuously adapting but never finding a C-agent to exploit.

The steady state found in the adaptive network corresponds to a stationary network structure and to a stationary configuration of strategies. When the system reaches a steady state, there are necessarily no links between D-agents, thus $p_{DD} = 0$ (see Fig. 1(b)), except in the unlikely case of having exactly the same aggregate payoff. Also, in general we have that $p_{CD} \neq 0$. Therefore, the steady state is composed by a collection of cooperators exploited by D-agents.

We define a *chain of cooperators*, as a connected component in $\mathcal{N}$ of sites occupied by cooperators in which each cooperator except the last one is linked to a neighboring cooperator with larger pay-off. All cooperators in the chain except the last one are actually unsatisfied, but as they imitate the same strategy they were playing on the previous step, they never change their relative payoff. When the system reaches a steady state ($p_{DD} = 0$), the only possibility for D-agents is to exploit the agents of a given chain and must necessarily be satisfied. In terms of payoff we say that D-agents are "passive" local maxima, in the sense that they have the maximum payoff in their neighborhood but nobody is imitating their strategy. In summary, every cooperator chain should satisfy the following ordering in terms of payoff:

$$\Pi_{l(i)} > \Pi_r \geq \Pi_i, \qquad r \in \text{neigh}(i) \tag{4}$$

where $i$ is a C-agent imitating from C-agent $l(i)$, and $r$ is an exploiting D-agent of $i$. It is clear that the number of different steady states which one can construct satisfying eq. (4) is huge for a fair number of agents $N$.

Finally, another salient feature of our cooperative network model is the inhomogeneous distribution of payoff for each subpopulation, which differs substantially from the non-adaptive network case. Figure 2 displays such distributions (normalized to the number of individuals in each subpopulation), and reveals that although defectors are outnumbered by cooperators, *on average they are wealthier*. This is an interesting result, which indicates that, in the long run, the searching capability of the D-agents rewards them. This behavior is observed systematically in the parameter regime $1 < b < 2$ for $K = 8$.

The above result seems surprising from the point of view of the traditional replication dynamics [18] used in evolutionary game theory, because one could conclude from Fig. 2 that D-agents should dominate the whole population. But our results indicate that the final highly cooperative state is not determined by average agents in the system, but rather by a small subset of those maximally connected C-agents. In the next section we discuss the important role played by these *cooperative leaders*, defined as those C-agents with the largest number of links.

# 4 Dynamics of network evolution and emergence of leaders

The network rule allows for the evolution of the connectivity of every agent and permits that the network reaches a steady state. The distribution of links in this steady state displays the heterogeneous structure of the network. Figure 1(c) shows the distribution of links between two C-, D-, and between D-C and C-D agents. Notice the broad band distribution of links between two C-agents, which may reach as 5 times the average connectivity of the network. Defectors, on the other hand, are shown to be connected only to C-agents, and have a narrow distribution centered at $K$.

The tail of the distributions of links identifies a very small number of C-agents with a large number of links to other C-agents We define the C-agent with the

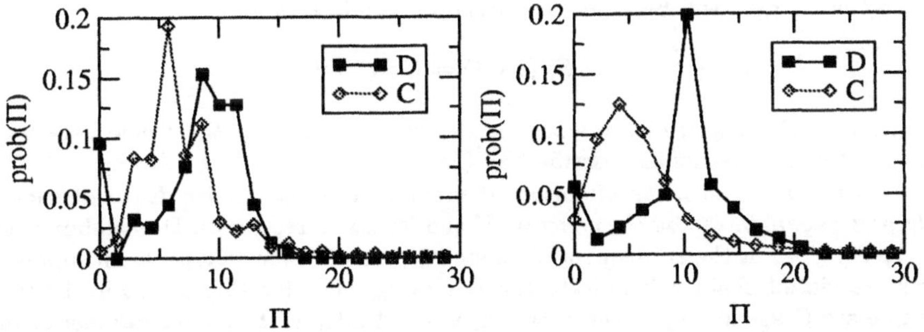


**Fig. 2.** Distribution of individual payoff for each subpopulation (C and D-agents) in a final equilibrium state (normalized to each subpopulation). (a) $b = 1.25$, (b) $b = 1.75$. ($p = 1.0$, $K = 8$).

maximum number of links as *the leader*. In a steady state this agent should be satisfied and leading a chain of cooperators. The number of C-agents which can fit on a given chain connected to the leader may be huge.

**Table 3.** Maximum number of links of the leader agent ($K_\alpha$), the D-agent with a largest number of links ($K_\beta$), and corresponding payoffs ($\Pi_\alpha$, $\Pi_\beta$) for different values of $b$ in a steady state. The results are averaged over 30 different initial conditions, $K = 8$ and $p = 1$.

| $b$ | $K_\alpha$ | $K_\beta$ | $\Pi_\alpha$ | $\Pi_\beta$ |
|---|---|---|---|---|
| 1.15 | 25.8 | 11.2 | 25.8 | 12.9 |
| 1.35 | 40.5 | 11.7 | 40.5 | 15.8 |
| 1.55 | 54.1 | 11.8 | 54.1 | 18.3 |
| 1.75 | 56.9 | 13.0 | 56.9 | 22.8 |
| 1.95 | 72.1 | 14.0 | 72.1 | 27.4 |

We have calculated the number of links of the leader agent for different values of the incentive to defect $b$ in a steady state. Table 3 gives a measure of the connectivity of the leader, labeled $\alpha$, compared with the number of links of the most connected D-agent labeled $\beta$ (recall that all connections are exploiting other C-agents), with their corresponding payoffs. Notice that for increasing $b$ the number of links of the leader increases while it remains essentially constant for the defector with largest number of links.

The leader agent leads the cooperative collective state of the system in several ways. On one hand, the leader favors the creation of chains of cooperators. On the other hand, and provided there are links between D-agents, leaders are selected through the dynamics of the adapting network, and are a direct consequence of the "searching" done by D-agents. Consider for example a D-agent $j$ which exchanges one of its D-neighbors with the leader $\alpha$. Assume that $\Pi_\alpha > \Pi_j$. In the next time step the D-agent will become a cooperator by imitation and

$$K_\alpha(t+1) = K_\alpha(t) + 1 \tag{5}$$

Whenever the payoff of the leader is the largest payoff in the whole network its connectivity increases as described by (5).

However, *an unstable situation occurs whenever the leader does not have the largest payoff in the whole network.* If the leader $\alpha$ receives a D-neighbor with a *larger* payoff, in the subsequent time steps a drastic event happens, for *alpha* and its associated chain will imitate the D-strategy. If after this cascading imitation there are C-agents left in the network, a new leader with a fewer number of links will be selected; otherwise a full defective network may be reached. This indicates the sensibility of the network structure to small perturbations on individual specially well-connected agents: A local event associated with a particular individual propagates in macrodynamical avalanches into the full network.

The phenomenon described above does not exclude the possibility that a D-agent $j$ selects a C-agent which would satisfy (4), allowing agent $j$ to become a passive local maximum and exploit the C-agent. In fact, it is by this mechanism that D-agents also increase their payoff.

We have performed several computer simulations to visualize the dynamic evolution of the network which we have just discussed. Each panel of Fig. 3 shows a comparison of the payoff of the leader $\alpha$ with the payoff of the D-agent with largest payoff (labeled $\gamma$[3]), together with the evolution of the fraction of cooperators in the network.

For small values of the adaptability parameter $p$, as shown in panel (a), the typical time for the network to reach a steady state is very large, and the leader $\alpha$ increase its connectivity via (5) very slowly. The fraction of cooperators increases on average as the leader increases its number of links. On the other hand for $p \approx 1$, the typical time to reach a steady state is very short (see Fig. 1), and in general the leader is always the wealthiest agent of the whole population.

An interesting intermediate regime occurs for $p \approx 0.05$ and high incentive to defect $b$, where a competition between the cooperative leader and the most wealthy defector arises. In this regime the number of links of the leader does not grow so fast and defectors may get a significant large payoff due to the high value of $b$. Panel (b) and (c) illustrates this situation. It is seen that there are time intervals in which the leader is not wealthier than the D-agent $\gamma$ with largest payoff. Whenever this event occurs, large global cascades involving a large fraction of the whole population are observed, with a significant fraction of the whole population being affected. The initial dropout of the fraction of cooperators affects immediately the payoff of exploiting D-agents and $f_C$ reaches a minimum approximately when the leader $\alpha$ becomes once again the wealthiest agent in the network. Once this stable situation is re-established, the leader may again increase its number of links until all D-agents have links exclusively with other C-agents ($p_{DD} = 0$).

An extreme example is observed in panel (c) at $t \approx 150$, where the network is composed of mostly D-agents together with a very wealthy leader. The leader is able

[3] We remark that at every time step a different D-agent may become the agent $\gamma$ with largest pay-off. This is a consequence of the competitive nature of D-agents in contrast to the conservative nature of C-agents.

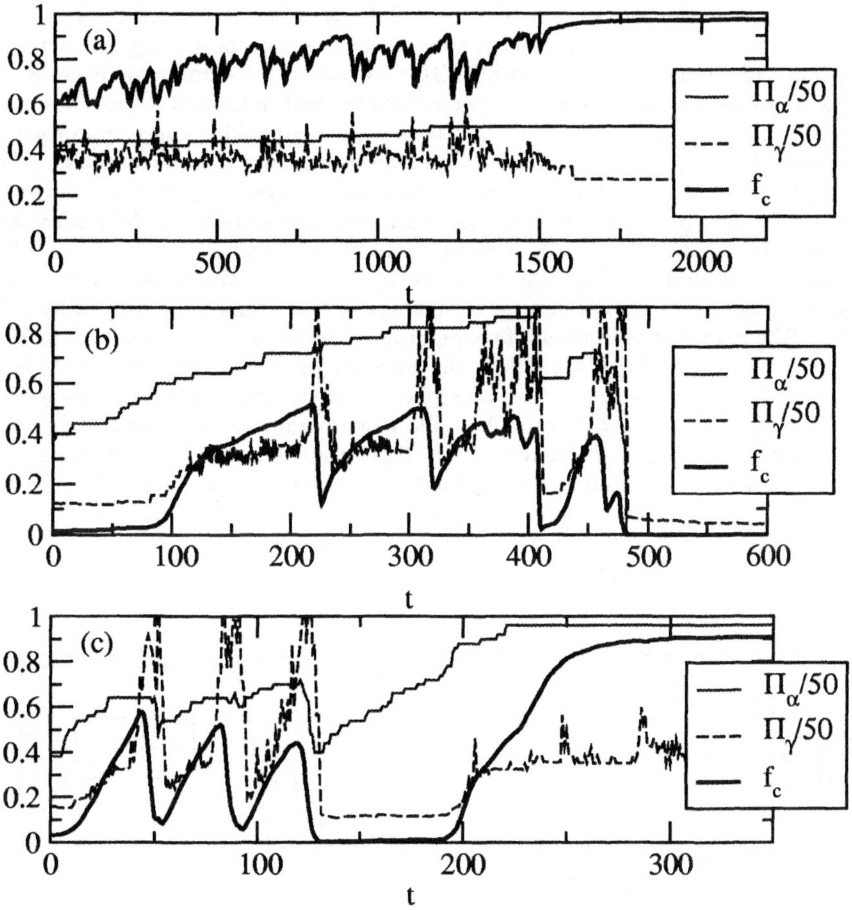


**Fig. 3.** Time series of $f_C$ and a re-scaled payoff of the cooperator leader ($\Pi_\alpha$) and the D-agent with maximum payoff ($\Pi_\gamma$). (a) $p = 0.005$, $b = 1.35$, (b) $p = 0.01$, $b = 1.75$, (c) $p = 0.05$, $b = 1.75$ ($K = 8$).

to increases its number of links by the intense (unsuccessful) "searching" done by D-agents, and by $t \approx 200$ the fraction of cooperators also increases. This recovery shows the importance of a wealthy leader, which enables a full cooperative final outcome. Another situation worth mentioning is that whenever strong competition between the leader and the wealthiest D-agent occurs, there is a possibility that the system ends in a full defective network. This is what is shown happens in panel (b) of Fig. 3. We remark that provided the initial fraction of C-agents is large, the full defective network is rarely reached for high enough $p$.

The above results show that the dynamic evolution of the network is intimately related to the fate of the leader. Another possible test, is to study how *noise* affects the network dynamics. If the noise is in the form of selecting a random agent and spontaneously changing its strategy, then the dominance of C-leaders is found to remain for nearly the whole range of $1 < b < 2$, for a sufficiently small noise intensity. However when the probability of a spontaneous change of strategy is increased, we find that a transition to the full defective network becomes more probable[4]. Clearly the leader may suffer such perturbations, and induce large cascades in the system. If such drastic perturbations have a small probability, the system has time to reach the cooperative outcome before the new leader is knocked down again. The simulation described in Fig. 4 illustrates how the system reaches a steady state at $t \approx 180$, and a spontaneous change of strategy was applied to the leader: large cascades results and a final cooperative outcome is again recovered. The transient oscillations observed in Fig. 4 at $t \approx 50$ before a steady state is reached, as well as the ones originated by the change of strategy of the leader constitute a Sysiphus effect. The drop in the fraction of cooperators comes together with a large increase of $p_{DD}$. Thanks to its adaptability, the network reacts creating more C-C links. However the attempts to build up a large cooperative behavior are not always successful and the system goes through oscillations in which non-cooperative behavior with large values of $p_{DD}$ is recovered. The frustrated attempts to build cooperation indicate that for cooperation to be robust, it has be built upon a specific networks of links. In the frustrated attempts to reach a collective stable cooperative state, the fraction of cooperators becomes large, but the spatial arrangement of links in the network is not the proper one.

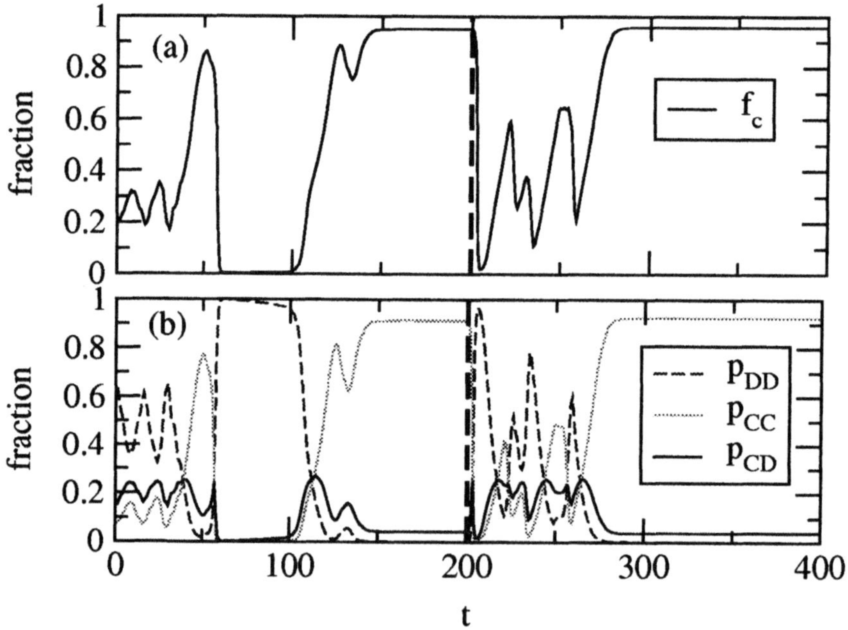


**Fig. 4.** Time series of $f_C$, where at $t = 200$ the leader agent changes strategy from C to D. Parameter values: $b = 1.75$, $K = 8$, $p = 0.05$.

[4] In fact a full defective network is reached for $b > b^{**}$, where the critical $b^{**}$ depends on the noise intensity.

## 5 Discussion

We have introduced a model of cooperation on an adaptive network, where cooperation is highly enhanced with respect to the situation of a fixed network. The network adaptation involves exclusively the D-agents, which in some sense are allowed to "search" for new neighbors, in the hope of finding C-agents to be exploited. Our study shows that this mechanism leads to a global cooperative state of the network. The asymptotic state reached by the system is a steady state in which the network structure and the average payoff per agent $\Pi$ remains stationary. However, most agents are unsatisfied, and continuously imitate the strategy of their neighbors with highest payoff (most of them C-agents). The structure of the network can be understood in terms of chains of cooperators with D-agents exploiting some cooperators.

The network adaptation that we have implemented in our spatial game gives rise to the emergence of a leader, a cooperator with maximum number of connections. Whenever the leader is the wealthiest agent in the network, and defectors are still "searching", the leader may increase its number of links with other cooperators. However, if some defector becomes wealthier than the leader, an unstable situation occurs and large global cascades may take place. Such cascades can also be induced if "noise" in the form of spontaneous change of the strategy is allowed.

We finally note that the dynamics described above cannot be explained in terms of average agents. In the final mostly cooperative state the average wealth of defectors is larger than the average cooperators wealth. The final collective state is dominated by rare individuals which build-up cooperation in the whole population because they have a number of links which is far from the average.

## Acknowledgments

MGZ acknowledges financial support from FOMEC-UBA. VME acknowledges financial support from the Danish Natural Research Council. MSM acknowledges financial support from the Spanish Ministerio de Ciencia y Tecnología project BFM2000-1108.

## References

1. D. Ashlock, M. D. Smucker, and L. Tesfatsion. Preferential partner selection in an evolutionary study of Prisoner's Dilemma. *BioSystems*, 37(1-2):99–125, 1996.
2. R. Axelrod. *The Evolution of Cooperation.* Basic Books, New York, 1984.
3. R. Axelrod and W. D. Hamilton. The evolution of cooperation. *Science*, 211:1390–1396, 1981.
4. M. Cohen, R. Riolo, and R. Axelrod. The emergence of social organization in the prisoner's dilemma: how context-preservation and other factors promote cooperation. Santa Fe Institute Working Paper 99-01-002, 1999.
5. S. Goyal and S. Joshi. Networks of collaboration in oligopoly. Mimeo, 1999.
6. B. A. Huberman and N. S. Glance. Evolutionary games and computer simulations. *Proc. Natl. Acad. Sci. USA*, 90:7716–7718, 1993.

7. M. O. Jackson and A. Watts. The evolution of social and economic networks. Vanderbilt University, Mimeo, 1999.
8. O. Kirchkamp. Spatial evolution of automata in the prisoners' dilemma. Discussion Paper B-330, Rheinische Friedrich Wilhelms Universität Bonn, Mimeo, 1995.
9. A. Kirman. Aggregate activity and economic organisation. *Revue Economique des sciences sociales*, 113:189–230, 1999.
10. K. Lindgren. Evolutionary dynamics in game-theoretic models. In Durlauf, Arthur and Lane, editors, *The Economy as an Evolving Complex System II*, volume XXVII, pages 337–367. SFI Studies in the Sciences of Complexity, 1997.
11. K. Lindgren and M. G. Nordahl. Evolutionary dynamics of spatial games. *Physica D*, 75:292–309, 1994.
12. A. Mukherji, V. Rajan, and J. R. Slagle. Robustness of cooperation. *Nature*, 379:125–126, 1996.
13. M. A. Nowak, S. Bonhoeffer, and R. M. May. Spatial games and the maintenance of cooperation. *Proc. Natl. Acad. Sci. USA*, 91:4877–4881, 1994.
14. M. A. Nowak and R. M. May. Evolutionary games and spatial chaos. *Nature*, 359:826–829, 1992.
15. M. A. Nowak and R. M. May. The spatial dilemmas of evolution. *Int. Jour. of Bif. and Chaos*, 3(1):35–78, 1993.
16. A. Watts. A dynamic model of network formation. Vanderbilt University, Mimeo, 1999.
17. D. J. Watts and S. H. Strogatz. Collective dynamics of small-world networks. *Nature*, 393:440–442, 1998.
18. J. Weibull. *Evolutionary Game Theory*. MIT University Press, 1996.
19. M. G. Zimmermann, V. M. Eguíluz, M. San Miguel, and A. Spadaro. Cooperation in an Adaptive Network. In Ballot and Weisbuch, editors, *Applications of Simulation to Social Sciences*, Hermes Science Publications (Paris, France), 2000.

# Part II.

# From Micro to Macro Behaviours

# Technology Diffusion, Stability and Decay: some Results

**J. Lizardo R. H. De ARAÚJO**

Instituto de Economia / Universidade Federal de Rio de Janeiro

**Abstract:** The proposition that history matters is not new in the economic debate, but growing recognition that technological change matters for economic growth has renewed and rephrased that proposition in more formal if narrower terms. Rosenberg, Paul David and Brian Arthur are names associated with path-dependence. Ebeling et alii used Markov processes as a random walk, and de Araújo used a birth-and-death model as a stochastic equivalent of a diffusion equation for finite markets. The present paper further explores this line, with new results. By examining similarities and differences between the author's results and those of others, we try to discern the implications of these assumptions for the results obtained on technology diffusion, and raise a few questions on the significance of stable equilibrium for analytic purposes. Finally, we discuss extensions to incorporate insights from other work aiming to arrive at a better understanding of the real processes of technology innovation and diffusion.

## Introduction

The proposition that history matters is not new in the economic debate. However, the growing recognition that technological change matters for economic growth has renewed and rephrased that proposition in more formal if narrower terms. As Rosenberg (1994) puts it, there is "a strong degree of path dependence, in the sense that one cannot demonstrate the direction or path in the growth of technological knowledge merely by reference to certain initial conditions". The concept of path-dependence in economic processes owes much to the work of Paul David, in both the eye-catching term "Economics of QWERTY" (Krugman, 1994) and its more precise definition as a non-ergodic stochastic process (Rosenberg, 1994).

Another name associated with path-dependence is that of Brian Arthur, for his stochastic modelling of diffusion processes [1], [2] as a random walk. Ebeling *et alii* [3], [8] also used Markov processes as a random walk, and de Araújo [5], [6] used a birth-and-death model as a stochastic equivalent of a diffusion equation for finite markets.

The present paper further explores this line, with new results. By examining similarities and differences between the author's results and those of others, we try to discern the implications of these assumptions for the results obtained on technology diffusion, and raise a few questions on the significance of stable equilibrium for analytic purposes. Finally, we discuss extensions to incorporate insights from other work aiming to arrive at a better understanding of the real processes of technology innovation and diffusion.

In a previous article [6] the author introduced a birth-and-death model for the competitive diffusion of two technologies in a finite market, with three variants: pure birth, birth and death without renewal, and birth and death with renewal. In the pure birth case, the process follows a pair of noisy logistic curves to randomly distributed final states. The state distribution depends on market size and on the relative diffusion rates, *as well as on the initial state*: the system exhibits path dependence. Moreover, experiments suggest that the market share needed to introduce a bias against a more attractive technology ***decreases*** with increasing market size.

Introducing 'deaths', that is to say consumer regrets, radically changes behaviour: the state (0,0), extinction for both technologies, becomes absorbing and all other states become transient. Despite this radical change, process realisations follow approximately the pure birth case until market saturation is approached, and exhibit a degree of stability around the saturated state reached. The combination of non-ergodicity with very slow decay towards the absorbing state suggests that transient behaviour is relevant.

By introducing a renewal to the initial state upon extinction all states become recurrent, the embedded chain is irreducible and one can calculate limit probabilities. These concentrate on states with only one technology (the lock-in effect found by Arthur [1], [2]), the probability of other states being positive but negligible. Also, the probability distribution of "lock-in" states depends both on diffusion rates and on the initial (renewal) state. These conclusions are robust regarding the shape of birth and death rates, even with diseconomies of adoption.

This paper examines in further detail these dynamics. The results not only confirm the existence of lock-in effects and state dependence. They also point to the fact that decay away from a saturated state is a complex phenomenon involving very large numbers of transitions and decay times, with both a substitution phase along near-saturated states and a bona fide decay phase after one technology becomes extinct.

These simulations pose an interesting problem for economic analysis and policy: Suppose a state is transient, but decay away from it is very slow compared with the time taken to reach this state. May not this state be considered as "stable" for practical purposes, specially if the likelihood of attaining such a state is non-negligible? In other words, should stability be considered in regard to a finite rather than an infinite time horizon? Since most agents take their decisions based on finite time frames, this is a pertinent issue.

In section 1 new results from the author's former work are presented and discussed with regard to those by Arthur and Ebeling. Section 2 analises the dynamics of the process without renewal showing that saturated and "lock-in" states are long-lived though transient, and discusses the implications of this fact. Section 3 discusses shortcomings of the approach and suggests extensions to improve its descriptive and predictive power.

# 1. Former work, new results

In an earlier paper (de Araújo, 1995) the author introduced a birth and death model for the competitive diffusion of two technologies in a finite market. This model has three variants: pure birth, birth and death without renewal, and birth and death with renewal. In the pure birth case the process follows a pair of noisy logistic curves attaining random final states. The distribution of these states depends on the market size and on the relative rates of diffusion, *as well as on the initial state*: the system exhibits path dependence. Moreover, experiments suggest that the market share necessary to introduce a bias against a more attractive technology ***decreases*** with increasing market size.

The complete model is as follows: if *two* technologies, **1** and **2**, are introduced at time 0 with diffusion coefficients $\alpha_1$ and $\alpha_2$, we have a two-dimensional process. This may be noted as $\{(N_1(t), N_2(t)), t \geq 0\}$ , $N_1(0) = 1$, $N_2(0) = b$, where transitions in $N_1(t)$ and $N_2(t)$ are independent but for the 'birth rates' $\lambda_1$ and $\lambda_2$ and 'death rates' $\theta_1$ and $\theta_2$. These are linked through the equation

Equation A
$$\begin{cases} \text{birth rate } \lambda_i = \alpha_i \cdot (n_i(t))^r \cdot (M - n_1(t) - n_2(t)), i = 1,2 \\ \text{death rate } \theta_i = \mu_i(n_1(t), n_2(t), M), i = 1,2 \end{cases}$$

This may be interpreted as follows.

During the interval $t, t + \Delta t)$ each remaining potential adopter has three options: wait and see, choose 1, choose 2. If $\Delta t$ is small enough, Pr{choose $i$} $\cong \alpha_i \cdot (N_i(t))^\tau \cdot \Delta t$, $i = 1,2$, where $\tau$ is a parameter that controls whether the probability of adoption is linear in the number of adopters ($\tau = 1$), concave ($0 < \tau < 1$), insensitive ($\tau = 0$) or decreasing ($\tau < 0$). Each agent chooses independently, and the probability of two simultaneous choices is zero. Finally, if at time $t$ technology $i$ has $n_i(t)$ adopters, the probability that an adopter will regret it and join the "potential adopters" set during interval $[t, t+h)$ tends to $\mu_i(n_1(t), n_2(t), M) \cdot h$ when $h \to 0^+$. In other words, the instantaneous death rate of each technology depends on the number of adopters for both technologies and on market size. Deaths follow the usual restriction that simultaneous events have zero probability. We also assume that $\theta_i$ is positive for all $n_i > 0$.

Introducing 'deaths', that is to say consumer regrets, radically changes behaviour: the state (0,0), extinction for both technologies, becomes absorbing and all other states become transient. Despite this radical change, process realisations follow approximately the pure birth case until market saturation is approached, and exhibit a degree of stability around the saturated state reached. The combination of non-ergodicity with very slow decay towards the absorbing state suggests that transient behaviour is relevant. Nevertheless, literature on transient behaviour in non-ergodic Markov processes is scant.

By introducing a renewal to the initial state upon extinction, all states become recurrent; the embedded chain is irreducible and it is possible to calculate limit probabilities. These concentrate on states with only one technology present (the lock-in effect found by Arthur [1], [2]), the probability of other states being positive but negligible. Furthermore, the probability distribution of "lock-in" states depends both on diffusion rates and on the initial (renewal) state. The above conclusions proved robust regarding the shape of birth and death rates, even with diseconomies of adoption.

Work with related motivation has been developed by Ebeling, Jiménez Montaño and others (1980, 1994). They use a random walk in which the sum of adopters in both technologies remains invariant; in other words, they are concerned with a substitution process in a finite market. They thus focus on the survival probability for a new technology competing against an established one. Survival probability is a function of market size, diffusion rates ratio between both technologies and the initial number of adopters of the new technology. Even though transition probabilities used in their model are not directly comparable with ours[1], their results are noteworthy. First, survival probability for a new technology decreases sharply with market size, i.e. with the established technology base. Second, survival grows only moderately with higher diffusion coefficients for the new technology. This prompts Ebeling *et alii* to speculate on the role of market niches as incubators for new technologies. They correctly note that the association of firms for development and use of a new technology acts as the creation of a relatively small protected niche. Once a technology dominates this niche, its chances for success in larger markets are enhanced.

The sum-up above raises a number of questions. First, what deductions may one draw from a comparison between the author's results and those by Arthur and Ebeling? Second, what can we learn about transient behaviour of competitive birth and death processes without renewal? We discuss the first point below. The second question will be viewed in the next section. The final section attempts an evaluation of the potential, implications and limits of the approach.

Several relevant differences exist between Arthur's hypotheses and results compared to ours. First, as noted in (de Araújo, 1995), Arthur uses an infinite stream of agents against a finite market in our case. Second, Arthur's agents are differentiated into two types (favouring technology A or B), while our 'potential adopters' crowd (which includes agents regretting former choices) is undifferentiated. Third, Arthur's lottery takes a random agent between the two types; this agent makes a choice according to his (or her) utility function, which takes into account adoptions already made. Against this, the lottery of the birth-and-death process chooses between adoptions or regrets, as well as for which technology; transition probabilities are a function of adoptions already made and of the number of potential adopters. Finally, Arthur postulates absorbing barriers when the difference between A and B adoptions reaches a predetermined threshold; in a birth and death model, extinction is the natural absorbing barrier.

The most relevant difference is that between a finite and an infinite market. The introduction of renewal in the birth and death model, which is akin to having an infinite stream of agents, leads to 'lock-in' effects whenever more adoption leads to

---

[1] In our model, substitution between saturated states leads to transition probabilities approximately of the form $\alpha_1\mu_2/(\alpha_1\mu_2+\alpha_2\mu_1)$ and $\alpha_2\mu_1/(\alpha_1\mu_2+\alpha_2\mu_1)$ from state $(n, M-n)$ to states $(n+1, M-n-1)$ and $(n-1, M-n+1)$ respectively. Ebeling *et al.*, on the other hand, use the forms $k\lambda_1 n^2(M-n)$ and $k\lambda_2 n(M-n)^2$ for the same transition probabilities, to stress acceleration and hyperselection. In other words, their model has "lock-in" effects built into its structure.

higher adoption probability. This is qualitatively similar to utility curves presenting 'economies of adoption' in Arthur. One should also note that dividing the market into two types of agents, as in Arthur, would not change the main results for the birth and death model (de Araújo, 1995) while complicating the treatment.

All resemblance ends, however, when 'diseconomies of adoption' are considered in a finite market. Contrasting with Arthur's results, which show convergence with probability one to an intermediate state, in a finite market results differ in degree only from those with economies of adoption (de Araújo, 1997)[2]. Figures 1 and 2, as well as Table 1, show that even with diseconomies of adoption the system tends to lock-in states and the initial (renewal) state is relevant. Also, if a technology (e.g. 1) has a larger diffusion coefficient, limit probabilities will be clustered near the state ($M$, 0) where $M$ is the size of the market. The net effect of diseconomies of adoption in birth and death processes with renewal is to increase variance, decrease the expected value of adoption for each technology and to slightly increase survival chances, in some cases, for the technology with smaller diffusion rate.

Figure 1

Limit probabilities for M=10, mu(i)=0.02,alpha(1)=0.04=2*alpha(2), tau= -1, renewal to (1,5)

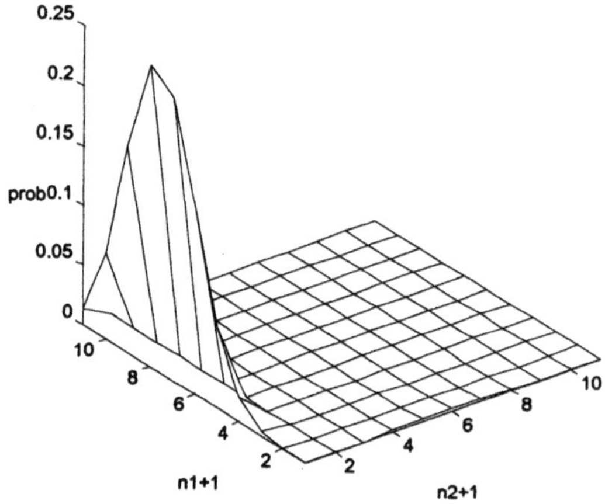


[2] A discontinuity found in (de Araújo, 1995) was due to the computer code used, which did not handle correctly diseconomies of adoption ($\tau \leq 0$). After correction, the discontinuity disappeared.

Figure 2

Limit probabilities for M=10, mu(i)=0.02,alpha(1)=0.04=2*alpha(2), tau= +1, renewal to (1,5)

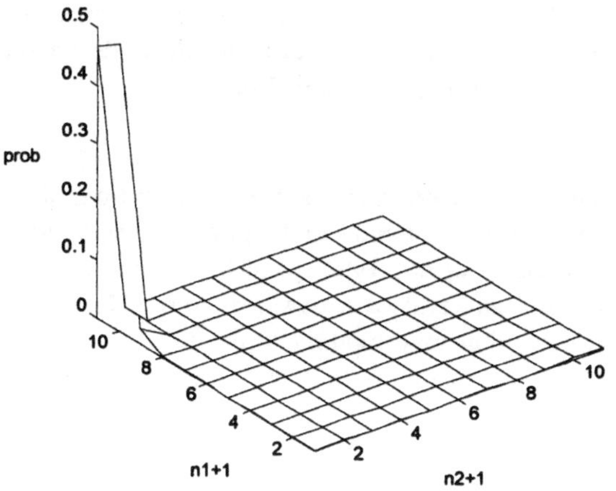


Table 1- Limit statistics for M=*10*, μ(i)=*0.02*, $\alpha_2$ = *0.02*, renewal to (*1*,$b_2$)

| $\alpha_1$ | $\tau$ | $b_2$ | $P(n_1{*}n_2>0)$ | $P(n_1>n_2)/P(n_2>n_1)$ | $P(n_2>0)$ | $E(n_1)$ | $E(n_2)$ |
|---|---|---|---|---|---|---|---|
| 0.02 | -1 | 1 | 0.0447 | 1.00 | 0.5220 | 2.5403 | 2.5403 |
| 0.02 | -1 | 5 | 0.0427 | 0.45 | 0.7043 | 1.6173 | 3.4789 |
| 0.02 | 0 | 1 | 1.9E-05 | 1.00 | 0.5000 | 4.2500 | 4.2500 |
| 0.02 | 0 | 5 | 1.6E-05 | 0.26 | 0.7955 | 1.7382 | 6.7618 |
| 0.02 | 1 | 1 | 6.1E-11 | 1.00 | 0.5000 | 4.6936 | 4.6936 |
| 0.02 | 1 | 5 | 3.5E-11 | 0.03 | 0.9688 | 0.2934 | 9.0939 |
| 0.04 | -1 | 1 | 0.0016 | 242.79 | 0.0054 | 6.4722 | 0.0236 |
| 0.04 | -1 | 5 | 0.0021 | 104.31 | 0.0109 | 6.4380 | 0.0515 |
| 0.04 | 0 | 1 | 5.5E-08 | 2677.00 | 0.00037 | 8.9966 | 0.0032 |
| 0.04 | 0 | 5 | 9.6E-08 | 631.15 | 0.0016 | 8.9858 | 0.0134 |
| 0.04 | 1 | 1 | 1.4E-13 | 2392.50 | 0.00042 | 9.4401 | 0.0039 |
| 0.04 | 1 | 5 | 5.8E-13 | 120.61 | 0.0082 | 9.3664 | 0.0072 |

Observe Figures 1 e 2, and Table 1: for $\alpha_1 = 0.04$ and $b_2 = 5$, one has $E(n_2) = 0.0515$ if $\tau = -1$, 0.0134 if $\tau=0$ e 0.0072 if $\tau = 1$ — in the figures $\tau$ is noted *tau*. For other parameter combinations behaviour appears paradoxical at first sight; e.g., when $b_2 = 1$ and $\beta = \alpha_1/\alpha_2 = 2$, $E(n_2)$ is smaller for $\tau=0$ than for $\tau=1$. Also, $P(n_2)$ for $\alpha_1 = 0.04$ and $b2 = 5$ is much less if $\tau=0$ than if $\tau=-1$ or $\tau=1$. This is due to the fact that a higher $\tau$ enhances birth over death rates, and also differences in diffusion rates, reducing the influence of the renewal state. The net result depends on the combination of $\tau$, birth and death coefficients and the renewal state.

As to Ebeling *et al.*, they are not concerned with diseconomies of adoption, since their model is purposefully designed to show *economies* of adoption as discussed in footnote 4 and acknowledged in (Bruckner *et al.*, 1994, page 86). They do raise relevant themes on the introduction and survival of a new technology facing an established one. However, it is not clear that investigating limit probabilities of a process with renewal, as they do, is adequate to that end. The next section investigates an alternative approach.

## 2. On the persistence of some transient states

The simulations shown in (de Araújo, 1995, 1997) pose an interesting problem for economic analysis and policy: if a state is transient, and therefore ***not*** a stable equilibrium, but decay from it is slow compared with the time taken to reach it, may not this state be considered as 'stable' for practical purposes, especially if the probability of attaining such a state is far from negligible? In other words, should stability be considered in terms of a finite rather than infinite time horizon? Since most agents make decisions based on finite time horizons, this is a pertinent issue and has far ranging theoretical implications.

This question raises another one: since (0,0) is an absorbing state, how does the state probability distribution evolve from a saturated state on the way to extinction? Is it possible to go beyond mere simulations and make probabilistic statements regarding transient behaviour? In the remainder of this section we shall see that it is effectively possible to obtain results on the state probability distribution for the decay period from an initial saturated state.

The propagation of state probability distributions of the market given the initial state $i_0$ follows the usual pattern. Given the transition matrix A and an initial probability vector $\pi_0 = [\delta(i, i_0)]$ where $\delta(\bullet,\bullet)$ is Dirac's delta, after $t$ transitions we have $\pi_t = \pi_{t-1}A = \pi_0 A^t$. in our case, a state is a pair ($n_1$, $n_2$) subject to $n_i \geq 0$, $i$ = 1,2 and $n_1 + n_2 \leq M$ (market size). A market size $M$ has thus $(M+1)(M+2)/2$ states.

Several measures may be proposed to analyse the evolution of $\pi_t$. One is the expected number of adopters for each technology, $E[n_i(t)]$, $i$ = 1,2. The information this gives is limited by the existence of 'lock-in' effects (the distribution tends to bimodality), but is relevant to analyse the dynamics of extinction. Another, complementary one is the probability of extinction for each technology, $P(n_i(t) = 0)$, $i$ = 1,2.

Finally, the state probability distribution itself

$\pi^t = \left[\pi_{n_1 n_2}(t)\right] = \left[\Pr\left(n_1(t) = n_1, n_2(t) = n_2 \middle| n_1(0), n_2(0)\right)\right]$ ($n_1 \geq 0$, $n_2 \geq 0$, $n_1+n_2 \leq 0$)

may be portrayed. Since states are two dimensional, one can either look at the distribution after a number of transitions or probe the evolution of $\Pr\left(d_M\left((n_1,n_2),(n_1(0),n_2(0))\right),t\right)$, where $d_M\left((n_1,n_2),(n_1(0),n_2(0))\right) = |n_1 - n_1(0)| + |n_2 - n_2(0)|$ is the Manhattan (or block) distance between states ($n_1$, $n_2$) and ($n_1(0)$, $n_2(0)$).

Analysis of these measures for several market sizes confirms the conjectures raised by de Araújo (1995) on the apparent stability of saturated or nearly saturated states, and also sheds more light on the dynamics of competitive diffusion. In what follows, we present some computer-generated results. Unless otherwise specified, $\tau$ = 1 in all of them.

One should note that decay goes through two distinct phases. In the first one the probability distribution concentrates on states saturating the market (or nearly so), with both technologies present. Decay in this phase is relatively fast, especially when both technologies start from equal shares and have unequal diffusion coefficients; but even in this case a number of transition several times larger than the market size is necessary in order to observe a noticeable displacement. After this, distribution concentrates on states with only one technology present, showing 'lock-in' and bimodality. Decay to extinction is very, very slow in this latter phase.

Figures 3 and 4 show state probabilities in a market size 30, after 200 transitions from the initial saturated state $n_1 = 15 = n_2$. In both cases $\mu_1 = \mu_2 = 0.01$, and $\alpha_2 = 0.02$; if we note $\beta = \alpha_1/\alpha_2$, Figure 3 pictures the case $\beta = 1$ and Figure 4 case $\beta = 2$. If the first case is extended to 500 transitions we obtain Figure 5, which marks the beginning of the second phase.

Figure 3

State probabilities for M=30 after 200 transitions from (15,15)

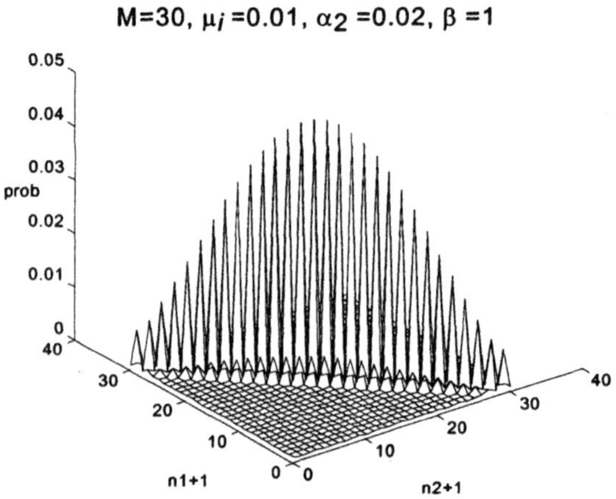


Figure 4

State probabilities after 200 transitions from initial state (15,15)

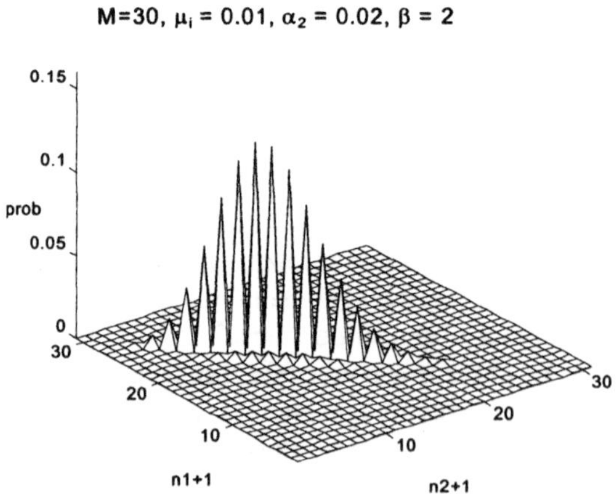

Figure 5

State probabilities after 500 transitions from initial state (15,15)

M=30, $\mu_i$ = 0.01, $\alpha_2$ = 0.02, $\beta$ =1

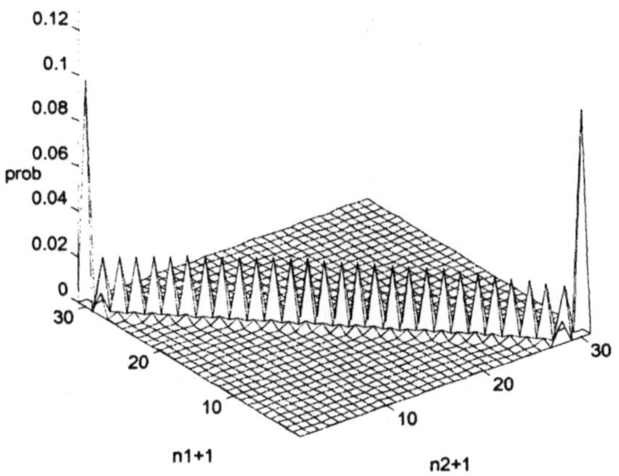


During the first phase E[$n_1$] remains constant if $\beta = 1$ or increases if $\beta >$ 1, while E[$n_1$+$n_2$] remains approximately constant; in the second phase both E[$n_1$] and E[$n$=2] appear nearly constant, although in fact they decay very slowly.

Figure 6

Evolution of E[$n_1$] from initial state (2,8)
M = 10, $\mu_i$ = 0.01, $\alpha_2$ = 0.02, $\beta$ = 2

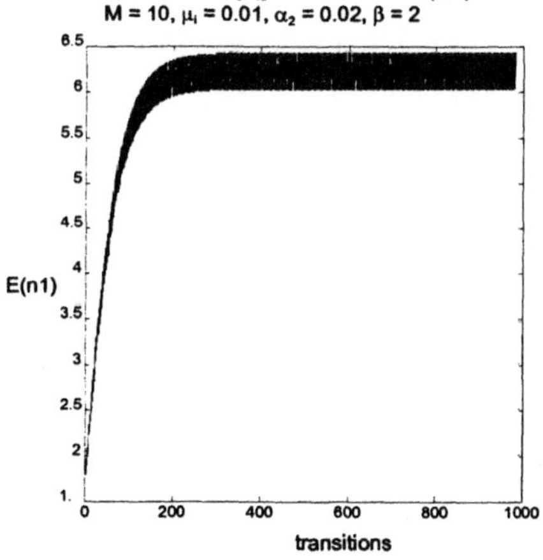

Figure 7

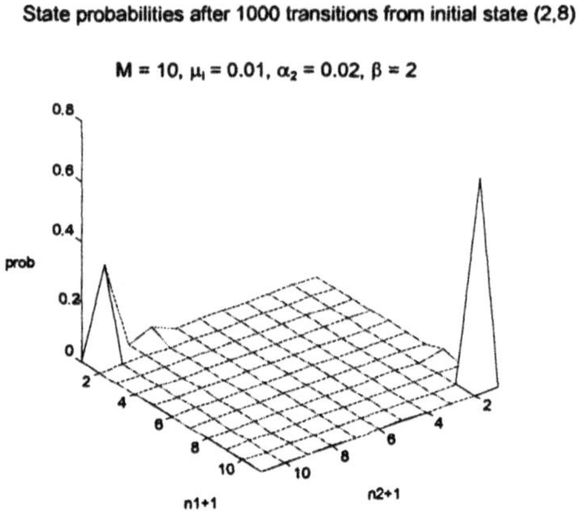


Figure 6 shows the evolution of E[$n_1$] in a market size 10 for 1,000 transitions from initial state (2,8) with the same parameters as above and $\beta = 2$, and Figure 7 shows the probability distribution after 1,000 transitions for the same case. Observe that in the second phase the distribution has two modes, and decay is directly linked to death rates; if these are reduced by an order of magnitude, the secondary peaks on axes ($n_1$, 0) and (0, $n_2$) in Figure 7 almost disappear. Death rates have in contrast a secondary role in the first phase, affecting only time between transitions: reducing $\mu_i$ a hundred times to 0.0001 in the cases of Figures 3 and 4 gives practically the same results, although the expected time to complete 200 transitions increases significantly.

Figure 8

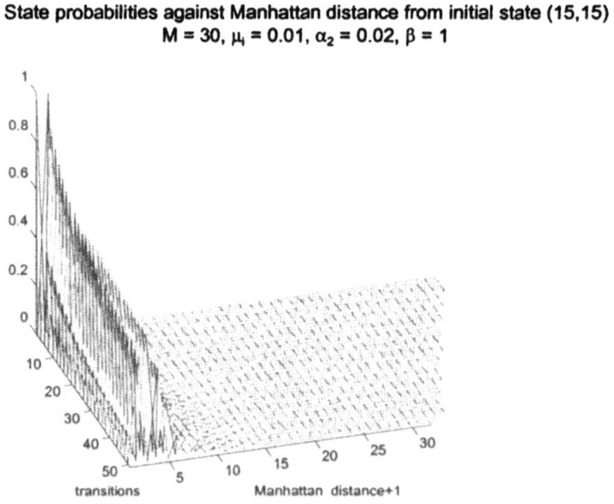

Another way of viewing the evolution of the probability distribution is through a graph of the probability distribution against the Manhattan distance from the initial state. Figures 8 and 9 show the evolution of two cases with the same parameters and initial conditions as Figures 3 and 4. Note that distribution functions are staggered in the transition axis, since return to a state requires an even number of transitions.

Figure 9

State probabilities against Manhattan distance from initial state (15,15)

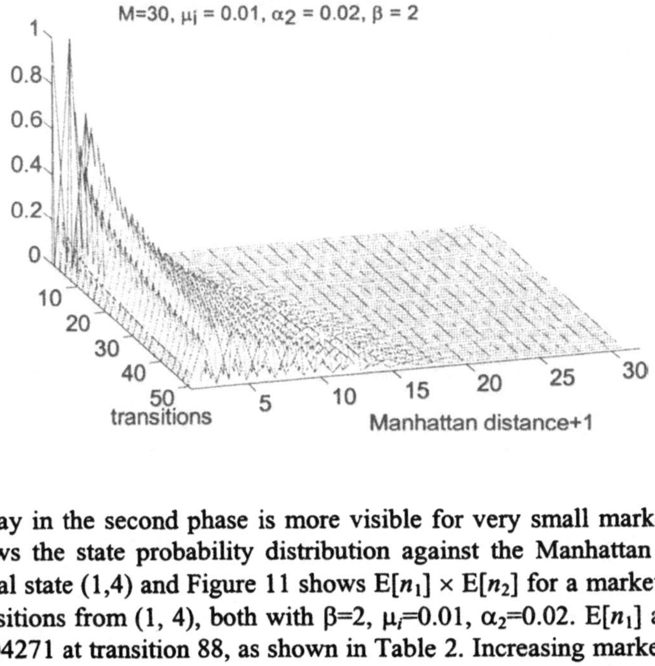


Decay in the second phase is more visible for very small market sizes. Figure 10 shows the state probability distribution against the Manhattan distance from the initial state (1,4) and Figure 11 shows $E[n_1] \times E[n_2]$ for a market size 5 during 500 transitions from (1, 4), both with $\beta$=2, $\mu_i$=0.01, $\alpha_2$=0.02. $E[n_1]$ attains a maximum 1,704271 at transition 88, as shown in Table 2. Increasing market size by 40% to 7 displaces the point of maximum approximately by a factor of four, as Table 3 shows.

Table 2 - Maximum of $E[n_1]$ for M=5, beta=2

| **Initial state** | (1,4) | (2,3) | (3,2) | (4,1) |
|---|---|---|---|---|
| **Max $E[n_1(t)]$** | 1.704271 | 2.888178 | 3.685168 | 4.205636 |
| **Transition *t*** | 88 | 84 | 80 | 70 |

Table 3 - Maximum E[$n_1$] for M=7, beta=2

| **Initial state** | (1,6) | (2,5) | (3,4) | (4,3) | (5,2) | (6,1) |
|---|---|---|---|---|---|---|
| **Max E[$n_1(t)$]** | 2.615056 | 4.259078 | 5.268505 | 5.874491 | 6.230764 | 6.436337 |
| **Transition *t*** | 334 | 330 | 324 | 316 | 306 | 286 |

Figure 10 *State probability distributions against transitions from initial state (1,4) for $M = 5$, $\mu_i = 0.01$, $\alpha_2 = 0.02$, $\beta = 2$*

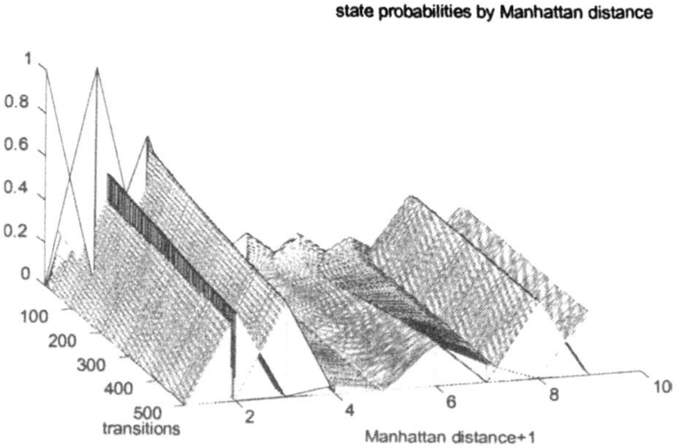


Figure 11 – E[$n_1$] × E[$n_2$] for 500 transitions from [1, 4] with $\alpha_1 = 2\,\alpha_2$

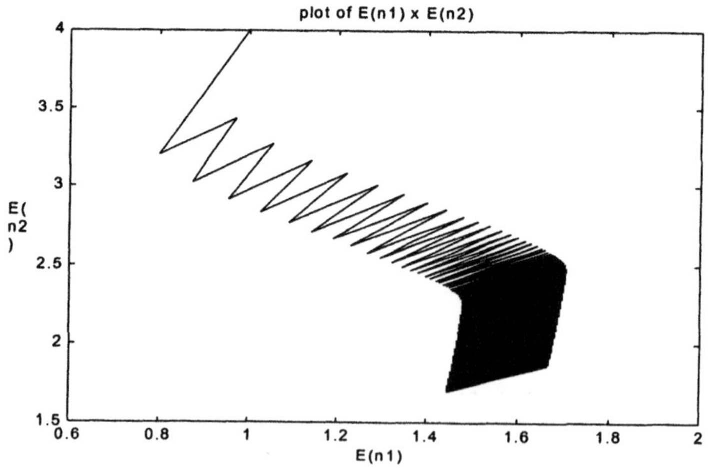

These two tables also confirm that the evolution of state probability distributions, and thus the probability of 'lock-in' into one technology in phase two, strongly depends on the initial state. A comparison of Figure 9 with Figure 12 (which portrays the evolution from state (1,29), all other parameters identical to those of Figure 9) and of Figure 4 with Figure 13 shows this dependence very graphically.

These results not only confirm the existence of 'lock-in' effects and path dependence. They also point out the fact that decay from a saturated state is a complex phenomenon involving large numbers of transitions and long decay times, with a first phase of substitution among quasi-saturated states and an effective decay phase after one technology becomes extinct.

Moreover, times involved in the processes of substitution and decay are large compared with the diffusion time to arrive at a saturated state. Evolution is faster in the substitution phase in saturated and near-saturated states, with transition rate $\nu = \mu_1 n_1 + \mu_2 (M-n_1)$ ($= \mu M$ if death rates are the same) for saturated states and $\nu = (\alpha_1+\mu_1)n_1 + (\alpha_2+\mu_2)(M-n_1)$ for states one transition away from saturation. If birth rates are equal for both technologies, and death rates likewise, $\nu = (\alpha + \mu)M$ in this latter case. This is **half** the transition rate at state (1,1), which means that the average time between transitions (i.e. $1/\nu$) along these states is double that at the beginning stages of diffusion, and much longer (by a factor of the order M/4) than that at middle stages. An idea of the periods involved may be given by a simple example: the average time <u>between</u> transitions for near-saturated states is roughly 10% the average time to go from 1 to 99% of saturation in a pure birth process[3].

For the decay phase, when a technology (e.g. technology 2) is extinct then we have a one-dimensional birth-and-death process with much larger birth rates than death rates (assuming $\alpha_i > \mu_i$). Thus, for $n_1 = M - k$, the probability of increasing to $M-k+1$ is $\alpha_1 \cdot k \cdot (M-k)$ while that of decreasing to $M-k-1$ is $\mu_1 \cdot (M-k)$. The birth/death probability ratio is $\alpha_1 \cdot k/\mu_1 > 1$ for all states in this phase, save boundaries, and ***increases linearly with distance from saturation***. This dominance of births over deaths leads to a high persistence for the surviving technology, especially for large markets, though it will eventually become extinct with probability one.

In other words, for analytical and decision-making purposes it is very difficult to distinguish empirically between such a transient but persistent state and a (quasi-

[3] The time from 1% to 99% of saturation in a logistic diffusion is $9.19/(\alpha M)$, and that from 0.1% to 99.9% is $13.81/(\alpha M)$. By comparison, average time between transitions in the substitution phase is $1/[M(\alpha+\mu)]$.

)stable equilibrium that moves with a changing world. One implication of this fact is that analyses should focus on decision-making ***processes*** rather than on compared properties of static equilibria, since there is no way of knowing *a priori* if these will be attained, and *ex post* analysis introduces biases.

Figure 12

*Evolution of state probabilities against Manhattan distance from initial state (1,29)*
*$M = 30$, $\mu_I = 0.01$, $\alpha_2 = 0.02$, $\beta = 2$*

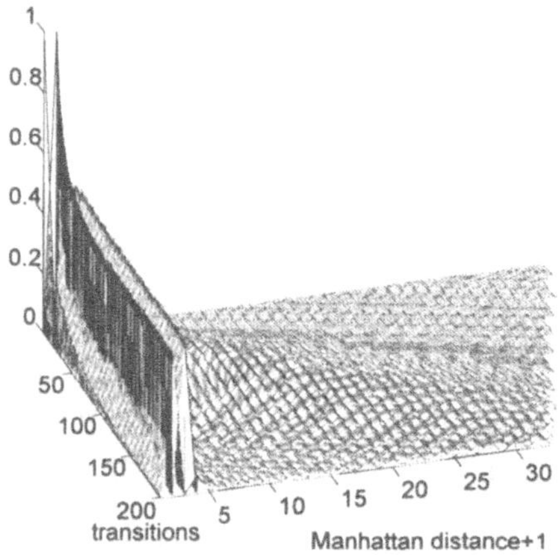


Figure 13

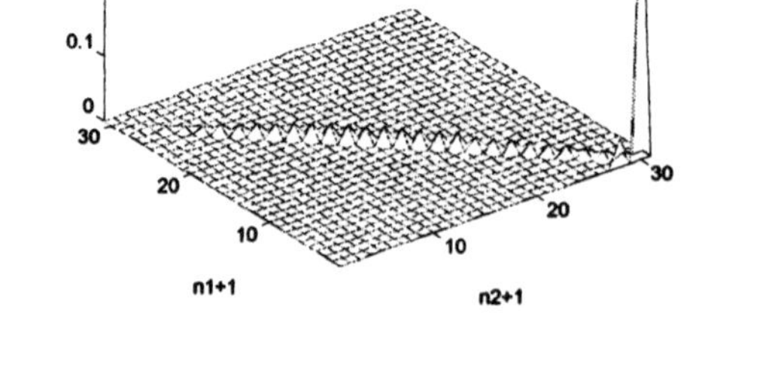

# 3. Extensions and implications

The above results give a three-phase view of the competitive diffusion process as a whole. In the first phase, two technologies with different initial endowments grow according to two noisy logistic curves, reaching a saturated state as described in (de Araújo, 1995). A substitution phase follows, until the market is locked in one technology; the probability for a given lock-in depends both on birth-and-death rates and on the saturated state reached in the first phase. Finally, the surviving technology exhibits a high persistence with low decay rates and approximately mimics a stable equilibrium. The persistence grows with market size.

This description does not depend on assumptions relative to economies or diseconomies of adoption. That is to say, lock-in behaviour appears to be intrinsic to the birth-and-death competitive diffusion model. As already mentioned, the existence of lock-in displaces the focus of analysis from static properties towards process dynamics. What is attractive about this model is its economy. Growth, competition, lock-in, near stability and decay, with path dependence, all come out of what is formally a very simple specification.

On the other hand, its limitations are equally clear. Chief among these is the fact that it offers no explanation of how innovations come about. For this, other families of models would have to be used, incorporating risk and interactions. Other limitations are easier to circumvent. For instance, the assumption of equal agents may be easily dropped without modifying the overall conclusions, as shown in (de Araújo, 1997). Indeed, when agents are unequal in size and in preferences, the model is readily extended as follows:

Formally, there are $N$ agents; agent $j$ controls market share $m_j$, and $m_1 + \cdots + m_N = M$. Moreover, let $I_i(t) \subseteq \{1, \ldots, N\}$, $i = 1,2$ e $I_1(t) \cap I_2(t) = \varnothing$ the sets of agents who have adopted technology $i$ at time $t$. Let also $M_i(t) = \sum_{j \in Ii(t)} m_j$ the market taken by technology $i$ at $t$. If we note $\{1, \ldots, N\} = EM$, the birth-and-death model may be rewritten as a ***many***-dimensional Markov process as follows.

(Births) If at time $t$ agent $j \in EM - I_1(t) - I_2(t)$, the probability he (she) will adopt technology $i$ in the interval $(t, t + h)$ is $\lambda_{ij} M_i^{\tau}(t)\, h + o(h)$, and Pr[2 or more adoptions in $(t, t + h)$] $= o(h)$. Here, $\tau$ is the parameter of economies or diseconomies of adoption; $\lambda_{ij}$ may vary among agents (one may have for instance $\lambda_{ij} = \alpha_{ij}\, m_j$, in which case larger agents tend to choose earlier and agents may be biased towards one technology). The probability for an increase in $M_i(t)$ during interval $(t, t+h)$ is then $M_i^{\tau}(t)\ \Sigma\{\lambda_{ij}|j \in EM - I_1(t) - I_2(t)\} + o(h)$. This reduces to $\alpha_i\ M_i^{\tau}(t)\ [M-M_1(t)-M_2(t)] + o(h)$ if $\lambda_{ij} = \alpha_i\ m_j$, as in the case of equal agents. Note only that now increases will be of unequal size.

(Deaths) If at time $t$ agent $j \in I_i(t)$, the probability he (she) will drop technology $i$ in the interval $(t, t+h)$ is $\mu_{ij}(m_j, M_i(t))h + o(h)$. If $\mu_{ij} = \mu_i\ m_j$, the model also reduces to a direct extension from the simple case of equal agents.

It is easy to see that the main conclusions taken from the basic model are valid for this extension, as long as all birth and death rates are positive. The main difference is the complexity that paths may take. In effect, the modified process has interesting features. First of all, path dependence and system complexity are ***enhanced*** since there are now $3^N$ states, instead of $(N+1)(N+2)/2$ for equal agents. On the other hand, coalitions of a small number of large agents are much more likely to occur than coalitions of a large number of small agents. *This suggests that an innovation is far more likely to succeed if a coalition of large actors (inter-firm linkages) endorses it.*

This leads to an interesting argument. Consider inter-sector linkages; in this case several markets exchange their products. If a coalition of large actors forms across sectors around an innovation affecting their markets, survival chances for that innovation are enhanced. This agrees with findings of empirical surveys such as those described in De Bresson *et al.* (1994), which noted a tendency of innovations to cluster around "bunches" linking firms and sectors.

Moreover, if such a coalition forms in a *new* market (i.e., one in which adoptions to date are a small fraction of the potential market) this may give an innovation a decisive advantage over a rival technology. Two examples come to mind: One is the launching of IBM PC, which effectively displaced the Apple II by the sheer weight of the Big Blue[4] (the Mac was a belated answer by Apple, which managed to carve a niche but never really threatened the PC standard). The other example was the struggle Betamax *versus* VHS, in which the coalition formed around the

---

[4] It is well to remember that when the IBM PC was launched, the former industry standard (Intel 8080 + CP/M) was being eroded by the upstart Apple. The points stressed in reviews of specialized magazines at the time were both the good points of the new machine ***and*** the weight of IBM. The launching of the Mackintosh (generally viewed as the better machine) by Apple came too late to turn the tables.

latter enhanced its chances and eventually gave it the victory, despite general consensus on the superior technical qualities of Betamax.

Another limitation of the model is the implicit assumption, by the standard birth-and-death process model, that information exchange is full and instantaneous. This makes speed of diffusion proportional to the size of potential market, so that for very large markets the time lag between innovation and saturation is almost null (Cf. footnote 3). A more reasonable specification may be easily made, however, by imposing a minimum lag between transitions and thus an upper bound to diffusion rates. This may be justified on psychological and economic grounds, and does not affect transition sequences.

# References

**Arthur, W.B.** (1989), Competing Technologies, Increasing Returns, and Lock-In by Historical Events, The Economic Journal, 99 , pp 116-131

—, **Ermoliev, Yu. and Kaniovski, Yu.** (1987), Path-Dependent Processes and the Emergence of Macro-Structure; European Journal of Operational Research, 30, pp 294-303

**Bruckner, E., Ebeling, W., Jiménez Montaño, M. A., & Scharnhorst, A. (1994),** Hyperselection and Innovation Described by a Stochastic Model of Technological Evolution, in **Evolutionary Economics and Chaos Theory**, L. Leydesdorff and P. Van den Besselaar, eds., Pinter, London.

**Cameron, H.M. & Metcalfe, J.S.** (1987), On the Economics of Technological Substitution, Technological Forecasting and Social Change, 32, pp 147-162

**De Araújo, J. L.** (1997), "Difusão de tecnologias e equilíbrios dependentes de trajetórias", Revista de Economia Contemporânea **1**(3).

**De Araújo, J. L.** (1995), "Are technology diffusion processes inherently historical?", Technological Forecasting and Social Change, April

**DeBresson, C.; Sirilli, G.; Hu, X.; and Kwan Luk, F. (1994),** "Structure and location of innovative activity in the Italian economy, 1981-85", Economic Systems Research **6**(2)

**Jiménez Montaño, M. A., and Ebeling, W.** (1980), A Stochastic Model of Technological Change, Collective Phenomena 3, 107-114

**Krugman, P.** (1994), **Peddling Prosperity: Economic Sense and Nonsense in the Age of Diminished Expectations**, W. W. Norton and Company, New York and London

**Rosenberg**, N. (1994), **Exploring the black box: Technology, economics and history**, Cambridge University Press, Cambridge UK

**Ross, S. M. (1983), Stochastic Processes**, John Wiley & Sons, New York

# On Dynamic Re-Specifications of Kiyotaki-Wright Model

Masanao Aoki*[1]

Department of Economics, University of California, Los Angeles, Fax Number 310-825-9528, e-mail aoki@econ.ucla.edu

**Abstract.** This paper examines the well-known Kiyotaki-Wright model for its dynamic specification. Because the authors are silent on the exact dynamic structure of their model, and on their model's behavior off the equilibrium states, the model dynamics can be introduced in several ways. One such interpretation focuses on the fact that the number of money traders on and off equilibria is fixed exogenously. One of our modifications introduces a possibility for the number of money traders also to evolve with time.

## 1 Introduction

The model of Kiyotaki and Wright (1993), henceforth the K-W model, deals only with equilibrium states, and pays no attention to dynamics of the model out of the equilibria. The model does not describe how the equilibria are supported, for example. Because of the interest generated by their model in the economic profession, we find it important to examine the dynamic behavior of their model off the equilibrium states.

We explicitly model the processes of exchanges by commodity traders and money traders as a jump Markov process, thereby introducing dynamics explicitly into the K-W model, as we did with the Diamond model in Aoki and Shirai (2000). Both the Diamond model and the K-W model share the feature that some parameters in the transition rates for the jump Markov processes are endogenously determined via value maximization processes by the agents in the model.[1] This is a feature that is absent in the discussion of jump Markov processes in probability textbooks. In this sense, our uses of jump Markov processes for economic dynamics introduce a new element to this field of search-theoretic models. Aoki (1996, 2001) have further examples of applcations of jump Markov process models to economic models.

Aoki and Shirai have used a finitary approach to re-examine the well-known search model by Diamond, Diamond (1982). That is, they start with a large, but a finite collection of agents, and describe interactions of agents in

* Prepared for the Wehia 2000 conference. The author wishes to thank Y. Shirai for his useful comments.

[1] In the K-W model they are Eq. (1) through (3).

the collection by specifying a set of transition rates between different types of agents. They then analyzed stationary and fluctuation behaviors of the model by means of the aggregate dynamics and the Fokker-Planck equations, which are deduced from the master equation for the jump Markov process. Then, the total number of agents is let go to infinity. A probabilistic criterion is proposed by them for selecting one of two locally stable equilibria by calculating probabilities for the state of the model to be in each of the two basins of attraction. Aoki and Shirai have contrasted their approach with the alternative and more commonly employed practice by economists, of considering models, from the beginning, composed of an infinitely many agents,! by using fractions of agents as state variables.[2] They have argued for the desirability of our alternative approach. See Aoki and Shirai (2000).

The same finitary approach is used in this paper to examine the dynamics of the adjustment processes in the K-W model. We will use the same notations as theirs as much as practical.

There are three types of agents in the K-W model; producers, commodity traders, and money traders. As in the original paper, we let $\Pi$ denote the probability that a randomly selected commodity trader accepts money, when they meet, and let $x$ be the probability that a randomly selected agent wants to trade when they meet. The probability that two randomly selected commodity traders exchange goods is denoted by $x^2$.[3]

In introducing a jump Markov process to examine the adjustment processes out of equilibrium, we immediately realize that the number of money traders is exogenously fixed in the K-W model. This is easy to see. To see this point clearly, let $m$ be the number of money traders, and $c$ be the number of commodity trader. Let $n$ be the total number of agents in the model. They keep this number fixed, so that, the state of the model is given by $(c, m)$, because the number of the producers is given by $n-c-m$. As we soon show, with $m$ fixed, $c$ serves as state variable. In the K-W model, the transition from state $(c, m)$ to state $(c-1, m+1)$ implies that one commodity trader exchanges his output for a unit of money with a money trader, and becomes a money trader. This changes $m$ to $m+1$. This former money trader, who has received good in exchange, however, immediately consumes it and becomes a producer. This reduces $m+1$ to back to $m$ again, instantly. Because of the transaction cost $\epsilon$, money traders always consume the goods they buy.

[2] Such models are often introduced by authors casually mentioning that an infinite number(uncountably many!) of agents are located on the unit interval or on a circle.

[3] Parameter $x$ plays no essential role in the dynamics. Its value merely affects the elements of the covariance matrice in the Fokker-Planck equation. These exogenously specified parameters should depend on the amount of money in the model. In the K-W model, the number of money traders remains constant and $m$ is also constant. For this reason this point does not come up. In some respecification of the model, the number of money traders are endogenous. Then this point is relevant.

In oher words, their Fig. 1 notwithstanding, these two events take place simultaneously, and are not two separate events. The transitions $(c, m) \to (c-1, m+1)$, and $(c, m+1) \to (c, m)$ always occur together, with the change in state vector $(-1, 1) + (0, -1) = (-1, 0)$. The other changes in the state vectorare (1,0),and (-2, 0) respectively.

Section 2 describes the transiiton rate specifications which fix the jump Markov processes involved. Sec. 3 describes the master equation for the state variable $c$. Its aggregate dynamics is discussed in Sec. 4, and the Fokker-Planck equation which governs fluctuations of the number of the commodity traders is described in Sec. 5. Our modification of the K-W model to allow for variable $m$ is discussed in Sec. 6.

We also take this opportunity to show in Sec. 7, as a somewhat of a side issue, that a slight modification of their value function (3) produces a model with their $\Pi$ values strictly inside the unit interval, not at 0, $x$ and 1 as the critical points of the dynamics of the K-W model.[4] The paper concludes with Sec. 8.

## 2 Transition rates

Although there are three groups, producers, commodity traders and money traders, there is only one state variable $c$, when we keep the total number of agents fixed at $n$.[5] Then, $n - c - m$ is the number of agents who are producers. The fractions of commodity and money traders in the model are $c/n$ and $m/n$ respectively.[6] Their equilibrium fractional values are denoted by $N_1$ and $N_m = M$. In our notation $M = m_s/n$ and $N_1 = c_s/n$, where subscript $s$ denotes stationary or equilibrium value.

There are three types of transition rates. They are denoted by $w$ with or without subscripts. The first is that a producer becomes a commodity trader at the rate $\alpha > 0$;

$$w_{0,1}(c) := w(c \to c+1) = \alpha(n - c - m) = n\alpha(1 - \frac{c}{n} - \frac{m}{n}),$$

for non-negative integers for $c, m$, such that $0 \le c + m \le n$. Subscripts to $w$ indicate the change in the type from type 0 (producer) to type 1 (commodity trader). The arrow in the argument of $w$ indicates transition in states. In this transition, the number of commodity traders increase by one. This transition rate specification implies that producers act independently.

The second one describes the rate of two commodity traders meeting randomly, trading their outputs, consuming the exchanged outputs, and re-

[4] This is intended to call the readers' attention to the fragility of the K-W model.

[5] This is merely a device to reduce the dimension of the state space from three to one.

[6] Kiyotaki and Wright assume an infinite number of agents, and use the fractions from the beginning.

turning to the production sector after trade. In this case, $c$ is reduced by 2.[7] Its transition rate is

$$w_{1,0}(c) := w(c \to c-2) = n\beta(1-\mu)x^2 \frac{c}{n},$$

for $2 \leq c \leq n$, where $\beta$ is a parameter for the arrival rate of the trading opportunity, and we use

$$\mu = \frac{m}{c+m},$$

as the definition at and off the equilibrium states.

The rate of a commodity trader becoming a money trader is specified by

$$w_{1,m}(c) := w(c \to c-1) = n\beta\mu\Pi xc/n,$$

for $0 \leq c \leq N$. Other specifications are possible.

## 3 The Master Equation

With these transition rates given, we first state the master equation, then solve it. For this purpose, we change the variable as shown in (1) below, and expand the master equation into the Taylor series in the power of $n^{-1/2}$ as explained in Aoki (1996, Sec. 5.2 and 5.3). Collecting terms proportional to $n^{1/2}$ we obtain the aggregate dynamic equation for the fraction of the commodity traders.

The master equation is given for positive integers of $c$, and $m$ where $m$ appears as parameter by

$$\frac{\partial}{\partial t}P(c) = P(c-1)w_{0,1}(c-1) + P(c+2)w_{1,0}(c+2)$$
$$+P(c+1)w_{1,m}(c+1) - P(c)[w_{0,1}(c)$$
$$+w_{1,0}(c) + w_{1,m}(c)].$$

At the boundary of the region, $c = 0$, $m > 0$, the boundary equation is specified.

At the boundaries such as $c \geq 1$, and $m = 0$, similar boundary equations need be specified. We omit the detail.

As in Aoki and Shirai (2000), and as exposited in Aoki (1996, Sec. 5.3), we solve this master equation by changing variables by

7 In Aoki and Shirai (2000), we discuss that it is better to respecify the transition rate as $\beta x^2 \frac{c}{2} b(c/N)$, with some positive function $b$ with $b' < 0$. This is due to the fact that the number $c$ may be reduced by 2 in two ways: either a given trader suceeds in finding a partner, or while he is searching for a partner, two other traders suceed in forming a partnership and return to the production sector. To keep the original framework as much as possible, however, we keep their specification.

$$c/n = \psi + n^{-1/2}\xi, \tag{1}$$

where the random variable $\xi$ is of the order $O(1)$, and by expanding the master equations in the inverse powers of $n$, and then equating terms of the same order of magnitude on both sides. We thus obtain the equations for $\psi$ as the aggregate or macroeconomic dynamic equation, and another equation for the probability density for $\xi$ as the Fokker-Planck equation.

# 4 Aggregate Dynamic Equation

When $m$ is fixed, the macroeconomic dynamic equation is given by

$$\frac{d\psi}{dt} = F(\psi),$$

with

$$F(\psi) = \alpha(1 - \psi - m/n) - 2\beta(1 - \mu)x^2\psi - \beta x \Pi \mu \psi, \tag{2}$$

where $F$ is the averages of the expected changes of the equilibrium value in the state variable, given by 1, -2, and -1, as mentioned in Sec. 1.

The equilibrium point of the macroeconomic dynamics is obtained by setting the right-hand side of this aggregate equation to zero. With $m = 0$, the critical point is given by $\psi = \alpha/(\alpha + 2\beta x^2)$. It is locally stable. With $m > 0$, there is a unique positive critical point betwee 0 and 1, given as the intersection of $1 + (m/n)/\psi$ and $(A\psi + B)/[\alpha(1 - m/n)]$, with $A = \alpha + 2\beta x^2$, and $B = (m/n)(\alpha + \beta x \Pi)$. This point is also locally stable.

# 5 The Fokker-Planck Equation

The corresponding Fokker-Planck equation for the probability density for the random variable $\xi$ is given by

$$\frac{\partial P(\xi, t)}{\partial t} = -\frac{\partial F}{\partial \psi}\frac{\partial}{\partial \xi}(\xi P) + \frac{1}{2}R\frac{\partial^2}{\partial \xi^2}P,$$

where

$$R = \alpha(1 - \psi - m/n) + 2\beta(1 - \mu)x^2\psi + \beta x \Pi \mu \psi.$$

Note that $\partial F/\partial \psi$ is negative, when the critical point is locally stable. Denote it by $-V$, where $V$ is positive.

In equilibrium or stationary state, the left-hand side is zero, and the equation is integrated.

We have $P$ as a mean zero Gaussian distribution with variance $2R/V$.

stationary probability density in the form of the above potential functions exist.

## 6 Use of Partition Vector as State Vector

There are several ways for introducing dynamics to the K-W model. We sketch two possibilities here.

We allow for a variable $m$ by introducing a different way of introducing states, called partition vector by Zabell (1992). It is called specification of the second kind by Sachikov (1996). It is not known in the economics literature but widely used in ecology or population genetics literature.

We allow money to be introduced exogenously, not only at the initial time, but throughout the process. We need to change the value function equations to induce agents to hold more than one unit of money. Without going to details, we assume that holding money stock of more than one unit is advantageous to money traders. More than one unit of money in stock may facilitate meeting with commodity traders, for example.

Let $a_i$ be the number of money traders with exactly $i$ units of money. Vector $\mathbf{a}$ with these $a_i$ as components is called partition vector. It describes how money stocks are held by various money traders. The total amount of money is denoted by $m$. It is given by

$$\sum_i i a_i = m,$$

and the number of money traders is given by

$$\sum_i a_i = K_m.$$

The variables $m$ and $K_m$ are both random variables. We may write the partition vector as

$$\mathbf{a} = \{a_1, a_2, \cdots\},$$

because only a finite number of elements of $\mathbf{a}$ is nonzero. This is shown by the Borel-Cantelli lemma, as in Shepp and Llyod (1966), and thus the set of all partition vectors is countable.

### 6.1 Modification 1

We adapt a simple Markovian dynamics expressed in partition vectors in Kelly (1976). See also Aoki (1996, Sec. 4.5.5) for detail.

Assume that the transition rates for the partition vector are specified by

$$w(\mathbf{a}, \mathbf{a} + \mathbf{e}_1) = \nu\lambda(m), \tag{3}$$

$$w(\mathbf{a}, \mathbf{a} - \mathbf{e}_j + \mathbf{e}_{j+1}) = j a_j \lambda(m), \tag{4}$$

and

$$w(\mathbf{a}, \mathbf{a} + \mathbf{e}_{j-1} - \mathbf{e}_j) = j a_j \mu(m), \tag{5}$$

$j \geq 2$, and its special case for $j = 1$

$$w(\mathbf{a}, \mathbf{a} - \mathbf{e}_1) = a_1 \mu(m), \tag{6}$$

where $\mathbf{e}_0$ is zero, and where $\mathbf{e}_i$ is a vector with the only nonzero entry of 1 at the $i$th position, and where $\nu$ is a constant and $\mu$, and $\lambda$ are fuctions of $m$, in general. The transition rate in (3) describes the rate at which a commodity trader becomes a money trader by trading his good for a unit of money. It corresponds to $w_{1,m}$ in the K-W model. Eq. (4) describes how a money trader with $j$ units of money acquires a unit of money to end up holding $j+1$ units of money. Eq. (5) shows how a trader with $j$ units of money buy commodity, and end up with $j-1$ units of money. In particular money traders with one unit of money, become producers after acquiring good. This exit rate of money traders to become producers is $a_1 \mu(m)$ of (6). For example, with more money traders commodity traders have easier time of exchanging commodities with money. Hence $\mu(m)$ may be increasing with $m$.

In this example, $m$ is also a random variable and has the transition rates

$$w(m, m+1) = (\nu + m)\lambda(m),$$

and

$$w(m, m-1) = m\mu(m).$$

These are obtained from (3) to (6) by summation over $j$.

In this case there is an equilibrium probability distribution for $m$. This is verified by the detailed balance condition. The stationary distribution is

$$P(m) = C^{-1} \binom{\nu + m - 1}{m} \prod_{k=1}^{m} \frac{\lambda(k-1)}{\mu(k)},$$

where we assume that

$$C = \sum_{m=0}^{\infty} \binom{\nu + m - 1}{m} \prod_{k \geq 1} \frac{\lambda(k-1)}{\mu(k)}$$

is finite. As an example suppose that there is a constant $\omega < 1$ such that the ratio $\lambda(k-1)/\mu(k) \leq \omega$ for $k \geq 1$. Then the product of the ratios is bounded from above by $\omega^m$, and $C \leq (1-\omega)^{-\nu} \leq \infty$. For simpler explanation we may assume that $\lambda$ and $\mu$ are constant and the ratio $\lambda/\mu$ is less than one.

The equilibrium probaility distribution for the partition vector is given by

$$\Pr(\mathbf{a}) = C^{-1} \prod_{j \geq 1} (\frac{\nu}{j})^{a_j} \frac{1}{a_j!} (\frac{\lambda}{\mu})^{j a_j},$$

where we use the relation $\sum_i i a_i = m$ to note that the last expression above becomes $(\lambda/\mu)^m$. That this is a bona fide probability distribution is verified by the sum becoming one. To see this we use the Kendall-Kelly identity

$$(1-x)^{-\nu} = \prod_{i\geq 1} \exp\left(\frac{\nu x^i}{j}\right).$$

On the left-hand side, the coefficient of $x^m$ is $C_{\nu+m-1,m}$. On the right-hand side, we have the sum over $a_j$, subject to the constraint $\sum ja_j m$ of the products $\prod_{j=1}^{\infty}(\nu/j)^{a_j}1/a_j!$.

This model is an example in which the equilibrium distributions have product forms. See Pollett (1986).

This probability has the form of the product of Poisson random variables with mean $\lambda_j = (\lambda/\mu)^j(\nu/j)$, $j = 1, 2, \ldots, m$ The products of Poisson distributions are not as special as they may first appear. Actually, distributions of Poisson random variables, conditional on their weighted sums arise in a fairly general context, as shown by Arratia and Tavaré (1992).

## 6.2 Ewens Sampling Formula: Modification 2

This section uses different specifications of transition rates for partition vectors. Suppose that we specify the rates for entry, exit, and changing of types as follows:

$$w(\mathbf{a}, \mathbf{a}+\mathbf{e}_i) = c_i(a_i + f), \tag{7}$$

$$w(\mathbf{a}, \mathbf{a}-\mathbf{e}_j+\mathbf{e}_k) = \lambda c_j a_j c_k(a_k + f), \tag{8}$$

and

$$w(\mathbf{a}, \mathbf{a}-\mathbf{e}_j) = d_j a_j, \tag{9}$$

with $c_j/d_j = g$ for all $j$, and for $k = j+1$ or $j-1$, as in (4) and (5). $\lambda$ is some positive parameter.

Eq. (7) is the rate at which a unit of money is exchanged for good by a commodity trader, (8) states the rate for a unit of money beging exchanged among money traders. Eq.(9) specifies the rate of money trader exchanging money for good.

The detailed balance conditions hold, and the stationary probability distribution is given by

$$\pi(\mathbf{a}) = \frac{K_m!}{a_1!a_2!\cdots a_m!} C^{-1}_{-Kf,m} C^{a_1}_{-f,1} C^{a_2}_{-f,2} \cdots C^{a_m}_{-f,\cdot}.$$

See Aoki (1996, 98). (Instead of $a$s here, $\beta$s are used.) If $K_m$ goes to infinity and $f$ to zero in such a way that the product $K_m f$ approach a non-negative constant $\theta$, then noting that $C_{-f,j} \to (f/j)(-1)^j$ the terms involving negative binomials approach

$$\binom{-\theta}{m}^{-1} \prod_j (f/j)^{a_j}.$$

Note that the factor $(-1)^m$ combined with $\prod_j (-1)^{ja_j}$ becomes one. and we obtain

$$\pi_m(\mathbf{a}) = C^{-1}_{\theta+m-1,m} \prod_{i=1}^{m} (\frac{\theta}{i})^{a_i} \frac{1}{a_i!}.$$

Alternatively, the right-hand side is expressible as

$$\pi_m(\mathbf{a}) = \frac{m!}{\theta^{[m]}} \prod_{i=1}^{m} (\frac{\theta}{i})^{a_i} \frac{1}{a_i!},$$

where $\theta^{[m]} := \theta(\theta+1)\cdots(\theta+m-1)$, called ascending factorial.

The first factor may also be written as $m!/[\theta]_m$. This is a heuristic way of deriving this justly famous formula in the population genetics. This form was conjectured by Ewens (1972), see Ewens (1990). Since then this distribution has been found in many different models. That this is a well-defined probability distribution can be verified by seeing that the expression sums to one. In verifying this we use the identity $-\ln(1-x) = \sum_j x^j/j$, or

$$(1-x)^{-\theta} = \sum_j e^{\theta x^j/j},$$

and comparing the coefficient of $x^m$ on both sides.

The exit rate of money traders becoming producers is $w(\mathbf{a}, \mathbf{a}-\mathbf{e_1})$ with $a_1 > 0$. A commodity trader may exchange his good with any one of money traders with positive units of money. The entry rate to money traders is thus $\sum_i w(\mathbf{a}, \mathbf{a}+\mathbf{e_i}) = \sum_i c_i(a_i + f)$.

This distribution has a single parameter $\theta$. This parameter represents correlations among money traders. It determines how likely the alternatives of $m$ money traders holding one unit of money each, and one or a few money traders holding most of money stocks. See Aoki (2000) for detailed discussions of this point. To obtain some ideas about the role of the parameter $\theta$, consider $m = 2$, with $a_1 = 0$, and $a_2 = 1$. This occurs with probability $\pi_2(a_1 = 0, a_2 = 1) = 1/(1+\theta)$ when partition vectors have the Ewens distribution. Therefore, the probability that a single money trader holds two units of money, rather than two money traders holding one unit each, approaches zero with large $\theta$. The smaller the value of $\theta$, the higher the probability that two money traders hold one unit each. More generally, it is easy to calculate the ratio of $\pi_m(a_j = 0, 1 \le j \le m-1, a_m = 1)$ over $\pi_m(a_1 = m, a_j = 0, j \ne 1)$, and to note that the probability of each of $m$ money traders holds a unit of money each is smaller, than one money trader holding all $m$ units, the smaller the parameter $\theta$ is. Although we do not discuss the dynamic process and the equilibrium distribution of money stocks, parameter $\theta$ of the Ewens formula influences and controls the nature of distributions of money stocks among money traders.

## 7 Aggregate Dynamic Equation

With $\mu$ as the fraction of money traders among all traders, the macroeconomic dynamic equations are:

$$\frac{d\psi}{dt} = E[G((\psi, m/n),$$

where the expectation is with respect to $m$, with

$$G(\psi, m/n) = \alpha(1 - \psi - m/n) - 2\beta(1 - \mu)x^2\psi - \rho(\psi, m/n)$$

, where $F$ is the averages of the expected changes of the equilibrium value in the state variable, that is 1,-2, and -1, and where $\rho$ is the entry rate to the money trader sector from the commodity sector.

In Aoki (2000), it is shown that

$$E(a_1) = \frac{m\theta}{\theta + m - 1},$$

$$E(m) = \sum_i iE(a_i),$$

and

$$E(K_m) = \sum_j e(a_j) = \int_\epsilon^1 \phi(p)dp,$$

where $\phi(p) = \theta p^{-1}(1-p)^{\theta-1}$ is called the frequency spectrum by Ewens.

The expression $x_j = ja_j/m$ is the fraction of money stock held by money holders each with $j$ units of money. The vector $(x_1, x_2, \ldots)$ is a point in the infinite dimensional simplex. Its order statistics $x_{(1)} \geq x_{(2)} \geq \cdots$ has distribution called Poisson-Dirichlet distribution, see Kingman (1993, 94).

The equilibrium point of the macroeconomic dynamics is obtained by setting the right-hand side of (3)to zero.

## 8 Value Functions

Denote the value function for the agents of type 0, 1, and $m$ by $V_j, j = 1, 2, m$. The first two relations are identical with those of the K-W model

$$rV_0 = \alpha(V_1 - V_0),$$

and

$$rV_1 = \beta(1-\mu)x^2(U + V_0 - V_1) + \beta\mu x \max_\pi[\pi(V_m - V_1)].$$

Here we write $U$ for $U - \epsilon$ in their paper since no information is lost by this short-hand notation. Denote by $\Pi$ the value of $\pi$ as the result of maximization.

In the third respecification, we change their Eq.(3) slightly as follows:

$$rV_m = \beta(1-\mu)\Pi x(U + V_0 - V - m) + \theta(m),$$

where $\theta(x)$ is some nonlinear function such that $\lim_{x\downarrow 0} \theta(x)/x^2 = \kappa$, where $\kappa$ is a positive constant, subject to a lower bound, to be given shortly. By setting $\theta$ to zero, we recover their original value function (3). This function $\theta$ replaces $-\beta(1-\mu)\Pi x\epsilon$ in their Eq.(3). We merely make the transaction term into something which favors holding money.

With this slight change, the maximization in their Eq.(2) no longer depends on the sign of $V_m - V_1$ only. The expression now has two terms: the first is proportional to $\pi - x$. This is the term in the K-W model which causes $V_m - V_1$ to be negative for $\pi < x$. The second term is positive,

$$V_m - V_1 = -\frac{A(x)}{\Delta}(x-\pi) + \frac{R(x)}{\Delta}\theta(x),$$

where $\Delta = D_0 + D_1\pi + D_2\pi^2/2$, where $D$s are coefficients dependent on $x$, where $A(x) = \beta(1-\mu)x(U-\epsilon)r$, and $R(x) = r[1 + \frac{\beta}{r+\alpha}(1-\mu)x^2]$.

This $\Delta$ is the determinant in the equations for $V_1$ and $V_m$, after substituting $V_0$ out by $V_0 = [\alpha/(r+\alpha)]V_1$. For small values of $x$ we have $\Delta \approx r^2$. Let $\phi(\pi) := \pi(V_m - V_1)$. Then, we verify easily that

$$\frac{d\phi(\pi)}{d\pi}|_0 > 0,$$

near $x = 0$,if $\kappa > \beta(1-\mu)(U-\epsilon)/r$.

We also verify that the derivative $(d\phi/d\pi)_{\pi=1}$ is negative near $x = 1$ if $\beta(1-\mu)\mu x^2/(r+\alpha) > r[1 + \beta(1-\mu)x^2/(r+\alpha)]$.

This implies the existence of a maximum of $\phi(\pi)$ inside the unit interval (0, 1). Thus, there is at least one $\Pi$ between 0 and 1. Because the derivative is cubic in $\pi$, there are at most three values of $\pi$ at which the derivative vanishes. In such cases, there are two local maxima and one local minimum of $\phi(\pi)$ in the interval (0 1).

# 9 Discussions

In the K-W model as originally posited, the number of money traders is exogenously fixed and does not change with time.

We have suggested several ways of modifying the K-W model. There are other possibilities for modifying the original K-W model, such as different specifications of some transition rates from the ones in this paper, or more explicit inclusions of externalities in choices, and transaction costs.

More fundamentally, by departing from the K-W framework, we could propose models in which agents are of the two types, employed and unemployed, both of whom hold one or more units of money. Only employed hold one unit of commodity each. Although this framework departs more radically from that of the K-W model, more interesting model behavior can be observed. This will be discussed in a separate paper.

## References


Aoki, M., *New Approaches to Macroeconomic Modeling: Evolutionary Stochastic Dynamics, Multiple Equilibria, and Externalities as Field Effects* , Cambridge Univ. Press, New York, 1996

——, *Modeling aggregate behavior and fluctuations in economics: Stochastic views of interacting agents* , forthcoming, Cambridge Univ. Press, 2001.

——, and Y. Shirai, "A new look at the Diamond model", forthcoming in *J. Macroeconomic Dynamics*, 2000.

——, "Cluster Size Distributions of Economic Agents of Many Types in Market", forthcoming, *J. Math. Analy. Appl*, Bellman Memorium issue, 2000.

Arratia, R., and S. Tavaré Independent process approximations for random combinatorial structures, *Advances Mathematics* **104** 90–154, 1994

Diamond, P., "Aggregate Demand Management in Search Equilibrium", *J. Political Economy*,**90**, 881–94, 1982

Ewens, W. J., "Population genetics theory–The past and the future" in *Mathematical and Statistical Developments of Evolutionary Theory* ed. S. Lessard, Kluwer Acad. Pub. London, 1990

Kingman, J. F. C., *Poisson Processes* , Clarendon Press, Oxford, 1993. Kiyotaki, N., and R. Wright, "A search-theoretic approach to monetary economics", *Amer. Econ. Review* **83**, 63–77, 1993.

Kelly, F. P. "On stochastic population models in genetics", *J.Appl.Prob.]* **13**,*127–31, 1976*

*———— Reversibility and Stochastic Networks Wiley, 1979*

*Gardiner, C. W., Handbook of Stochastic Methods for physics, chemistry and the natural sciences , Second Edition, Springer-Verlag, New York, 1994*

*Pollett, P. K., "Connecting reversible Markov processes", Adv. Appl. Probab.* **18**,*880–900*

*Sachikov, V., Combinatoiral Methods in Discrete Mathematics ,Cambridge Univ. Press, New York, 1996.*

*Shepp, L. A. and S. P. Lloyd Ordered cycle lengths in a random permutaion, Trans. Amer. Math. Soc.* **121** *340–57, 1966*

*van Kampen, N., Stochastic Processes in Physics and Chemistry , North Holland, 1992*

# Industrial Dynamics with Quasi-Zero Intelligence Interacting Firms

**Roberto Leombruni**
**Domenico Delli Gatti**
**Mauro Gallegati**

ABSTRACT This paper presents some artificial stylised facts emerging in a simulated contestable market where firms interact with each other in taking their stay or go decision. We use nearly zero-intelligence firms: no optimisation is considered, and all the firms sell at a fixed price an equal quantity of the good. The entry of new firms is triggered by the overall profitability of the market, measured with the spread between the average rate of profit and the interest rate. The exit decision is modelled *via* a mean field effect, to take into account in the decision process both the performance of the individual firm, and the information about the profitability of the market that can be abduced looking at the stay or go decision of the other firms. Financial requirements of production are considered, with a spread between creditor and debtor interest rates. The model is simulated with the ACE approach, using the Swarm libraries released by the Santa Fe Institute.

## 1 Introduction

Empirical literature in industrial dynamics has underlined since a long time many stylised facts relating firms distribution characteristics - such as different dynamic path followed by large and small firms, a right skewed size distribution, and the presence of a positive firms turnover -, that cannot hence be adequately tackled within the Representative Agent (RA) framework[1]. As it seems, real markets are characterised by a "tremendous within-industry heterogeneity" that is not cancelled out by the 'selection of the fittest', and that requires a specific attention (Haltiwanger 1997 [Hal97]).

Putting the focus on the dynamics of firm size, theoretical literature has tried to better fit the data putting heterogeneity in some firm level characteristics, mainly in the productivity level (since Jovanovic 1982 [Jov82]),

[1] See Caves, 1998 [Cav98], for a survey on recent empirical findings on industrial dynamics.

and in the managerial ability (since Lucas 1978 [REL78]). For what concerns that component of the overall market dynamics, shaped by the entry and exit fluxes, its explicit consideration dates back to Brock (1972 [Bro72]). A sum-up of these two branches of the literature has been put forth by Hopenhayn (1992 [Hop92]), who derives a limit distribution of firms size in which there is both heterogeneity in productivity and a positive equilibrium firms turnover. The dynamic behaviour of the model is limited to a comparative statics analysis.

In the last decades, however, the RA framework has been questioned not only in its "strongest" version, i.e. when the assumption is that the macrobehaviour can be grasped looking at a single representative agent; but also when the heterogeneity, although considered, is managed via a massive set of unrealistic assumptions imposed to derive an exact microfoundation to the aggregate behaviour; in some way, cancelling out most of the potential effects that heterogeneity can have on the dynamics (cf. Martel 1996 [Mar96], Kirman 1992 [Kir92]). The relaxing of this (apparently) "weak" RA hypothesis has started a growing literature on the effects that heterogeneity and interaction among individuals can have on the macrodynamics in presence of market imperfections and strategic complementarities (Gallegati and Kirman 1999 [GK99]).

In this paper, we tackle this latter research program adopting an agent based computational approach. The focus of the model presented is on the partial equilibrium dynamics emerging in a market with free entry exit of firms endowed with heterogeneous equity, in presence of financial imperfections and local knowledge on the market profitability. As we'll see, we choose to specify these various hypotheses in somewhat a *naïve* way, for two reasons. The first is bound to the kind of methodology we adopted, the so-called bottom-up, or generative, approach (see Epstein and Axtell 1996 [EA96]). One of the main tenets of this approach is the effort of emulating some macrobehaviours of interests, building artificial societies populated with agents as simple as possible. In Epstein's words: "The issue is not how much rationality there is (at the micro level), but how little is enough to generate the macroequilibrium" (Epstein 1999 [Eps99]).

The seconde exigence that guided us, was that of having an analytical benchmark with which to interpret the outcomes of the simulations. In particular, the adoption of this "kiss principle"[2], has permitted us to analyse the interaction among the firms in an efficient way, *via* a mean field effect, and to concentrate the simulations on the effects of the heterogeneity.

The simulations have been conducted with the Swarm Simulation Toolkit, and produced some artificial stylised facts of interest: a right skewed distribution for the equity base, attributable mainly to a composition effect;

---

[2]The reappraisal proposed by Axelrod of the old army slogan: "Keep it simple, stupid!" [Axe97].

a long run positive firms turnover, due, as in Hopenheyn and in Jovanovic, to a positive probability of exiting the market also during expansions; large cycles, due to the interaction component of the decision taken by the firms.

In next section, we present the algebra of the model, and a qualitative analysis of its dynamics with no heterogeneity among firms. Some other results on the dynamics emerging are given in the following section, where we simulate the same base model. The effects of heterogeneity are then studied, putting idiosyncratic shocks on the price at which firms are selling their good. Some concluding remarks will follow.

# 2 The model

The model we are presenting extends a previous work (see Leombruni et al. 2001 [LGG01]), where we introduced nearly zero-intelligence (ZI) firms to study industry dynamics. The use of ZI-agents, allowed us to derive a sort of benchmark for the behaviour of a market with free entry exit, linear production costs, and equity rationing[3].

There, the entry exit decision were essentially external to the firms: given a positive (negative) profitability of the market - measured by the spread between the interest rate of a secure asset and the average profit rate - a certain number of firms were driven into (out of) the market.

Here, we give back to the firms a bit of intelligence, to model the presence of local and dispersed knowledge on the market profitability.

On one side, potential entrants have access just to price signals, so that entries are still driven by the presence of more than normal profits.

On the other side, incumbents firms have at their disposal local knowledge on the market profitability. This information, however, is dispersed, and has to be extracted looking at the behaviour of the other competitors. The stay or go decision, then, is modelled assuming non-price interaction among firms: each of them will make its choice considering both their own performance, an the signals on the market profitability that they have looking at the stay or go decision taken by its competitors.

## *2.1 Firms and households*

The demand side of the model consists of a constant and exogenous income $Y$, which is entirely spent by households to buy equal quantities of the goods produced by the incumbent firms. The price too is given, and is normalised

[3]The dropping of the first "s" in the Axelrod's KISS principle and the straightforward adoption of a ZI assumption dates back at least to the works of Gode and Sunder (1993 [GS93]), and has a specific interest in the simulation literature (see Terna 2000 [Ter00] for a discussion).

to one plus an idiosyncratic shock identically and indipendently distributed across time and firms.

The $N$ incumbent firms face fixed unitary (production) costs $\alpha$. The $i$th firm's production is financed by means of its equity base $A_i$, and the eventual negative slack between equity and the financial requirements yields a cost of $r$ times the slack. This is equivalent to assuming a spread between creditor and debtor $r$, the former being set to zero.

Each firm then will be characterised by the following profit equation:

$$\Pi_i = \frac{(1+\epsilon_i-\alpha)Y}{N} - r\left(\frac{\alpha Y}{N} - A_i\right) \qquad \text{if } \frac{\alpha Y}{N} - A_i > 0 \qquad (1.1a)$$

$$\Pi_i = \frac{(1+\epsilon_i-\alpha)Y}{N} \qquad \text{otherwise} \qquad (1.1b)$$

Firms remunerate their shareholders at the rate $r$, while the excess (shortfall) of the profit over the dividend increases (decreases) the equity base. Therefore, the motion equation of the equity of $i$th firm will be the following:

$$\begin{aligned}\Delta A_i &= \Pi_i - rA_i \\ &= A_i(\pi_i - r)\end{aligned} \qquad (1.2)$$

with $\pi_i \equiv \Pi_i/A_i$.

To study the overall behaviour of our market, we will neglect for a while the idiosyncratic shocks on prices, and consider separately the dynamics of $A$ and $N$. For the sake of simplicity, we'll also treat $N$ as continuous.

## 2.2 *Equity base equilibrium*

Holding $N$ constant, the equilibrium condition $\Delta A_i = 0$ is reached when

$$\pi_i = r \qquad (1.3)$$

It can be shown that, if the condition $\alpha(1+r) < 1$ holds[4], equation (1.3) implies also no liability for the firm, so we can substitute (1.1b) into it to obtain the following equilibrium value of $A$:

$$A = \frac{Y(1-\alpha)}{rN} \qquad (1.4)$$

[4]The condition states that the unitary production costs, augmented for the costs of their financing, must be lower than the price. It is a sort of minimum requirement for the market to exist, and we'll assume it always verified.

If we draw equation (1.4) against $N$, we have the hyperbolic "Normal profits curve" (NPC) that identifies all the infinite couples $(A, N)$ in which $\Delta A = 0$ (cp. figure 2). When any firm has an equity base lower than that of the NPC, the excess profit will accumulate until the equilibrium value is reached, and similarly in the opposite case. As a consequence, in equilibrium any initial heterogeneity in the equity wipes out.

## *2.3 Firms' turnover equilibrium*

Now let's hold $A$ constant and equal for all firms, and let us specify separately the entry and the exit mechanisms.

Given the partial equilibrium nature of the model, we assumed the existence out of our market of an unbounded set of potential entrants. We assumed also no entry barriers (pure contestable market hypothesis), so that the inflow of new firms is triggered simply by the difference between the average rate of profit and the interest rate.

Entries, then, will "happen" only when the market is not too crowded for the given aggregated demand, that is when actual $N$ is lower than the value that satisfies $\pi = r$.

The condition is the same than (1.3), but the passage to (the inverse of) equation (1.4), necessary to compute an $N^*$ of equilibrium, is less direct. In fact, here we are considering only the entries in the market, so that the $N^*$ that we can compute from (1.4) is just a ***benchmark*** with which to compare the actual number of firms: it is the maximum number of firms compatible, given $A$, with at least normal profits. Whether or not that value will be a feasible equilibrium depends also on the exit fluxes; and as we will show, the answer is no.

As a consequence, the equilibrium in $N$ given $A$ does not imply no liability for the firms, and to solve $\pi = r$ we must consider both profit equations (1.1a-1.1b).

The switch between the two profit equations is the "No liability line" (NLL) we drawn on figure 2, defined by the condition $\alpha Y/N = A$. In the region lying over it, firms have no liability, the "active" profit equation will be (1.1b), and the $N^*$ satisfying $\pi = r$ can again be found on the NPC.

In the region below it, the "active" profit equation will be (1.1a). It is easy to see that if $\alpha(1 + r) < 1$ is verified, with any couple $(A, N$ falling in this area firms are earning more than normal profits. Hence, new firms will continue to entry until $N$ goes over the NLL.

In conclusion, given $A$, the unique $N$ compatible with the condition $\pi = r$ is that read on the NPC. To determine the entity of the inflow, then, we calculate an equilibrium $N^*$ *via* equation (1.4), and let the number of entrant $I$ be equal to the difference - if positive - between actual $N$ and $N^*$, times an adjustment coefficient $\gamma$. Writing it in relative terms, we'll have

$$i \equiv \frac{I}{N} = \max\left\{\gamma\left[\frac{Y(1-a)}{rAN} - 1\right], 0\right\} \tag{1.5}$$

Turning to the incumbent firms, the decision they have to take is whether to stay in the market or to leave it. As said, the information they have to base on to take their choice is twofold: their own performance, and the profitability of the market.

To evaluate the latter, they try to extract some information looking at the ***behaviour*** of their competitors. When they observe that some of them are leaving, they interpret this fact as a signal that the market profitability is getting worse. ***Via*** this interaction effect, then, a firm has a positive probability of exiting even if it is making more than normal profits.

Formally, this is equivalent to the presence of strategic complementarities: firms will judge convenient to make a stay or go choice of the same sign of that of their competitors.

A convenient way of tackling this social interaction effect, is to slightly modify a mean field effect model as those proposed by Brock and Durlauf (see for instance [BD00]), to let the external field be determined endogenously.

Calling $\omega_i$ the choice of the $i$th firm, and $\omega_{\sim i}$ the average choice of its competitors, we write the expected benefit of $i$th firm as:

$$V(\omega_i) = h_i\omega_i + J\omega_i E_i(\omega_{\sim i}) + \eta(\omega_i) \tag{1.6}$$

where:
$\omega_. = -1$ stands for the "go" choice, and $\omega_. = 1$ for the "stay" choice;
$h_i$, the external field, is a measure of the firm's own profitability;
$J$ is a parameter measuring the strength of the interaction;
$E_i(.)$ is the operator that gives the expectations of firm $i$;
$\eta(\omega_i)$ is a random term whose realisation depends on the decision taken, independent across individuals.

Assuming the difference $\eta(1) - \eta(-1)$ to be logistically distributed, we'll have that the probability that $i$th firm will stay in the market is proportional to:

$$Pr[\omega_i = 1] = Pr[V(1) > V(-1)] \propto \exp\{\beta[h_i + JE_i(\omega_{\sim i})]\} \tag{1.7}$$

where $\beta$ is the inverse of the scale parameter of the logistic distribution. Note that the introduction of a random term in the expected benefit $V$ implies a positive exit probability not only with $h$ positive, but also when the firm observes none of its competitors exiting.

Using (1.7), we can compute the mathematical expectations of the individual behaviour $E(\omega_i)$, $i = 1 \ldots N$, and aggregate them to compute the expected average choice $\omega$ in the population. Imposing a coherence between this latter and the expectations of the individuals, i.e. letting $E(\omega) = E_i(\omega_{\sim i})\forall i$, and remembering that the equity base, the market share, and hence the firms' profitability situations as expressed by $h_i$ are the same for all firms, we can write the condition for the average choice $\omega^*$ to be an expectational equilibrium:

$$\omega^* = \tanh\left\{\beta\left[h + J\omega^*\right]\right\} \tag{1.8}$$

Since tanh(.) is continuous, and is a contraction of $[-1;1]$ into itself, there is at least one solution to (1.8); when $J\beta < 1$, it can be shown that the solution is unique. Assuming this condition verified, we can re-scale $\omega^*$ to obtain a unique value for the rate $o$ of exiters which gives a self-consistent equilibrium in the expectations:

$$o = \frac{1-\omega^*}{2} \tag{1.9}$$

Now that we have defined the entry exit rules, we can put them together to write the equilibrium condition $\Delta N = 0 \Leftrightarrow i = o$. We can proceed as follows.

Firstly, we can observe that for $h$ to be a measure of the firm's profitability it must be increasing in $\pi$, which in turn is a continuous decreasing function of $N$, defined in $(0, \infty)$ and there differentiable almost everywhere.

If we let also $h(N)$ be continuous, and observe that equations (1.8-1.9) define implicitly $o$ as a decreasing function of $h$, with values in (0; 1), we'll have that

$$o(N) : \Re^+ \mapsto (0;1) \tag{1.10}$$

is a continuous and increasing function of $N$.

By a simple analysis of the two curves $i = i(N) - o = o(N)$, descends that there is a value $N$ strictly lower than $\frac{Y(1-\alpha)}{rA}$ in which entries are equal to exits; that is, firms turnover will be in equilibrium in a point lying strictly under the NPC.

A convenient way to better qualify this equilibrium is to consider both $i$ and $o$ as functions of the product $AN$. In figure 1 we plotted them with values of the parameters satisfying the hypotheses adopted till now, and specifying $h$ as:

$$h = \frac{\pi - r}{r}$$

If we read this figure "forgetting" for a while the $A$ on the X-axes, we can see the $N$ of equilibrium for the turnover derived above.

The key observation, is that the function $o(.)$ is separable in $AN$. If we assume the same separability for the function $i(.)$, we'll have that the equilibrium condition for the firms' turnover will depend only on the ***product*** of

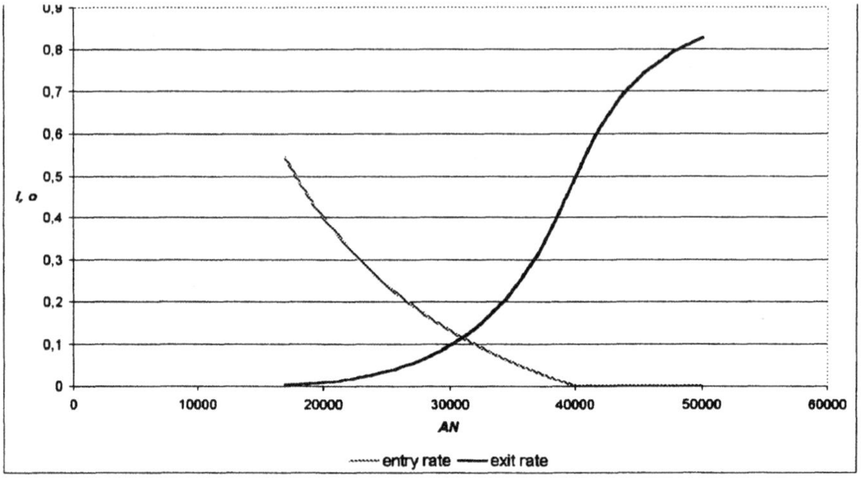


FIGURE 1. Firms' turnover equilibrium.

$A$ and $N$. In the figure, a change $\Delta N = 1/\Delta A$ will not change the position of the entry/exit rates curves, and hence the position of the equilibrium.

If we call $k$ the value of the product $AN$ at which the two curves intersect, we can write the map that defines the turnover equilibrium in the space $(N, A)$ as:

$$N = \frac{k(.)}{A} \tag{1.11}$$

where $k(.)$ doesn't depends on $A$, but depends on all the other parameters of the model.

This turnover equilibrium (TE) curve acts as an attractor for $N$ (cp. figure 2). Given $A$, for values of $N$ lower than that identified by equation (1.11), we'll be at the left of the equilibrium of figure 1: entries will be higher then exits, and $N$ will increase. Similarly in the opposite case.

## *2.4 Market dynamics and heterogeneity*

In the last two paragraphs, we derived the equilibrium condition for the equity base as a function of $N$, and the equilibrium condition for the firms' turnover as a function of $A$.

In figure 2 we draw the two equilibrium maps, together with the no liability line. The direction of the trajectories around these maps is also reported: according to what stated in paragraph 2.2, in the area above the NPC the equity will decrease; while in paragraph 2.3 we saw that in the area above the TE curve there will be an excess of exits on entries, and $N$ will decrease - similarly in the opposite cases.

We also saw that the latter lies strictly under the former, so that between the two we have an attractor-basin where the system will fall with certainty.

The dynamics inside the basin is that depicted: with agents all equal, the market will go towards higher equity and lower $N$.

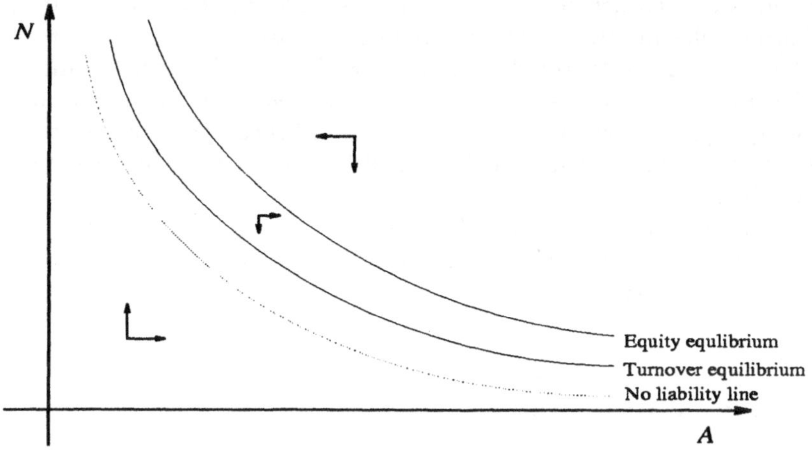


FIGURE 2. $(A, N)$ phase diagram - no heterogeneity.

We must consider two more factors to better assess the behaviour of our market: the role of liabilities, and that of heterogeneity.

In the figure, we draw the no liability line under the two other curves, but that is not always the case. The position of the NLL and of the NPC depends only on the parameters $r$ and $\alpha$, and on the level of the aggregated demand $Y$, while the TE curve will depend also on the parameters $\beta$ and $J$, with which we modelled the strategic interaction among firms, and on $\gamma$. For some configurations of these three latter parameters, then, the TE curve can go under the NLL. This will change slightly the trajectories around the equilibrium maps, since both the equity and the turnover will evolve according to a different profit equation.

Actually, the TE curve itself will have a different formulation, since the implicit function $k(.)$ of the model parameters will change. Without getting into details, it can be shown that it will lie under the TE curve as calculated ignoring financial costs. With no heterogeneity, anyhow, this will not change the qualitative dynamics depicted so far.

When we allow for heterogeneity in the firms' equity base, the portrait can change sensibly.

On one side, the equilibrium condition we derived for the firms turnover is again no more directly applicable. While the entries are driven only by the average measure of the market profitability, the exit rule has a nonlinearity that makes the aggregation sensible to the dispersion of the distribution.

Experimentally, as we'll see, the heterogeneity implies a downward shift in the TE.

In addition, if the equity dispersion goes over a threshold, a not-empty subset of firms will lie under the NLL. This will change both the benefit function on which they base their stay or go choice on, and the motion equation of their equity. In other words, we can have two non empty subset of firms following two different dynamic regimes.

On the other side, the direction followed inside the basin attractor can change its sign, provided that the new firms entering the market have an average equity lower than the incumbents'. This composition effect, in fact, countervails the tendency in the rise of the average $A$ when we are under the NPC.

Both these effects are easily handled building an agent based simulation of our model, with which we can also have some hints on the dynamics around these maps.

# 3 The simulations

In this section we first report some technical details concerning the simulations conducted. We then proceed to show the artificial time series generated, first reproducing the results above derived assuming no heterogeneity among firms and a unique equilibrium in the stay or go choice, then relaxing both of them.

## *3.1 Technical details*

The simulations have been conducted in Swarm, a set of software libraries developed at the Santa Fe Institute in New Mexico - starting from 1995 -, to help simulating complex systems. The underlying programming languages of Swarm are Objective-C and Java, which object-oriented architecture is particularly suited to run agent-based simulations[5]. In a nutshell, a simulation in Swarm is built up putting together a population of artificial agents - independent "pieces" of software containing the agents' behavioural algorithms and their vector of state variables, and a schedule with the ordered list and the timing of all the actions each agent will make.

The core of our simulation has been built up with three kinds of agents: firms and households (whose state variables and behaviour are a straightforward translation of the model characteristics described above), and an

[5]For a discussion of the specificities of object-oriented simulations see Joines and Roberts (1999 [JR99]). For economic simulations in Swarm see Luna and Stefansson (2001 [LS01]). Technical details on the Swarm Toolkit can be found on the web, at http://www.swarm.org.

environment, whose role is to collect and distribute statistics, to co-ordinate entries and exits, and to behave as an interface between the other agents.

For every time-step of the simulation, the main items of the schedule are the following:

1. Households: do shopping;
2. Environment: check for bankruptcies;
3. Firms: take the stay or go decision;
4. Environment: create new firms.

During the first step, a representative household spends entirely its yearly income. It divides it equally among the incumbent firms, except for a random multiplicative shock on the price payed to each seller. The shock is extracted from a uniform distribution with mean one, whose support is a simulation parameter modifiable at start up. The eventual slack between yearly income and actual spending is added to the following period income.

In the second step of the schedule the environment checks if there have been "hard" bankruptcies, i.e. if any firm's equity base has fallen under a threshold - that we set to zero. Afterwards, it asks to all firms to take their stay or go decision, providing them with the macro variables values they need in order to take it.

To implement the stay or go choice, we had to better specify equation (1.6). For the external field we used the following:

$$h_i = \frac{1}{b}\sum_{i=1}^{b} L^i(\pi_i - r),$$

where $L$ is the lag operator, and $b$ is the number of periods considered by the firms to evaluate their own profitability. For what concerns the expectations on the behaviour of their competitors, we assumed again an adaptive mechanism: they have been put equal to the lagged proportion of exiters re-scaled to the interval [-1, 1]:

$$E_i(\omega_{\sim i}) = 1 - 2L(o)$$

Finally, the entry of new firms is driven by the Environment object. If there's a positive slack between the overall rate of profit and the interest rate, the environment will create a number of new firms according to equation (1.5). In accordance with empirical findings, we let new entrants have an average equity base lower than the entrants. However, not to put a bias towards a predetermined equity level, we extracted it from its distribution among the first quantiles of the incumbents.

### 3.2 Simulation runs

To get started, we simulated the model with no shocks on prices, and a unique solution for the stay or go choice - the condition being $\beta J < 1$ [6].

As we can see in figure 3, the qualitative outcomes we obtain are in line with those predicted by the formal study of the model. Once the trajectory goes over the TE curve[7], the market starts moving towards low $N$ and high $A$; the path seems quite close to the TE curve, showing faster adjustments in the number of firms.

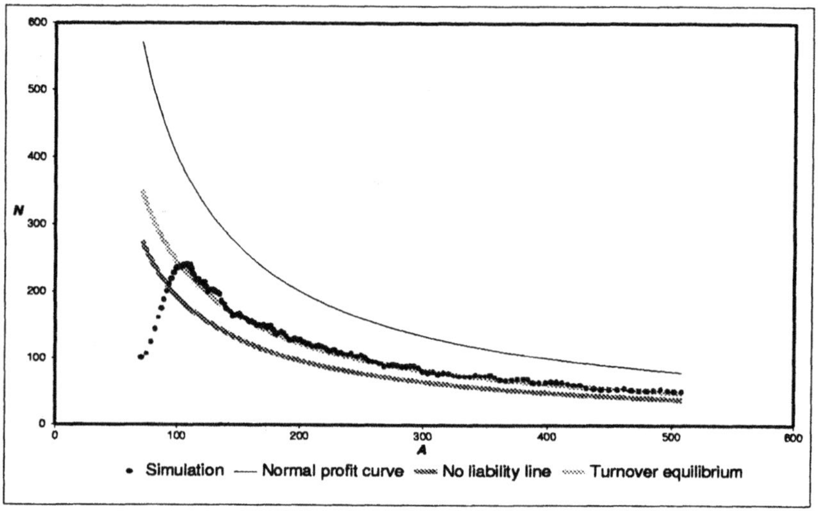


FIGURE 3.

Putting an idiosyncratic shock on prices, this portrait changes as follow.

First of all, the dynamic of the average equity base will be dumped by a composition effect, since new firms will be endowed on average of an equity base lower than the incumbents. As a consequence, the equilibrium condition $\Delta A = 0$ will be satisfied somewhere ***under*** the NPC.

The movement towards low $N$ and high $A$, then, will be either slower, or even reversed.

As a first step towards this second case, we calibrated the simulation parameters so to obtain the almost coincidence of the two curves. In figures 4 and 5 we can see two recurrent stylised facts that characterise the dynamics also in this quasi-equilibrium case, namely: a right skewed distribution

---

[6] Assuming, as we are doing, adaptive expectations, the equilibrium is also stable (cp. Brock and Durlauf [BD00]).

[7] The turnover equilibrium curve has been plotted solving numerically $k(.)$ in function of the parameters.

of the equity base, and irregular fluctuations of the number of firms' time series[8].

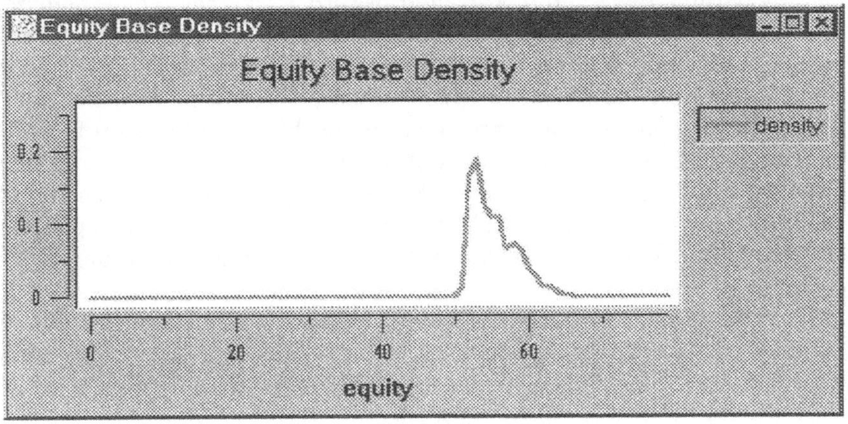


FIGURE 4.

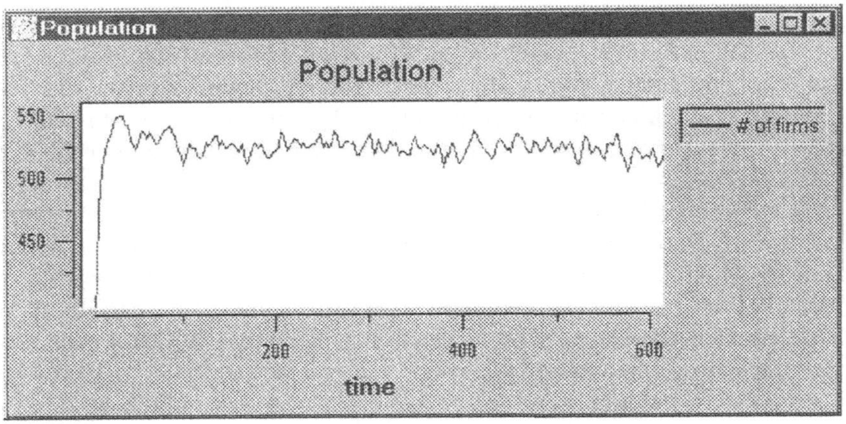


FIGURE 5.

The former is surely related to the composition effect due to the low equity level of the entrants; so as the latter are obviously fired by the stochastic components of the simulation.

[8]The (most "crucial") parameters values were the following: $r = 0.02$, $\alpha = 0.96$, $\beta = 2.2$, $J = 0.6$. In this case, so as in some following simulations, we have $\beta J > 1$, so that we can have multiple equilibria in the stay or go decision. Anyway, it can be shown that a jump from one equilibrium to another can only occur either with a very large exogenous shock, or with a downward shift in the $h$, towards negative values. The latter in our model cannot occur, since $h$ is endogenous, and anneals rapidly to positive values, the former seemed to us out of the scope of our analysis.

Both of them, however, are strictly related also to the interaction term ruling the exits. As a matter of fact, the relation between the asymmetry and the parameters $J$ and $\beta$ is straightforward, since they influence directly the entity of the exit fluxes; and hence also the way in which the turnover changes the composition of the population. For some set of parameters, the skewness can also turn negative: when all the firms face no liability costs, "little is better": a lower $A$ brings an higher rate of profit, which in turn is sufficient to countervail the signal received by the exit fluxes, strongly amplified by the high $J$. While the right tail of the distribution can be almost be cancelled out. With heterogeneity, and a non empty quota of firms facing financial constraints, this somewhat counterintuitive effect disappears.

The strenght of interaction has a quite direct relationalso with business fluctuations, *via* two effects: it can stuck an agent to a behaviour consistent with the others' one, hence stabilising it; and it can amplify random shocks that wouldn't elsewhere have great effect on the aggregate.

Letting the heterogeneity be higher - relatively to the distance between the NPC and the NLL - we can observe a change in the dynamic regime.

We already saw how the NPC, because of the composition effect, can be lower than in the base model. It happens something similar to the TE curve, that is, the curves that we draw as a function of the *average* value of the equity, overestimates the level at which there is equilibrium in the firms turnover. This is a common result of all the simulations conducted, and maybe it was predictable, studying more deeply the nonlinear specification of the entry-exit rules.

A little bit harder to predict, was the fact that the two curves can change their relative position. In figure 6, we report the outcomes of two simulations, conducted with all the structural parameters equal, but with a different variance of the random shock. As a benchmark, we draw only the NLL together with the series, since it is the only curve whose position is not changed by the heterogeneity.

Apparently, the higher dispersion in the equity base caused the "empiric" firms' turnover equilibrium curve to go above the line that stabilise the average $A$. This fact, changed the direction followed by the system inside the attractor basin, and caused the second experiment end towards higher values of $N$. In the figure, actually, we reported three hundred time steps for the series with the lower noise, and six hundreds for the higher variance one.

This latter, in fact, for the first three hundred time steps remains about the same levels of $N$. Afterwards, "something happens", and the series starts moving.

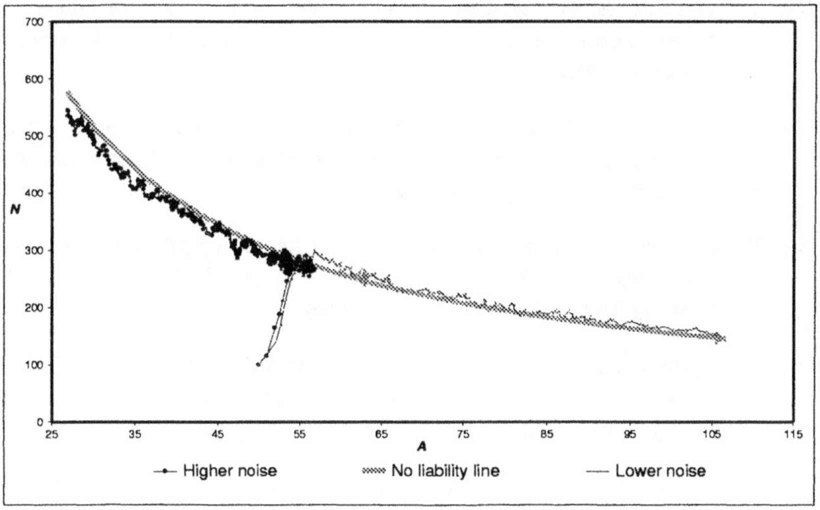


FIGURE 6.

# 4 Conclusions

Perfect competition is usually associated with the idea that firms take for exogenous the behaviour of their competitors. We studied a case in which this idea is not connected with the assumption of no interaction at all between agents. Namely, we considered the non price interaction between incumbent firms, due to the process of knowledge extraction - about the market profitability - in which they are involved in.

This assumption has been modelled with a mean field effect, in which the external field - the firms' own profitability - is endogenous, since it depends also on the entry fluxes and on the dynamics of their equity base.

Allowing for idiosyncratic shocks on prices, we simulated some artificial stylised facts of interest: a right skewed distribution for the equity base, attributable mainly to a composition effect; a long run positive firms turnover, which is related, among the others, to the dispersion of the equity distribution; business cycles, triggered by the stochastic components of the model and magnified by the interaction component of the decision taken by the firms.

Finally, we showed how the equity heterogeneity can have a role in switching the regime ruling the long run dynamics of the market.

# 5 References

[Axe97] Robert Axelrod. *The Complexity of Cooperation: Agent-Based Models of Competition and Collaboration.* Princeton University Press, Princeton, NJ, 1997.

[BD00] William A. Brock and Steven N. Durlauf. *Discrete Choice with Social Interaction.* working paper, University of Wisconsin - Madison, 2000.

[Bro72] William A. Brock. On models of expectations that arise from maximizing behaviour of economic agents over time. *Journal of Economic Theory*, 5:348–376, 1972.

[Cav98] Richard E. Caves. Industrial organization and new findings on the turnover and mobility of firms. *Journal of Economic Literature*, 36:1947–1982, December 1998.

[EA96] Joshua M. Epstein and Robert L. Axtell. *Growing Artificial Societies - Social Science from the Bottom Up.* MIT Press, Cambridge, MA, 1996.

[Eps99] Joshua M. Epstein. Agent based models and generative social science. *Complexity*, pages 55–77, May/June 1999.

[GK99] Mauro Gallegati and Alan Kirman. *Beyond The Representative Agent.* Edward Elgar, 1999.

[GS93] Dhananjay K. Gode and Shyam Sunder. Allocative efficiency of markets with zero-intelligence traders. *Journal of Political Economy*, 101:119–137, 1993.

[Hal97] John C. Haltiwanger. Measuring and analyzing aggregate fluctuations: The importance of building from microeconomic evidence. *Review of the Federal Reserve Bank of St. Louis*, pages 55–77, May/June 1997.

[Hop92] Hugo A Hopenhayn. Entry, exit, and firm dynamics in long run equilibrium. *Econometrica*, 60:1127–1150, 1992.

[Jov82] Boyan Jovanovic. Selection and the evolution of industry. *Econometrica*, 50:649–670, 1982.

[JR99] Jeffrey A. Joines and Stephen D. Roberts. Simulation in an object-oriented world. In *Proceedings of the 1999 Winter Simulation Conference*, 1999.

[Kir92] Alan P. Kirman. Whom or what does the representative individual represent. *Journal of Economic Perspective*, 6:117–136, 1992.

[LGG01] Roberto Leombruni, Domenico Delli Gatti, and Mauro Gallegati. Building theories from the bottom-up: A simple entry-exit model. In Francesco Luna and Benedikt Stefansson, editors, *Economic*

*Simulations in Swarm: Agent-Based Modelling and Object Oriented Programming.* Kluwer Academic Publishers, Cambridge, 2001.

[LS01] Francesco Luna and Benedikt Stefansson, editors. *Economic Simulations in Swarm: Agent-Based Modelling and Object Oriented Programming.* Kluwer Academic Publishers, 2001.

[Mar96] R. J. Martel. Heterogeneity aggregation and meaningfull macroeconomics. In D. Colander, editor, *Beyond Microfoundation.* Cambridge University Press, Cambridge, 1996.

[REL78] Jr. Robert E. Lucas. On the size distribution of business firms. *Bell Journal of Economics*, 9(2):508–23, May/June 1978.

[Ter00] Pietro Terna. Mind no-mind dilemma in agents for social science simulations. In G. Ballot and G Weisbuch, editors, *Applications of simulation to social sciences.* HERMES Science Publishing Ltd, Oxford, 2000.

# Part III.

# Finance

# Stability of Pareto-Zipf Law in Non-stationary Economies

**Sorin Solomon* and Peter Richmond****

*Racah Institute of Physics, Hebrew University of Jerusalem

**Department of Physics, Trinity College Dublin 2, Ireland

**Abstract:** Generalized Lotka-Volterra (GLV) models extending the (70 years old) logistic equation to stochastic systems consisting of a multitude of competing auto-catalytic components lead to power distribution laws of the (100 years old) Pareto-Zipf type. In particular, when applied to economic systems, GLV leads to power laws in the relative individual wealth distribution and in the market returns. These power laws and their exponent $\alpha$ are invariant to arbitrary variations in the total wealth of the system and to other endogenous and exogenous factors. The measured value of the exponent $\alpha = 1.4$ is related to built-in human social and biological constraints.

## 1. Background on logistic equations and on power laws

Aoki [1998] and Aoki and Yoshikawa [1999] have emphasized the importance of the logistic equation

1.1 $$d w / dt = A w - B w^2$$

in generic economic systems. They interpret w as the total product demand in a market. The linear term $Aw$ represents the fact that the emergence of new products is proportional to the present size of the market. The nonlinear term $- B w^2$ expresses the fact that the products have to compete with one another for a finite total potential market. Equation 1.1. was used for a very long time under the name of Bass' formula (for a review see Mahajan et al 1990) to parametrize the spread of new methods/ products/ concepts in a market (see Solomon et al 2000 for a multi-agent spatial generalization). However, Aoki and Yoshikawa emphasize that the logistic equation 1.1. admits a much wider range of interpretations. Solomon and Levy [1996] have suggested that w can represent the total capital within a financial system. In this interpretation, the first term represents the average returns that the system offers, while the term $-B w^2$ represents the effects of competition and other growth limiting factors. Hofbauer and Sigmund [1998] have studied similar systems in the context of evolutionary games and genetic dynamics. In fact, equations of the type 1.1 were introduced long ago in population biology by Lotka [1925] and Volterra [1926]. In this context, w represents the population size, Aw the aggregated effects of birth and natural death, while $- B w^2$ represents the effects of the competition for limited resources. In its discrete form, the logistic equation had a crucial role in the study of chaos (May 1974, Feigenbaum 1981). Aoki and Yoshikawa quote Montroll [1978] to the effect that "almost all the social phenomena, except in their relatively brief abnormal times obey the logistic growth".

A non less universal (and until recently un-related) property, spanning a wide range of disciplines from linguistics to economics and to biology, is the presence of scale-invariant probability distributions [see Stanley et al. 1998 for a review].

This property was initially observed by Pareto [1897 sic! ] in the context of individual wealth distribution: in each economy, the fraction P(w) of people owning a wealth w is proportional to a power of w:

$$1.2 \quad P(w) \sim w^{-1-\alpha}$$

For the last hundred years the value of $\alpha \sim 3/2$ changed little in time and across the various capitalist economies.

The emergence of Pareto power laws 1.2 in dynamical systems with random multiplicative dynamics was known since a long lime in a variety of fields: the frequency of words in texts [Yule 1924], economic growth [Gibrat 1931, Champernowne 1953], cities populations [Zipf 1949], wealth distribution [Ijiri and Simon 1977], renewal stochastic processes [Kesten 1973] etc.

Power laws are a crucial phenomenon in the emergence of macroscopic features in systems consisting of many sub-components. Indeed, the power laws imply that the sizes of the sub-components span many orders of magnitude. Therefore, the presence of the power laws constitutes a bridge between the microscopic structure of a system and its macroscopic emergent features (see book by Levy, Levy and Solomon 2000). The fact that the sizes of the sub-components are so different invalidates the usual "mean field" approach that aggregates all the sub-components into a "representative" component governed by a single aggregate equation.

It was shown [Solomon and Levy 1996, Levy and Solomon 1996, Malcai et. al 1999] that in fact, systems of the type 1.1, when studied at the microscopic agents level rather than in the aggregate form 1.1, lead to power law distributions of the form 1.2. These Generalized Lotka-Volterra (GLV) systems [Solomon 1998, 2000] treat each component of the system individually while taking into account their non-linear interactions. GLV explains not only the ubiquitous emergence of the power laws in many fields but also their stability in generic systems with non-stationary dynamics and arbitrarily varying total size [Biham et al 1998, Blank and Solomon 2000]. In particular, GLV explains the measured values of the exponent of the Pareto wealth distribution in terms of the social and biological constraints on the economy [Solomon and Levy 2000].

One can therefore say that the careful reconsideration of the system 1.1., led to the solution of a 100 years old puzzle by a 75 years old equation.

In the next section we introduce the GLV model, its various interpretations and show how it reduces to a set of decoupled stationary linear stochastic equations.

In section 3 we derive analytically the Pareto law for the relative wealth distribution in the GLV model.

In section 4 we discuss the stability of the Pareto exponent based on biological and social constraints and on the analytic probability distribution found at section 3.

In section 5 we relate via the GLV dynamics the Pareto individual wealth distribution to the market returns distribution.

Section 6 discusses the generic features that GLV implies for economic systems.

In the Appendix we relate the properties of the economic systems as described by GLV to the properties of statistical mechanics systems.

## 2. The generalized Lotka -Volterra model

The dynamics of the GLV system in the discrete formulation is defined by the recursive formula [Solomon and Levy 96, Solomon 98, Biham et al 98, Solomon 2000]:

2.1 $w_i(t+\tau) = r_i(t)\, w_i(t) + a\, w(t) - c(w,t)\, w_i(t)$

where $r_i(t)$'s are random numbers distributed with the same probability distribution (independent on i) with a square standard deviation D of order $\tau$. The functions a and c(w,t) are of order $\tau$ too in order to insure a meaningful "continuum limit" $\tau \rightarrow 0$.

- If one considers $w_i(t)$ as the individual wealth of the agent i,
- the random multiplicative factor $r_i(t)$ represents the random part of the returns that its capital $w_i(t)$ produces during the time between t and $t + \tau$.
- The coefficient a expresses the auto-catalytic property of wealth at the social level, i.e. it represents the wealth the individuals receive as members of the society in subsidies, services and social benefits. This is the reason it is proportional to the average wealth. This term prevents, as we shall show, the individual wealth falling below a certain minimum fraction of the average. The exact mechanism by which this happens (subsidies, minimal insurance or wage, elimination of the weak and their substitution by the more fit) is not, at this level of description, important.
- The coefficient c (w,t) controls the overall growth of the wealth in the system. It represents external limiting factors: finite amount of resources and money in the economy, technological inventions, wars, disasters etc. It also includes internal market effects: competition between investors, adverse influence of bids on prices (such as when large investors sell assets to realize their profits and cause thereby prices/ profits to fall). This term has the effect of limiting the growth of w (t) to values sustainable for the current conditions and resources.

c (w,t) parametrizes the general state of the economy. Time periods during which -c(w,t) is large and positive correspond to boom periods during which the wealth is on average increasing. Periods during which -c(w,t) is negative correspond to recessions, when typically the investments lead to negative or small returns. The surprising fact (proven in section 3) is that as long as the term c(w,t) and the distribution of the $r_i(t)$'s are common for all the equations 2.1 (for all i's), the Pareto power law 1.2 holds and its exponent is independent of the nature of c(w,t).

One can also look to the term c(w,t) as an expression of the inflation. If one thinks of $w_i$ as the real (as opposed to numerary) wealth of each individual, an increase of the total numerary in the system w(t) means that an agent with individual wealth $w_i$ will loose due to inflation an amount proportional to the increase in w and proportional to ones own wealth: $-c\, w(t)\, w_i(t)$.

A different interpretation of GLV may consider the market as a set of companies i=1,.....,N whose shares are traded at variable prices $w_i(t)$. The price of each stock $w_i(t)$ is proportional to the capitalization of the corresponding company i ( the total wealth of all the market shares of the company).

In this case,

- $r_i(t)$ represents the fluctuations in the market worth of the company. For a fixed total number of market shares, $r_i(t)$ also measures the relative changes in the individual share prices. These changes are typically fractions of the nominal share price (measured in percents or in points).
- a represents the correlation between the worth of each company $w_i$ and the market index w(t).
- c represents the competition between the companies for the finite amount of money in the market (and limits their worth). Time variation in the global resources may lead to lower or higher values of c that in turn lead to increases or decreases in the total (or average) wealth w(t).

Yet another interpretation of the GLV equation 2.1 is in the context of the investors herding behavior:

- $w_i(t)$. is the number of traders adopting a similar investment policy / position (they constitute a "herd" i).
- one assume that the sizes of these sets vary auto-catalytically according to the random factor $r_i(t)$. This can be justified by the fact that the visibility and social connections of a herd are proportional to its size.
- the a term represents the diffusion of traders between the herds.
- The nonlinear term c (w,t) represents the general status of popularity of the stock market as a whole. This term also includes the competition between various herds in attracting individual traders as members.

Very unexpectedly, many of the properties of the nonlinear system of coupled differential equations with time-depenent (and variable-dependent) coefficients 2.1 can be studied analytically.

To do this, let us first take the average in both members of 2.1 and get:

2.2 $w(t+\tau) = (r(t) + a)\, w(t) - c(w,t)\, w(t)$

where w and r are averages over i.

The equation 2.2 reduces in the continuum limit to a differential equation of the form 1.1. Therefore, at the aggregate level, the system 2.1 represents the same system as 1.1. However, the detailed representation 2.1 allows one to uncover properties that would be impossible to guess from contemplating 1.1.

The equation 2.1 can be written:

2.3 $w_i(t+\tau) - w_i(t) = [r_i(t) - 1]\ w_i(t) + a\, w(t) - c(w,t)\ w_i(t)$

Let us assume further that the average

2.4 $s = (r(t) - 1) = \langle r_i(t) - 1 \rangle$

is of the same order of magnitude $\tau$ with

2.5 $D = \langle r_i^{\,2} \rangle - r^2 \sim \langle (r_i - r)^2 \rangle = \langle [\varepsilon_i(t)]^2 \rangle$

where

2.6 $\varepsilon_i(t) = r_i(t) - r$

Note that

2.7 $\langle \varepsilon_i(t) \rangle = 0$

With these notations, 2.3 becomes:

2.8 $w_i(t+\tau) - w_i(t) = [\varepsilon_i(t) + s]\, w_i(t) + a\, w(t) - c(w,t)\ w_i(t)$

and consequently (assuming that in the N -> infinity limit the random fluctuations cancel according 2.7 [see however Huang and Solomon 2000, Malcai et al 1999 and Blank and Solomon 2000]):

2.9 $w(t+\tau) - w(t) = [s + a]\, w(t) - c(w,t)\, w(t)$

Note that 2.9 shows explicitly that when aggregated, the system 2.1 reduces in the $\tau \to 0$ limit to the logistic form 1.1 (with the identifications $A = [s+a]/\tau$ and $c(w,t) = Bw\tau$).

By introducing the new variable

2.10 $x_i(t) = w_i(t) / w(t)$

and applying the chain rule for the differentials $dx_i = x_i(t+\tau) - x_i(t)$, $dw_i = w_i(t+\tau) - w_i(t)$ and $dw = w(t+\tau) - w(t)$, the equation 2.8 becomes (considering 2.9):

$$
\begin{aligned}
2.11 \quad dx_i(t) &= 1/w(t)\, dw_i - w_i(t)/w(t)\, dw(t) \\
&= [\varepsilon_i(t) + s]\, x_i(t) + a - c(w,t)\ x_i(t) \\
&\quad - x_i(t)\, [s + a - c(w,t)] \\
&= [\varepsilon_i(t) - a]\, x_i + a
\end{aligned}
$$

Up to here there is no assumption that the system of $w_i$ is in a steady state, yet we were able to show that the stochastic dynamics of the relative individual wealths $x_i$ reduces to a set of identical decoupled equations 2.11 which are independent on c(w,t). In fact the dynamics of the relative wealths depends only on D and a (since the average r(t) is substracted in 2.11, 2.6) and not on the details of the interactions c(w,t) or on the average growth r(t).

In particular, even for large variations of c(w,t)) and w(t), if one keeps a/D constant, the dynamics of the relative wealth will be time independent. Consequently, even in a very nonstationary system, after some time, the relative wealth distribution will eventually converge to a time invariant form that we compute analytically in the next section. The approach of the asymptotic distribution by the $x_i$'s is governed by the equations 2.11 and therefore is independent on the global non-stationary dynamics induced by c (w,t) (and/or r(t)) on w(t).

# 3. The Pareto wealth distribution in GLV

Let us write the generic equation:

3.1 $x(t+\tau) - x(t) = \varepsilon(t)\, g(x(t)) + f(x(t))$

Without loss of generality, we can assume

3.2 $< \varepsilon(t) > = 0$

since the non-random part of $\varepsilon(\tau)$ can be absorbed in a redefinition of

$f \rightarrow f + < \varepsilon(t) > g$.

In order for the noise $\varepsilon(t)$ to be relevant as one takes "the continuum limit" $\tau \rightarrow 0$ we assume the square standard deviation:

3.3 $D = < \varepsilon(t)^2 >$

to be of order $\tau$.

As a consequence, we will have to keep in the computations below terms of order $\varepsilon(t)^2$ and therefore will have to keep occasionally terms of the second order in the differential

$dx = x(t+\tau) - x(t)$.

For a meaningful "continuum limit", the function f (x) is taken of order $\tau$ while g(x) is of order 1.

In order to find the asymptotic probability distribution corresponding to the dynamics 3.1 one will try to reduce it by performing an appropriate change of variables:

3.4 $y(t) = y\,(x(t))$

to a Langevin process [Richmond 2001] with constant (unit) coefficient for the random term:

3.5 $y(t+\tau) - y(t) = \varepsilon\,(t) + j\,(y(t))$

Such an equation is known to lead to the (Maxwell-Boltzmann) stationary distribution [McQuarrie 2000]:

3.6 $P(y)\,dy = \exp\left[\,2/D \int^{y} j\,(z)\,dz\,\right] dy$

(note that the formula 3.6 makes sense in the "continuum limit" since both D and j are of order $\tau$).

The time evolution equation for the new variable y(t) is obtained from 3.1 using the chain differential rule up to second order (in order to keep the terms of order D):

3.7 $y(x(t+\tau)) - y(x(t)) = dy = (dy\,/d\,x)\;\; dx + 1/2\; d^2y/dx^2\;\, (dx)^2 + \text{etc.}$
$= dy\,/dx\, [x(t+\tau\,) - x(t)] + 1/2\; d^2y/dx^2\;\, [x(t+\tau\,) - x(t)]^2 +$ etc.
$= dy\,/dx\, [\,\varepsilon\,(t)\, g\,(x(t)) + f\,(x(t)] + Dg^2/2\;\, d^2y\,/dx^2 + \text{etc.}$

Where we denoted by etc. the terms in the r.h.s. that vanish faster than $\tau$ in the continuum limit $\tau \to 0$.

In the classical particular case g(x) = x and f(x) = 0 one has:

3.8 $x(t+\tau) - x(t) = \varepsilon\,(t)\, x(t)$

which transforms through y(x) = ln x into:

3.9 $y(t+\tau) - y(t) = \varepsilon\,(t) - D/2$

rather than just naively

3.10 $y(t+\tau) - y(t) = \varepsilon\,(t)$

(see also Maslov, Marsili and Zhang 98, Sornette and Cont 97, Bouchaud and Mezard 2000 which parametrize the stochastic term using the form of an exponential:

$x(t+\tau) - x(t) = [\exp \varepsilon\,(t) - 1]\, x(t) \sim [\varepsilon\,(t) + D/2]\, x(t)$

which therefore transforms into

$y(t+\tau) - y(t) = \varepsilon\,(t)$

We choose here to parameterize the stochastic terms by the simple form 3.8, 3.1, 2.11 which is more directly related to the parameters used in the discrete numerical simulations of GLV [Biham et al 98, Huang and Solomon 2000]).

Obviously, in order to bring 3.7 to the form 3.5 one needs to make the change of variables:

3.11 $$dy = 1/g\, dx$$

With this change, the equation 3.7 becomes:

3.12 $$y(t+\tau) - y(t) = \varepsilon(t) + f(x(y))/g(x(y)) - D/2\, dg/dx$$

According to 3.6, this means:

3.13 $$P(y)\, dy = \exp\left[ 2/D \int^{y} (f(x(z))/g(x(z)) - D/2\, dg/dx)dz \right] dy$$

One can use 3.11 to change the variables in the integrals and obtain

3.14 $$P(x)\, dx = P(y)\, dy$$

$$= \exp\left[ 2/D \int^{x} (f(u)/g^2(u) - D/2(dg/dx)/g)\, du \right] 1/x\, dx$$

$$= \exp\left[ 2/D \int^{x} (f(u)/g^2(u)]du \right] 1/x^2\, dx$$

In order to find the stationary distribution of $x_i(t) = w_i(t)/w(t)$ all one has to do is to apply 3.14 to 2.11 i.e. to take:

3.15 $$f = a(1 - x)$$

and

3.16 $$g = x$$

and obtain therefore, according 3.14:

3.17 $$P(x)\, dx = \exp\left[ 2/D \int^{x} (f(u)/g^2(u)]du \right.\quad 1/x^2\, dx$$

$$= \exp\left[ 2/D \int (a-x)/x^2\, dx\right] 1/x^2\, dx$$

$$= \exp\left[ 2/D \int (a-x)/x^2\, dx\right] 1/x^2\, dx$$

$$= x^{-1-\alpha} \exp\left[-2a/(xD)\right]$$

with

3.18 $$\alpha = 1 + 2a/D$$

The distribution P(x) has a peak at $x_0 = 1/(1 + D/a)$. Above $x_0$, the relative wealth distribution P(x) behaves like a power law while below it P(x) vanishes very fast (all the derivatives diverge at x=0).

# 4. The social and biological constraints: $\alpha$ stability

Until now, we have explained the survival of the power law 3.17-3.18 in the presence of large exogenous and endogenous changes in the total wealth. We now relate the constant value of $\alpha \sim 3/2$ measured over the last 100 years (and for all the major capitalist economies) to the social and biological constraints that any society is submitted to.

The main idea is to exploit the particularities of the wealth distribution shape 3.17 in order to relate the power decay of the probability distribution at large wealth to the wealth distribution of the poorest. This is obviously possible since both the exponent $\alpha = 1 + 2a/D$ of the large wealth power law and the coefficient 2a/D in the exponential of -1/x which dominates the low wealth behavior, depend on the single parameter a/D.

Consequently, the constraints on how poor the poor are allowed (or can afford) to be, determine the power low distribution of the upper society wealth.

This relation is not limited to the GLV dynamics and takes place for any dynamics that leads to an asymptotic power law distribution for large relative wealth and to a very sharp decay at low values.

In our case note that the decay of the probability density 3.17 as x -> 0 is extremely fast: all the derivatives diverge!

In these conditions, the lowest relative wealth $x_m$ is estimated roughly by assuming that there are no individuals below it and that above it, the power law is fulfilled. Then one gets $x_m$ from the identity

$\langle x\rangle = \langle x_i(t)\rangle = \langle w_i(t)/w(t)\rangle = w/w = 1$ which implies:

4.1 $$\langle x\rangle = 1 = {}_{x_m} x^{-\alpha}\, dx \,/\, {}_{x_m} x^{-1-\alpha}\, dx$$
$$= [-1/(1-\alpha)\, x_m^{\,1-\alpha}]/[-1/(-\alpha)\, x_m^{\,-\alpha}]$$
$$= x_m\, \alpha/(\alpha-1)$$

or:

4.2 $\alpha = 1/(1 - x_m)$

(Malcai et al 99)

And according 3.18:

4.3 $x_m = 1 - 1/\alpha = 1/(1 + 1/2\ D/a)$

Which is a reasonable value for $x_m$ considering that the peak of P(x) 3.17 is at $x_0 = 1/(1+D/a)$ and that the decay below this value as x->0 is extremely sharp.

Based on 4.2-4.3 one can now give a general scenario of how the internal interests and constraints within society lead to the actual value of $\alpha \sim 3/2$ measured repeatedly in various economies in the last 100 years.

Suppose that in a given economy the wealth necessary to keep a person alive is K. Certainly, anybody having less than that will have a very destabilizing effect on the society, so the number of people with wealth less then K should be negligible if that society is to survive. Let us now look at the average family supported by an average wealth. Let us assume there are in average L dependents per one wage earner. Certainly they will need a wealth of order KL, otherwise the wage earners will try to correct the situation by strikes, negotiations, elections or revolts. Note that in a sense, KL is the **definition** of the minimal amount for supporting L dependents, since the prices of the prime necessities will always adjust to it: if the average wealth increases so will do the prices of housing, services, etc.

In short, while the poorest persons (which cannot even afford a family) will make sure they do not get less than 1/L of the average, the average will almost by definition take care that their income is at least L times the minimal wealth necessary for supporting one person. All in all, we are lead to the prediction that $x_m \sim 1/L$ and thus (according 4.2) $\alpha = 1/(1 - x_m) \sim L/(L-1)$.

These relations fit well the known numbers in the typical capitalist economies in the last century: family size L ~ 3-4, poverty line (below which people get subsidized) $x_m \sim 1/4 - 1/3$ and $\alpha \sim 1.33 - 1.5$.

The key result we obtain is, therefore, that the relative poverty lower bound totally governs the overall relative wealth distribution. The dynamical details by which this distribution arises are of course complex and depend on the interactions in the system.

The low birth rate in some of today's societies might suggest higher values for $x_m$ with associated higher values for leading to greater equality and stability. On the other hand, if the fluctuations D in the economy are large, the social subsidies, as measured by the coefficient a need to be increased in order to ensure that $x_m = 1/(1 + 1/2\ D/a)$ (and $\alpha = 1 + 2a/D$) remains constant. For example, energetic stock markets combined with stagnant social security or pensions may lead to a decrease in $\alpha$. This is a well known effect: a period of large financial fluctuations leads to a significant number of "nouveau riches" which may leave many others far behind financially.

# 5. Heavy tails of market returns in GLV

We will discuss here some of the financial market implications of the formal results obtained in the previous sections. Let us first discuss the fluctuations induced in w(t) by the dynamics 2.1.

The fluctuations in w(t) are important because one can think of w(t) as a measure of the total worth (capitalization) of the stock market. Therefore w(t) is proportional to the market index and its time variation is related to the market returns:

5.1 $R(t) = \ln[w(t+1) / w(t)]$

Or, assuming

5.2 $\Delta w(t) = w(t+1) - w(t) << w(t)$,

and expanding the logarithm in 5.1:

5.3 $R(t) \sim \ln[(w(t) + \Delta w(t)) / w(t)] \sim \Delta w(t) / w(t)$

This quantity measures the wealth at time t+1 of an agent that invested 1 Dollar in the stock at time t.

In order to estimate the probability distribution of the returns R(t) as resulting from the GLV model 2.1 let us consider here the discrete GLV dynamics in which the individual wealths $w_i$ are updated sequentially. More precisely, at each time t, a random integer i between 1 and N is extracted and the corresponding wealth $w_i(t)$ is updated according to 2.1. (Updating one agent at a time means effectively that the time is rescaled $\tau \rightarrow N\tau$ and therefore, in order to describe the same continuum process, one has to rescale also a, D ,c in 2.1 by a factor N).

The change in w(t) effectuated by the updating of a single $w_i$ at time t will be:

5.4 $\Delta w(t) = [w_i(t+1) - w_i(t)]/N$

since the changes due to the contributions to w(t) due to all the other agents are null ($w_k(t+1) = w_k(t)$ if k is different from i).

Using 2.1 in 5.4 one gets

5.5 $\Delta w(t) = [(r_i(t) - 1) w_i(t) + a\, w(t) - c(w,t)\; w_i(t)]/N$

Substituting 5.5 in 5.3 and using 2.10 $x_i(t) = w_i(t)/w(t)$ one gets:

5.6 $R(t) \sim \Delta w(t) / w(t) \sim [(r_i(t) - 1) x_i(t) + a - c(w,t)\; x_i(t)]/N$

One sees that the returns consist of 2 components:

- one deterministic $[a - c(w,t)]\, x_i(t)\; /N$ depending on the social security policy a and the state of the economy c(w,t) and
- one stochastic which dominates the short time fluctuations:

5.7 $R(t) \sim (r_i(t) - 1)\, x_i(t)/N$

The stochastic part 5.7 is seen to be proportional to the $x_i$'s and therefore it inherits the stochastic properties of the probability distribution P(x) 3.17. In particular, in a wide range of parameters, the variations R(t) have a power law distribution:

5.8 $P(R) \sim R^{-1-\alpha}$

A random walk with steps of sizes distributed by the power law probability distribution 5.8 is called a Levy walk of index α. Τηε

The sum of many such steps does not converge to a Gaussian distribution as expected (by the central limit theorem) from a random walk with steps of fixed scale. Rather, the sum converges to a universal shape called a Levy distribution of index α denoted by the symbol $L_\alpha(w)$. In a certain range of w the function $L_\alpha(w)$ itself behaves as a power law 5.8.

Accordingly, GLV predicts that the market returns will be distributed (in a certain R range) by a (truncated) Levy distribution $L_\alpha(w)$ of index α given by Eq. 3.18 (Solomon 1998).

This unexpected relation between the wealth distribution and the market returns [Levy and Solomon 1997] turns out to be in accordance with the actual experimental data [Mantegna and Stanley 1996].

# 6. Conclusions

It is well known and sometimes over emphasized that ill-willed or incapable politicians may influence economics in the negative way by preventing people from working and trading or simply by stealing. A less clear issue is whether good-willed capable politicians can do anything positive to improve the economic and social welfare of the citizens.

By analyzing the economic dynamics from a very general point of view we extracted in this paper, features which are common to most economies and which put generic limits on how much (and at which price) one can improve the financial and social realities. Even from weak generic assumptions on the capital dynamics, one was able obtain very specific predictions on the way the social wealth is distributed. A crucial assumption was that the capital market is fair, i.e. equal capitals have equal opportunities. E.g. by investing twice 100 USD in the same asset one is likely to obtain the same output as from investing once 200 USD (independently on the investor's identity). Mathematically, this was expressed by our assumption that there is a unique probability distribution, independent on i for all the random factors $r_i(t)$ and that the same function c(w,t) appears in all the equations 2.1 (for all i's).

We showed that in such a market, the wealth distribution among the individual investors fulfills a power law 3.17.

The exponent $\alpha$ 3.18 has been measured repeatedly in the last hundred years and found to be a constant of order 3/2. This means that in a system with say 250 million people, the poorest one will have approximately 400000 times less than the richest one. The average individual will have roughly 100000 less than the wealthiest. These numbers are in agreement with the actual ones in the US economy. Social security initiatives cannot change the Pareto "power law", they can only seek to change the value of the exponent $\alpha$. For instance, if one subsidizes the poorest citizens in order to prevent the last one to fall below a certain "poverty line" (say a fraction $x_m$ of the average wealth) one is lead to a value of the power law exponent 4.2. $\alpha = 1/(1-x_m)$.

This connection between the relative wealth of the poorest and the wealth hierarchy among of the richest [Anderson 1995] emphasizes the subtle connections that make financial management of the social ecology [Levy et al 1996] very difficult to control and predict.

The value $\alpha$ =3/2 above is common to most capitalist economies over most of their history. As discussed in section 4, this indicates that having a ratio $x_m$ = 1/4-1/3 is not the result of the policies/ actions of the various governments but rather a result of more basic biological constraints.

The balance between "fair play" for the capital and minimal socio-biological needs of the humans seems to trap the world economy into a power law wealth distribution which determines much of its dynamical and equilibrium properties.

One sees now that without underestimating the responsibility of the governments to pursue fair, humane and efficient policies, one cannot expect them to change in a very dramatic way the above economic/ financial realities.

Let us remark that low $x_m$ values that lead to $\alpha \sim 1$ have cf. 4.2 a dramatic influence on the stock markets stability: $\alpha \sim 1$ means all of the wealth belongs to just a few individuals. This in turn leads cf. 5.8 [Biham et al 98, Huang and Solomon 2000] to macroscopic fluctuations in the financial indices. Having all the wealth concentrated in just a few hands, implies chaotic instability in the markets ( in contrast to the case in which the wealth is distributed among many individuals and their various fluctuations average smoothly). One sees that beyond the humanistic arguments, a judicious social security policy is a requirement of the capital markets stability as well.

Mechanisms similar to the described above apply in appropriately modified ways to companies and countries [Solomon 2000, Solomon 2001] and establish severe limits to how equalitarian (or how unequal) one can expect/afford the world economy to be.

# APPENDIX Econodynamics vs Thermodynamics; Market Efficiency vs. Thermal Equilibrium; Pareto vs Boltzmann laws

We have used intensively in this paper the formal equivalence between the non-stationary systems 2.1 of interacting $w_i$'s and the equilibrium statistical mechanics systems governed by the universal Boltzmann distribution 3.6. One can take seriously/ literally this formal equivalence and construct a series of analogies between the two systems. This leads to new connections between known economic and financial facts. E.g one can relate the Pareto distribution to the efficient market hypothesis:

We have seen that in order to obtain a Pareto power law wealth distribution it is sufficient that the relative returns of the agents are stochastically equivalent, i.e. there are no investors or strategies that can obtain "abnormal" returns. This is usually the claim of the believers in the **efficient market hypothesis**. By definition an *efficient market* is a market in which the market pricing mechanism is so efficient that it reaches the "right price" before any of the agents can take systematic advantage (arbitrage) of the mis-pricing of one item vs. another.

Therefore, the presence of a Pareto wealth distribution is a sign of "market efficiency" in analogy to the Boltzmann distribution in statistical mechanics systems whose presence is a sign of thermal equilibrium.

Indeed physical systems which are **not** in thermal equilibrium (e.g. are forced by some external field - say by laser pumping) do **not** fulfill the Boltzmann law.

Similarly, markets that are not efficient (e.g. when some groups of investors make systematically more profit than others) do not yield power laws [Solomon and Levy 2000].

Market efficiency and power laws can then be thought as the short time and long time faces of the same medal/phenomenon.

This analogy is consistent with the interpretation of market efficiency as an analog to the Second law of Thermodynamics:

- one can extract energy (only) from systems that are not in thermal equilibrium

- one can extract wealth (only) from markets that are not efficient.

- by extracting energy from a non-equilibrium thermal system one gets it closer to an equilibrium one.

- by extracting wealth from a non-efficient market one brings it closer to an efficient one

- in the process of approaching thermal equilibrium, one also approaches the Boltzmann energy distribution
- in the process of approaching the efficient market one also approaches the Pareto wealth distribution.

- by having microscopic information on the state of the system (beyond the knowledge of the macroscopic thermodynamic measurables), one can extract additional energy from a systems in thermal equilibrium (e.g.Maxwell demons "gedanken experiment" [Leff and Rex 1990]).
- by having detailed private information on a financial market, (beyond the publicly available data), one can extract excess profits even if the market pricing is efficient.

# References


P. W. Anderson in The Economy as an Evolving Complex System II (Redwood City, Calif.: Addison-Wesley, 1995), eds. W. B. Arthur, S. N. Durlauf, and D. A. Lane.

M. Aoki, New Approaches to Macroeconomic Modeling : Evolutionary Stochastic Dynamics, Multiple Equilibria, and Externalities As Field Effects, Cambridge Univ Pr 1998

M. Aoki and H. Yoshikawa, Demand creation and economic growth, U. of Tokio , Ctr. for Int'l. Research on the Japanese Econ. 1999.

C.H. Bennett, "Demons, Engines, and the Second Law." Scientific American Nov. 1987: 108-116.

O. Biham, O. Malcai, M. Levy, S. Solomon, Phys. Rev. E 58, 1352 (1998)

A. Blank and S. Solomon "Power laws in cities population, financial markets and internet sites (scaling in systems with a variable number of components)", Physica A 287 (1-2) (2000) pp. 279-288.

J. P. Bouchaud and M. M\'ezard, Physica A 282, 536 (2000)

D.G. Champenowne, Economic Journal 63 (1953) 318.

M. J. Feigenbaum, Universal behavior in nonlinear systems, Los Alamos Science, 1, 4-27 (1981).

R. Gibrat, Les in'egalite's 'economiques (1931 Paris, Sirey).

J. Hofbauer, K. Sigmund, Evolutionary Games and Population Dynamics Cambridge Univ Pr 1998.

Z. F. Huang and S. Solomon, e-print, cond-mat/0008026, and to appear in Eur. Phys. J. B.

Y. Ijiri and H. A. Simon, Skew Distributions and the Sizes of Business Firms (North-Holland, Amsterdam, 1977).

H. Kesten, Acta Math. 131 (1973) 207.

H.S. Leff and A. F. Rex. Maxwell's Demon: Entropy, Information,Computing. Princeton University Press 1990.

M. Levy, S. Solomon Power Laws are Logarithmic Boltzmann Laws International Journal of Modern Physics C , Vol. 7, No. 4 (1996) 595; adap-org/9607001

M. Levy M, and S. Solomon (1997), Physica A 242, 90.

M. Levy, H. Levy and S. Solomon, "Microscopic Simulation of Financial Markets; from Investor Behavior to Market Phenomena" Academic Press, New York, 2000.

M. Levy, N. Persky, and S. Solomon, "The Complex Dynamics of a Simple Stock Market Model" International Journal of High Speed Computing, 8, 1996 http://www.ge.infm.it/econophysics/papers/solomonpapers/stock_ex_model.ps.gz .

A.J. Lotka, (editor) Elements of Physical Biology, Williams and Wilkins, Baltimore, 1925;

V. Mahajan, E. Muller, and F.M. Bass, New product diffusion models in marketing: A review and directions for research. Journal of Marketing 54, 1 (January 1990), 1-26.

D. A. McQuarrie, Statistical Mechanics, University Science Books 2000.

O. Malcai, O. Biham and S. Solomon, Phys. Rev. E, 60, 1299, (1999).

R. Mantegna and H. E. Stanley, An Introduction to Econophysics: Correlations and Complexity in Finance Cambridge University Press, Cambridge, 1999.

M. Marsili, S. Maslov and Y-C. Zhang, Physica A 253,(1998) 403

E. W. Montroll and M. F. Shlesinger, Proc. Nat. Acad. Sci. USA 79, 3380 (1982).

R. May, Biological populations with nonoverlapping generations: stable

points, stable cycles, and chaos. Science, 186, 645-47, (1974). .

V. Pareto, Cours d"economie politique. Reprinted as a volume of Oeuvres Compl`etes (Droz, Geneva, 1896-1965). V. Pareto, Cours d'Economique Politique (Macmillan, Paris, 1897), Vol. 2. V. Pareto, Le Cours d' ' Economie Politique (Macmillan, London, 1897).

S. Redner, Am. J. Phys. 58, 267 (1990) ; Eur. Phys. J. B4, 131 (1998).

P. Richmond, Power Law Distributions and Dynamic behaviour of Stock Markets to appear in Eur. J. Phys 2001.

M. F. Shlesinger and E. W. Montroll, Proc. Nat. Acad. Sci. USA (Appl. Math. Sci.) 79, 3380 (1982)

H. A. Simon and C. P. Bonini, Amer. Econ. Rev. 48, 607 (1958)

S. Solomon and M. Levy, adap-org/9609002 , Int. J. Mod. Phys. C7, (1996) 745

S. Solomon, in Decision Technologies for Computational Finance, edited by A.-P. Refenes, A. N. Burgess, and J. E. Moody (Kluwer Academic Publishers, 1998).

S. Solomon, Generalized Lotka-Volterra (GLV) Models and Generic Emergence of Scaling Laws in Stock Markets, in "Applications of Simulation to Social Sciences" ,Eds: G Ballot and G. Weisbuch; Hermes Science Publications 2000.

S. Solomon, Why Do Impossible Things Always Happen?, to appear Princeton U. Press 2001.

S. Solomon and M. Levy, Market Ecology, Pareto Wealth Distribution and Leptokurtic Returns in Microscopic Simulation of the LLS Stock Market Model, http://arXiv.org/abs/cond-mat/0005416 ; To appear in the Proceedings of "Complex behavior in economics: Aix en Provence (Marseille), France, 2000".

S. Solomon, G. Weisbuch L. de Arcangelis, N. Jan, D. Stauffer, Social Percolation Models, Physica A 277, 239 (2000)

D. Sornette and R. Cont ,in J. Phys. I France 7 (1997) 431

H.E. Stanley, L.A.N. Amaral, J.S. Andrade, S.V. Buldyrev, S. Havlin, H.A. Makse, C.K. Peng, B. Suki and G. Viswanathan, Scale-Invariant Correlations in the Biological and Social Sciences. Phil. Mag. B, vol. 77, 1998, p. 1373.

V. Volterra [1926], Nature, 118, 558.

S. Moss de Oliveira , H. de Oliveira and D.Stauffer, Evolution, Money, War and Computers, B.G. Teubner Stuttgart-Leipzig 1999.

U. G. Yule, Phil. Trans. B. 213, 21 (1924)

G. K. Zipf, Human Behavior and the Principle of Least Effort (Addison-Wesley Press, Cambridge, MA, 1949).

# Toy Models of Markets with Heterogeneous Interacting Agents

Matteo Marsili[1]

Istituto Nazionale per la Fisica della Materia (*INFM*), Unitá di Trieste-SISSA, I-34014 Trieste

**Abstract.** Simple models of financial markets with heterogeneous adaptive agents have been recently investigated using tools of statistical mechanics of disordered systems. We review and discuss the main results of this approach.

## 1 Introduction

The need of going beyond the representative agent approach, faces the difficulties of accounting mathematically for a complex reality: The description of a system of $N$ truly heterogeneous interacting agents requires a number of variables proportional to $N$ whose behavior cannot be captured by few "representative" degrees of freedom.

Theoretical physics has faced quite similar problems when disordered materials, such as glasses or disordered alloys, have started to attract attention. At odds with fluids, the atoms of disordered materials are not free to move. But they are neither placed on a regular grid, as in a crystal. Hence the elegant methods devised for studying homogeneous or periodic systems – such as fluids or regular crystals – are of little use.

The approach of theoretical physics to disordered systems [1] is based on a statistical approach: A typical disordered system is a realization of a random system in which the interactions between degrees of freedom are drawn from a given probability distribution. As long as the system is large, i.e. composed of a large number $N$ of interacting units, the macroscopic behavior is subject to laws of large numbers (under generic assumptions of regularity of the distribution of interactions) which constrain its behavior. In other words, for a typical realization of the random system, with a probability which converges to one as $N \to \infty$, all macroscopic quantities attain a value which is independent of the particular realization of the interactions. This property, known as *self-averaging*, means that the description of a typical realization of the random system is also valid for a whole populations – or *ensembles* – of systems which share the same distribution of interactions. The implication is that the aggregate macroscopic behavior is therefore independent of the detailed microscopic interaction.

Financial markets are an ideal prototype example of a system for which this approach can be applied. Indeed they are composed of a large number of interacting units – the agents. Agents are heterogeneous in their perception of the world around them and in the way they react to it[1]. Furthermore their interaction, mediated by market prices, has the same structure of mean field interaction in physics, for which exact results are possible. Because of the statistical nature of the laws which govern the macroscopic behavior, one expects that collective properties enjoy a high degree of independence of microscopic details. Common sense then suggests to look for highly simplified models which, on one side, preserve the whole complexity of the interaction and, on the other allow for an analytic approach.

The aim of this contribution is to review recent results on this line of research. The models we shall discuss describe systems of heterogeneous agents interacting via a market mechanism. We first discuss the El Farol bar problem [2], which has inspired much of the research in this field. We mention how this can be formalized in an asset market model and finally we discuss its later refinement, the Minority Game [3,4]. We focus on four aspects:

1. A key issue in financial markets is the interplay between agents and information. In few words, agents receive pieces of information and they process it in order to take their trading decisions. This process transmits information into prices. The first issue that these models allow to address is then market informational efficiency: How efficiently is an information on the future state of the market incorporated into market prices?
2. The second issue concerns market impact. Agents may trade as *price takers*, i.e. as if they were gambling against the market price process. In reality agents have an impact on prices because market prices depend on how much agents invest. What is the consequence of price taking behavior and to what extent is market impact important?
3. Market can be regarded as ecosystems where different types of traders interact [4,5]. While some trader may use the markets as an instruments for exchanging goods – which is why traditionally markets exist – some other may trade for pure speculation. What is the interplay between these two populations?
4. High frequency data on financial markets has shown quite remarkably stable non-trivial empirical laws – sometimes called the "stylized facts" [6]. Can one reproduce stylized facts within such simple models and, if so, what is the theoretical insight which agent based models offer?

The simplicity of the models allows one to give clear and detailed answers to most of these as well as to other questions. In particular recent research has shown that

1. markets typically become more and more efficient as the number of participants increases [7,8].

---

[1] Many different sources of heterogeneity are possible: beliefs, expectations, asymmetric information, past experience, differences in agent's environment and capabilities, etc.

2. The assumption of price taking behavior is not at all an innocent one. One may naively expect that, when the number of agents gets large, each single agent becomes negligible and hence whether agents take into account their market impact or not makes a vanishingly small difference. Within the models we discuss, this argument is wrong. In particular price taking behavior is responsible for inefficiencies such as excess volatility [9].
3. Different types of traders may coexist in a market, with one species helping the other [10]
4. the main stylized facts can be recovered and their origin can be partially understood within this framework [11–15].

While both the El Farol bar problem and the Asset market model originates in the economic literature – with the former representing a provocative departure from the mainstream trend on which the second is rooted – the Minority Game has been introduced and studied mostly by physicists up to now. Still we shall see that these models belong to one and the same generic class of models and they share the same generic behavior.

Our principal goal is to present the physicists' approach to these complex systems to an audience of economists and social scientists, with the hope that this may encourage further collaboration between the two communities. We shall try to cover this material in a descriptive way. The reader interested in more detailed discussion is referred to original papers. It is finally worth pointing out that, in spite of the author's effort for completeness, the exposition shall inevitably be biased towards his own contributions and the work he knows best.

In the following we shall let $N$ be the number of agents and we shall be interested in the limit $N \to \infty$. The market interaction is repeated over an infinite number of periods. Our reference framework is agent based models. Also we shall think in terms of dynamic stochastic processes and their stationary states rather than in terms of equilibria in the economic interpretation.

## 2 El Farol bar problem and Asset markets

The El Farol bar problem [2] deals apparently with a non-market context. In each period $t$, $N$ agents have to decide whether to go ($a_i(t) = 1$) or not ($a_i(t) = 0$) to the El Farol bar. The problem is that the bar is enjoyable only if it is uncrowded. In Ref. [2] $N = 100$ and there are only $S = 60$ seats. So the bar is enjoyable only if $A(t) = \sum_{i \in N} a_i(t)$ does not exceeds $S$.

Arthur [2] points out the frustration inherent in such a situation: If everybody thinks that the bar will be crowded, no one will go and the bar will be empty. Conversely all agents may attend at once the bar, if they all expect it to be uncrowded. Hence he argues that this is a situation which forces expectations of different agents to diverge. We expect that, if agents start with different expectation models and revise them according to their performance, agents expectations will never converge and agents' heterogeneity will be preserved forever.

Arthur shows by computer experiment[2] that inductively rational agents are able to self-organize so that the attendance (demand) $A(t)$ fluctuates around the comfort level (supply) $S$.

The focus of Ref. [2] is mostly on the distinction between deductive and inductive rationality. The El Farol problem has also been regarded as a paradigm for congestion problems in transportation and information networks [16]. The case of a population of agents who face the choice between two roads is quite similar: everyone will try to avoid the crowded one. The El Farol bar problem has acquired considerable popularity and is regarded as a paradigm of complex systems [17,18]. The reformulation of the ideas of Ref. [2] in a well defined mathematical model – the Minority Game [3] – has allowed for a considerable progress in the understanding of the properties of such complex systems.

### 2.1 The El Farol bar as an asset market

The El Farol bar can be regarded as a very simple market: $S$ units of an asset are offered to $N$ agents who may decide to invest 1 \$ to buy it ($a_i(t) = 1$) or not to invest ($a_i(t) = 0$). Each unit delivers a return of $R = 1$ \$ to his owner at the end of the period. The demand $A(t) = \sum_i a_i(t)$ fixes the price $p(t) = A(t)/S$ of the asset, via the market clearing condition. Hence an agent who invests $a_i(t)$ \$ in the asset, receives $a_i(t)/p(t)$ units of it. These will be worth $a_i(t)/p(t)$ \$ at the end of the period. If $p(t) > 1$, which occurs if $A(t) > S$ (crowded bar), it is not convenient to invest (attend). If $p(t) < 1$ it is convenient to invest (attend). A market model of this kind has been recently discussed in detail by Berg *et al.* [8]. They study a well established market model [19] in which the return of the asset fluctuates and agents receive some asymmetric information on the return. This information is heterogeneous across agents and the issue that Ref. [8] addresses is whether the market mechanism is able to aggregate the dispersed information into a single market price. In more precise terms, this means to understand to what extent market prices follow returns. The outcome depends crucially on the assumption of price taking behavior: If agents behave as price takers, which means that they neglect their impact on the price, the market converges to a competitive equilibrium. As the number of agents in the market increases, prices approach more and more returns. At and beyond a critical number of agents $N_c$ prices become equal to returns: In other words, all information transmitted to agents is incorporated into market prices, i.e. the market is efficient[3]. The critical number $N_c$ is proportional to the number of states of the world, which measures the number of market relevant events. This marks

[2] We refer to the original reference [2] for details on Arthur's model.

[3] We refer here to the strong form of efficiency which states that a market is efficient if prices do not change when all private information is made public.

a phase transition in the statistical mechanics formalism, which separates an efficient from an inefficient phase.

If agents are sophisticated and account for their market impact, market prices differ from competitive prices and the phase transition disappears: Still prices approach returns as the number of agents increases but they do so gradually. Prices only converge to returns in the limit $N \to \infty$ of infinitely many agents. This is a surprising result: The contribution of each single agent to prices is of order $1/N$, which seems negligible for $N \to \infty$. But if all agents account for it, the collective behavior changes radically.

## 3 Minority Game

The minority game (MG) has been proposed to model speculative behavior in financial markets. Agents sell and buy asset shares or currencies with the only goal of profiting from market's fluctuations. Hence, at odds with previous models, both demand and supply fluctuate in time. The MG is a highly simplified picture of this context.

In the MG agents can take, in each period $t$, one of two actions $a_i(t) = \pm 1$. The agents who take the minority action win, whereas the majority looses. This is captured by the payoffs[4]

$$u_i(t) = -a_i(t)\frac{A(t)}{N}, \qquad A(t) = \sum_{i \in N} a_i(t) \tag{1}$$

Refs. [3,24–26] deal with a model where $u_i(t) = -a_i(t)\text{sign}\, A(t)$. Ref. [10] has shown that the difference between the two models is however immaterial as far as the qualitative collective behavior is concerned. The connection with financial markets is established naively by observing that markets are instruments for reallocating goods. No arbitrage implies that no gain can be made, in principle, by pure trading. Hence the market should be a zero sum game. Transaction costs and other frictions make it an unfavorable game, on average, i.e. a minority game. Ref. [14] expand these types of arguments in much more details.

Having not even introduced a price, it may hard to be satisfied with these arguments and accept the MG as a model of a market. For this reason, we give below a derivation of the interaction (1) from a market mechanism. Those

[4] Our reference framework is not Game Theory, but rather agent based models. Agents are adaptive and their learning dynamics shall be defined in terms of reinforcements or "attractions" [20–23]. Then $u_i(t)$ should better be called "perceived" payoff (specially because we are going to assume later that agents consider counter-factually also the outcomes of strategies which they did not actually play). We shall borrow some terminology from game theory and speak simply of payoffs. Also, our adaptive agents chose among different modes of behavior which we shall call simply strategies.

who find the previous naive arguments sufficient and prefer to rush to the results, may skip the next subsection.

### 3.1 Minority Game as a Market model

A more detailed derivation of this interaction goes along the following lines: $a_i > 0$ means that agent $i$ contributes $a_i\$$ to the demand for the asset. $a_i < 0$ means instead that $i$ sells $q_i(t) = -a_i/p(t-1)$ units of asset, which is the current equivalent of $|a_i|\$$. With $a_i(t) = \pm 1$, the demand is given by $D(t) = \frac{N+A(t)}{2}$, whereas the supply is $S(t) = \frac{N-A(t)}{2p(t-1)}$. Price is fixed by the market clearing condition, $p(t) = D(t)/S(t)$, i.e.

$$p(t) = p(t-1)\frac{N+A(t)}{N-A(t)}. \tag{2}$$

Consider an agent who buys 1\$ of asset at time $t$ (i.e. $a_i(t) = 1$): He loses 1\$, which he invests, and gets $1/p(t)$ units of asset. These will be worth $p(t+1)/p(t)$ dollars if sold tomorrow. Hence the payoff is

$$u_i(t) = \frac{p(t+1)-p(t)}{p(t)}, \qquad \text{if } a_i(t) = +1 \tag{3}$$

The price $p(t+1)$ at the next time is not known. We can replace it by the expectation of agent $i$ at time $t$, denoted by $E_t^{(i)}[p(t+1)]$, which we express by a linear formula:

$$E_t^{(i)}[p(t+1)] = (1-\psi_i)p(t) + \psi_i p(t-1) \tag{4}$$

Then $E_t^{(i)}[u_i(t)|a_i(t) = +1] = -\psi_i[p(t)-p(t-1)]/p(t)$ and, using Eq. (2) we find $E_t^{(i)}[u_i(t)|a_i(t) = +1] = -2\psi_i A(t)/[N+A(t)]$.

Likewise, if agent $i$ sells $1/p(t-1)$ units of assets at time $t$ (i.e. $a_i(t) = -1$) and buys it back at the expected price $E_t^{(i)}[p(t+1)]$, elementary algebra leads to $E_t^{(i)}[u_i(t)|a_i(t) = -1] = 2\psi_i A(t)/[N-A(t)]$. This means that generically:

$$E_t^{(i)}[u_i(t)] = -2\psi_i a_i(t)\frac{A(t)}{N+a_i(t)A(t)}. \tag{5}$$

When $A(t) \ll N$ we can neglect the term $a_i(t)A(t)$ in the denominator and, with $\psi_i = 1/2$, we recover Eq. (1). This assumption is justified, for $N \gg 1$, because typically $A(t) \sim \sqrt{N}$ in the MG.

Note that $\psi_i > 0$ means that agent $i$ expects that future prices is an average of past and present prices. Naively speaking, agent $i$ is a "fundamentalist" who thinks that the market is close to a stationary equilibrium: Future price increments tend to compensate past fluctuations out of the equilibrium. On the contrary, $\psi_i < 0$ corresponds to "trend followers" who believes that the next price change $\Delta p' = p(t+1) - p(t)$ is in the same direction of the previous one. Indeed with $\psi_i = -1$ Eq. (4) is a linear extrapolation. The interaction between fundamentalists and trend followers has been analyzed by several authors [27,28,5]. Typically trend followers cause instabilities and are responsible for bubbles, i.e. for market phases where actual prices diverge from fundamentals.

In the present context, trend followers ($\psi_i < 0$) are described by a majority game, whereas fundamentalists are described by the minority game. The minority game has been studied in many details recently [29], as we shall review below. Much less work [30] has been done on the majority game. The mixed model, where fundamentalists and trend followers interact, has not yet been studied.

It is worth to remark that agents would revise and calibrate their expectations in the real world. In other words $\psi_i$ should reflect self-fulfilling expectations. This can be checked *a posteriori* and we shall see that expectations can be self-fulfilling in both cases.

### 3.2 Global efficiency

How smaller will the minority side be with respect to the majority? If agents are rational at all, we expect that they will drive the system to a state where none of the two actions $a_i = \pm 1$ identifies systematically the minority side. Hence, we expect that agents reduce

$$\langle A \rangle = \lim_{t_0, T \to \infty} \frac{1}{T} \sum_{t=t_0+1}^{t_0+T} A(t) \tag{6}$$

as close as possible to zero. Note that both a state where $A(t) = 0\ \forall t$ and a state where $A(t) = (-1)^t N$ have $\langle A \rangle = 0$. But these states are quite different: in the former no agent loses whereas in the latter all agents lose. In order to distinguish these two cases, we introduce the quantity

$$\sigma^2 = \langle A^2 \rangle = \lim_{t_0, T \to \infty} \frac{1}{T} \sum_{t=t_0+1}^{t_0+T} A^2(t) \tag{7}$$

which measures global efficiency or the extent to which agent coordinate to decrease wastes due to fluctuations.

### 3.3 Plain Minority Game with adaptive agents

In order to see the relevance of these quantities, it is preferable to consider a simple version of the MG. Several simple and analytically tractable versions of the MG have been proposed [31–34]. In order to make connection with our later discussion, we focus on Ref. [9] and give the basic intuition and the key elements of the mathematical approach. The interested reader is referred to Ref. [9] for more detailed account of the results.

Agents learn from past experience which action $a_i(t)$ is the best one. The learning dynamics is the one used in general in minority games and it is well rooted in the economic literature [20–23]. The past experience of agent $i$ is stored in the "score" $\Delta_i(t)$: $\Delta_i(t) > 0$ means that the action $a_i = +1$ is (perceived as) more successful than $a_i = -1$ and *vice-versa*. Agents use the information accumulated in $\Delta_i(t)$ to take decisions[5]:

$$\mathrm{Prob}\{a_i(t) = \pm 1\} \equiv p_i(t) = \frac{e^{\Delta_i(t)}}{e^{\Delta_i(t)} + e^{-\Delta_i(t)}} \tag{8}$$

and they update $\Delta_i(t)$ by

$$\Delta_i(t+1) = \Delta_i(t) - \frac{\Gamma}{N}[A(t) - \eta a_i(t)]. \tag{9}$$

This learning dynamics is easily understood for $\eta = 0$: if $A(t) < 0$ agents observe that the best action was $+1$ at time $t$. Hence they increase $\Delta_i$ and the probability $p_i$ of playing $a_i = +1$.

The $\eta$ term in Eq. (9) describes the fact that agent $i$ may account for his own contribution to $A(t)$. For $\eta = 1$ indeed, agent $i$ considers only the behavior of other agents $A(t) - a_i(t)$ and does not react to his own action $a_i(t)$. In other words, $\eta$ measures the extent to which agents account for their "market impact". Finally the factor $\Gamma/N$ modulates the strength of the response in the behavior of agents to the stimulus $A(t) - \eta a_i(t)$. How does the collective behavior depends on the parameters $\eta$ and $\Gamma$?

Let us start by considering *naive* agents – i.e. $\eta = 0$ – who do not account for their impact on the aggregate. If at time $t = 0$, $\Delta_i(0) = \Delta^{(0)}$ for all $i$, then $\Delta_i(t)$ and $p_i(t)$ are independent of $i$ for all times, hence we may suppress the index $i$. For $N \gg 1$, the law of large numbers implies that $A(t)/N \simeq \tanh \Delta(t)$[6] so that

$$\Delta(t+1) \simeq \Delta(t) - \Gamma \tanh \Delta(t) \tag{10}$$

This is a dynamical system with a fixed point $\Delta^* = 0$. However this fixed point is stable only if $\Gamma < \Gamma_c = 2$. For $\Gamma > 2$ a bifurcation to orbits of period two occurs, i.e. $\Delta(t) = (-1)^t \Delta^\dagger$ where $\Delta^\dagger(\Gamma)$ is a solution of $2x = \Gamma \tanh x$. This has dramatic effects on the collective behavior: For $\Gamma < 2$ the system converges to a state where agents draw their actions at random ($p_i = 1/2$). Hence $A(t)$ has Gaussian fluctuations of magnitude $\sqrt{N}$ around zero and

---

[5] The exponential form, which results from a Logit discrete choice model, is taken here for simplicity. Any increasing continuous function $\chi_i(x)$, with $0 \le \chi_i(x) \le 1$ for all real $x$, $\chi(x) \to 0$ as $x \to -\infty$ and $\chi(x) \to 1$ as $x \to \infty$, leads to the same results [9].

[6] The distribution of $K(t) = [N + A(t)]/2$ is indeed binomial with parameters $N, p = 1/(1 + e^{-2\Delta})$. Hence the fluctuations of $A(t)/N$ around its average are of order $1/\sqrt{N}$.

$\sigma^2 \simeq N$. On the contrary, for $\Gamma > 2$, the attendance oscillates between the the two values $\pm N \tanh \Delta^{\dagger}$, which implies that $\sigma^2 \propto N^2$ is much larger than in the previous case. Note that in both cases $\langle A \rangle = 0$, but global efficiency is very different!

The transition from a state where $\sigma^2 \propto N$ to a state with $\sigma^2 \propto N^2$ is generic in the minority game, and it has been been discussed by several authors [24,35,7,36].

For $\eta = 0$, the asymptotic state of the dynamics depends on the initial conditions. Indeed the above argument generalizes by observing that $\Delta_i(t) - \Delta_i(0)$ does not depend on $i$. Hence the initial heterogeneity across agents is preserved by the dynamics and the asymptotic state will depend on the initial conditions. The dependence on initial conditions is also generic. It occurs in the efficient phase of the asset market model [8] as well as in the Minority Game in its full complexity [37,38].

The situation changes drastically as soon as agents start to account for their market impact. To see this, let us again follow Ref. [9]. Take the average of Eq. (9) in the long time limit and define $m_i = \langle a_i \rangle$. We note that

$$\langle \Delta_i(t+1) \rangle - \langle \Delta_i(t) \rangle = -\frac{\Gamma}{N} \left[ \sum_{j \in N} m_j - \eta m_i \right] = -\frac{\Gamma}{N} \frac{\partial H_\eta}{\partial \eta} \tag{11}$$

where

$$H_\eta = \frac{1}{2} \left( \sum_{i \in N} m_i \right)^2 - \frac{\eta}{2} \sum_{i \in N} m_i^2. \tag{12}$$

A close inspection[7] of these equations implies that $m_i$ are given by the minima of $H_\eta$.

Note that $H_1$ is an Harmonic function of $m_i$'s. Hence it attains its minima (and maxima) on the boundary of the hypercube $[-1, 1]^N$. So for $\eta = 1$ all agents take always the same actions $a_i(t) = m_i = +1$ or $-1$ and the waste of resources is as small as possible: Indeed $\sigma^2 = 0$ or $1$ if $N$ is even or odd, which is a tremendous improvement with respect to the case $\eta = 0$ (where $\sigma^2 \sim N$ or $N^2$). These states are indeed Nash equilibria [9] of the associated $N$ persons minority game. These are the stationary states of the learning process for any $\eta > 0$[8].

[7] The first order conditions on $H_\eta$ imply that if $-1 < m_i < 1$ then $\langle \Delta_i(t+1) \rangle = \langle \Delta_i(t) \rangle$, i.e. the process $\Delta_i(t)$ is stationary. Else if $m_i = +1$ (or $-1$) one should have $\Delta_i(t) \to +\infty$ (or $-\infty$), which is precisely what constrained minimization and Eq. (11) say.

[8] The proof of this statement goes as follows: Suppose that $m^*$ is an equilibrium with $-1 < m_k^* < 1$ for $k = 1, \ldots, n$ and $m_i^* = \pm 1$ for $i > n$. The conditions for a minimum requires that $H_\eta$ is locally positive definite around $m^*$. At least

Again the asymptotic state is not unique and it is selected by the initial conditions. However now the set of equilibria is discrete and the system jumps discontinuously from an equilibrium to another, as the initial conditions $\Delta_i(0)$ vary. This contrasts with the $\eta = 0$ case, where the equilibrium shifts continuously as a function of the initial conditions [9].

While with $\eta = 0$ the dynamics preserves the initial diversity, with $\eta > 0$ the heterogeneity of agents increases in time and agents polarizes on the two extreme positions $m_i = \pm 1$. This shows that the intuition of Arthur [2], that market interaction forces heterogeneous expectations, crucially depends on whether agents account for their market impact ($\eta > 0$) or not ($\eta = 0$).

Going back to the market derivation of section 3.1, it is worth to observe that expectations are generically fulfilled for $\eta = 0$: a linear fit of the $p(t+1)-p(t) = cA(t+1)$ versus $p(t)-p(t-1) = cA(t)$ reveals a negative slope $-\psi$, whose magnitude increases with $\Gamma$. With $\eta > 0$ both price increments are zero: However if the equilibrium is disturbed with a positive price increment it will be restored by a negative one, hence $\psi > 0$ also in this case. Expectations are also self fulfilled in the majority game, i.e. when we assume $\psi < 0$. It is not hard to find that the dynamics is described by Eq. (10) with $\Gamma < 0$. Hence, depending on initial conditions, $\Delta_i(t) \to \pm\infty$ linearly in time and all agents will either buy ($a_i(t) = +1$) or sell ($a_i(t) = -1$) all the time. Of course then successive price increments will be positively correlated, indeed $A(t+1) = A(t)$, i.e. $\psi = -1$ is the correct expectation model. Of course this behavior is typical of a market bubble: In real situations such rallies never last forever.

## 3.4 Information and Trading Strategies

In the El Farol bar problem and in the minority game agents are heterogeneous because they react differently to the state of the market $\omega$. The state $\omega$ is one of $\Omega$ possible states[9] and it is randomly drawn uniformly and independently in each period.

Agents resort to simple rules or *schemata* [2] to cope with the information $\omega$. These are simple information processing devices which prescribe an action $\rho^\omega$ as a function of $\omega$. If there are only two possible actions (buy +1 or sell

---

$n$ eigenvalues of the matrix $\frac{\partial^2 H_\eta}{\partial m_i \partial m_j}$ must be non-negative. But this matrix has only one positive eigenvalue $\lambda = N - \eta$ and $N - 1$ negative eigenvalues $\lambda = -\eta$. Hence $n$ can at most be 1, which can occur for $N$ odd.

[9] The original formulation [3] of the MG aimed at describing the behavior of *technical traders*, who look for predictable patterns in historical prices. In Ref. [3] $\omega$ encodes the outcome of past market interactions, i.e. it is an *endogenously* generated information. It was later found [26] that the same collective behavior holds for $\eta = 0$ for the simpler case where $\omega$ is an exogenously generated random process. This is only qualitatively true for $\eta = 0$ [39] and it is wrong for $\eta > 0$ [38].

$-1$), there are $2^{\Omega}$ such functions, which we call ***trading strategies***. Each agent is assigned a number $S$ of such trading strategies, each of which is randomly drawn from the set of all possible functions. We label by $i$ and $s = 1, \ldots, S$ the strategies of agent $i$ so that $a_i = \rho^{\omega}_{s,i}$ is the action taken by agent $i$ if he takes strategy $s$ when the state is $\omega$. We define $s_i(t)$ as the strategy chosen by agent $i$ at time $t$, so that $a_i(t) = \rho^{\omega}_{s_i(t),i}$.

Agents then face the problem of deciding which strategy to use, irrespective of what the state $\omega$ is. Each agent $i$ assigns a score $U_{s,i}(t)$ to each of his strategy $s = 1, \ldots, S$ and draws $s_i(t)$ from the distribution

$$\mathrm{Prob}\{s_i(t) = s\} = Z^{-1} e^{\Gamma U_{s,i}(t)}, \qquad Z = \sum_{s'=1}^{S} e^{\Gamma U_{s',i}(t)}. \tag{13}$$

Agents adapt their behavior by updating the scores of their strategies. In the most general case, this takes the form

$$U_{s,i}(t+1) = U_{s,i}(t) - \rho^{\omega}_{s,i} \frac{A(t)}{N} + \eta \rho^{\omega}_{s,i} \frac{a_i(t) - \rho^{\omega}_{s,i}}{N}. \tag{14}$$

The reinforcement for $\eta = 0$ is the virtual payoff agent $i$ would have received if he had played strategy $s$. We call this virtual because the real payoff must account for the fact that, if agent $i$ had really played strategy $s$, $A(t)$ would have changed into $A'(t) = A(t) - a_i(t) + \rho^{\omega}_{s,i}$. With $\eta = 0$ Eq. (14) models agents who neglect their market impact, i.e. who behave as price takers. The last term[10] corrects this behavior either approximately ($\eta < 1$) or exactly ($\eta = 1$).

### 3.5 Early results

The MG was introduced [3], and is often discussed [24–26,35,7,30] with $\eta = 0$, $\Gamma = \infty$ – which means that agents use their "best" strategy – and with Eq. (14) replaced by $U_{s,i}(t+1) = U_{s,i}(t) - \rho^{\omega}_{s,i} \mathrm{sign}\, A(t)$. In this setting, numerical studies [3,24,40] have uncovered a remarkably rich phenomenology as a function of the the number of states $\Omega$ – or information complexity – and the number of agents $N$. Savit *et al.* [24] have found that the collective behavior does not depend independently on $\Omega$ and $N$ but only on their ratio $\alpha \equiv \Omega/N$. This is illustrated in Fig. 1, which reports the value of $\sigma^2/N$ as a function of $\alpha$ for different values of $\Omega$. When $\alpha \gg 1$ information is too complex and agents essentially behave randomly. Indeed $\sigma^2/N$ converges to one – the value it would take if agents were choosing $a_i(t)$ by coin tossing. As $\alpha$ decreases – which means that $\Omega$ decreases or the number of agents increases – $\sigma^2/N$ decreases suggesting that agents manage to exploit the information in order to coordinate. But when agents become too numerous,

---

[10] This term is equivalent to a term $-\eta\delta_{s_i(t),s}$, which is $-\eta$ when $s_i(t) = s$ and 0 otherwise. This latter term is the one discussed in Refs. [37,38].

$\sigma^2/N$ starts increasing with $N$. Savit *et al.* [24] found that, at $\Omega$ fixed, $\sigma^2$ increases linearly with $N$ as long as $N \ll \Omega$ but with a quadratic law $\sigma^2 \sim N^2$ for $N \gg \Omega$, which implies $\sigma^2 \sim 1/\alpha$ for $\alpha \ll 1$. The behavior for $\alpha \ll 1$ has been attributed to the occurrence of "crowd effects", and it has been studied in many details both numerically and by approximate methods [35,4,40]. Ref. [24] found evidence for a phase transition. Ref. [7] showed that the transition occurs where $\sigma^2/N$ attains its minimum and that it separates a symmetric ($\alpha < \alpha_c$) from an asymmetric phase ($\alpha > \alpha_c$). The symmetry which is broken is that of the average of $A(t)$ conditional on the state $\omega$

$$\langle A|\omega\rangle = \lim_{t_0\to\infty}\lim_{T\to\infty}\frac{1}{T_\omega}\sum_{t=t_0+1}^{t_0+T} A(t)\delta_{\omega(t),\omega} \quad \text{where } T_\omega = \sum_{t=t_0+1}^{t_0+T}\delta_{\omega(t),\omega}. \tag{15}$$

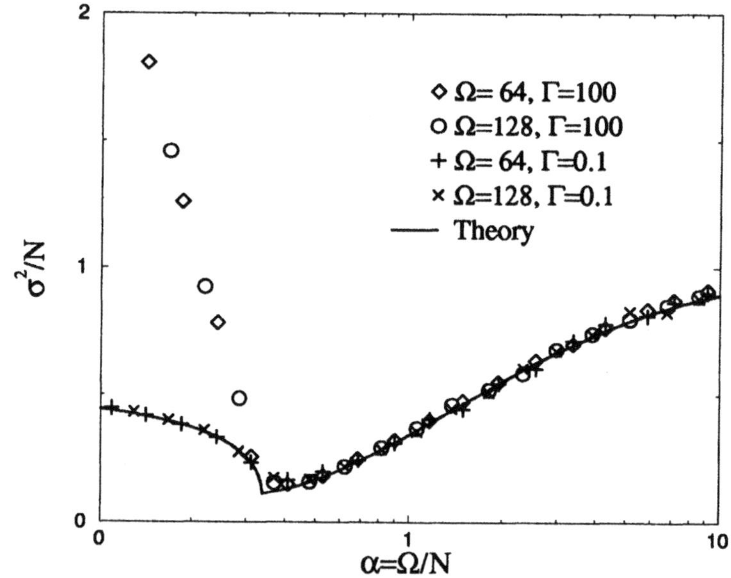


**Fig. 1.** Global efficiency $\sigma^2/N$ for $\Omega = 64$ and $\Omega = 128$ and $N$ ranging from $0.1\Omega$ to $10\Omega$. Agents have two trading strategies. Each point is obtained as the average over 100 random systems in the asymptotic stationary state of the dynamics. The collapse of the two curves implies that for large $N$ the collective properties only depend on the ratio $\alpha = \Omega/N$. For small values of $\alpha$, $\sigma^2$ is larger for fast learning (∘, ⋄ correspond to $\Gamma = 100$) than for slow learning (×, + correspond to $\Gamma = 0.1$). The full line are the results of the theory based on the statistical mechanics approach, which is valid in the limit $\Omega \to \infty$ and for small $\Gamma$.

In the asymmetric phase, $\langle A|\omega\rangle \neq 0$ for at least one $\omega$. Hence the sign of $A(t)$ is predictable, to some extent. A measure of the degree of predictability is given by the function

$$H = \frac{1}{\Omega}\sum_{\omega=1}^{\Omega}\langle A|\omega\rangle^2. \tag{16}$$

In the symmetric phase $\langle A|\omega\rangle = 0$ for all $\omega$ and hence $H = 0$.

Ref. [7] found that $H$ is a decreasing function of the number $N$ of agents (at fixed $\Omega$): This means that the agents who enter the market exploit its predictability and hence reduce it.

Cavagna *et al.* [36] have proposed a further generalization of the model where

$$a_i(t) = \sum_{\omega=1}^{\Omega} \rho^{\omega}_{s_i,i}\eta^{\omega}(t) \tag{17}$$

where $\eta^{\omega}(t)$ is white noise (i.e. a Gaussian variable with zero average and unit variance, independent for each $\omega$ and $t$). In Ref. [36] trading strategies $\rho^{\omega}_{s_i,i}$ are "continuous", i.e. they are drawn from a continuous distribution – a Gaussian – rather than from the bimodal. The idea is that the exogenous process which drives the market, or the news arrival process, is an $\Omega$ dimensional vector $\eta(t)$. Agents respond to it with linear strategies, which are also $\Omega$ dimensional vectors $\rho_{s,i}$. This model reduces to the previous one if we assume that $\eta(t) = (\ldots, 0, 1, 0, \ldots)$ can only be one of the $\Omega$ versors of the orthogonal basis and that $\rho^{\omega}_{s,i}$ is drawn from the bimodal distribution. Challet *et al.* [41] have remarked that this model has the same collective behavior as the one discussed above. Though more elegant, the model of Ref. [36] is much more demanding for numerical simulations: indeed it needs $N \times \Omega$ elementary operations, in each time step, whereas the "binary" model only requires $N$ operations.

Cavagna *et al.* [36] found that, for $\alpha < \alpha_c$, global efficiency increases if $\Gamma$ is reduced (see figure 1). We already saw in section 3.3 how, following Ref. [9], one can understand this effect in the much simpler setting of the MG without information (and $\eta = 0$). That case corresponds to the limit $\alpha \to 0$. The same instability of fixed point solutions occurs in whole $\alpha < \alpha_c$ phase as $\Gamma$ exceeds a critical learning rate $\Gamma_c(\alpha)$ which can also be computed within the statistical mechanics approach [9]. For $\Gamma > \Gamma_c(\alpha)$ the systems is in a turbulent dynamical phase and $\sigma^2$ is correspondingly much larger[11].

[11] It must be said that the behavior of agents in this regions of parameters raises serious doubts on their assumed rationality, however bounded. Indeed typically agents statistically switch from a strategy to the other, using the wrong one, most of the time. This results in a periodic behavior of period $2\Omega$ [7] which causes the increased inefficiency. Ref. [38] also comments about this.

### 3.6 Statistical Mechanics of Minority Games

A considerable progress in our understanding of the properties of the MG was made in Refs. [37,38]. There the properties of the stationary state of the MG were shown to be related to the properties of disordered spin models. Statistical mechanics makes a systematic study of these systems possible and it eventually leads to exact results. We refer the interested reader to Refs. [37,38,10] for a detailed account of the approach. Here we just give an outline of this approach and the basic results.

The relevant variables are the frequencies $\pi_{s,i}$ with which each agent plays a given strategy in the stationary state. This approach is based on the observation that the stationary conditions of the dynamics correspond to the first order condition for the constrained minimization of the function

$$H_\eta = \frac{1}{\Omega}\sum_{\omega=1}^{\Omega}\langle A|\omega\rangle^2 - \eta\sum_{i=1}^{N}\pi_{s,i}^2, \tag{18}$$

subject to $\pi_{s,i} \geq 0$ and $\sum_s \pi_{s,i} = 1$. Here

$$\langle A|\omega\rangle = \sum_{i=1}^{N}\sum_{s=1}^{S}\pi_{s,i}\rho_{s,i}^{\omega}. \tag{19}$$

The argument is analogous to what happens in the simpler case discussed in section 3.3. However the characterization of the equilibria is more complex in this case and it requires sophisticated techniques of statistical mechanics [1,38].

Without entering into too many details, we can give a simple explanation, based on linear algebra, of why a phase transition occurs for $\eta = 0$. This not only gives a flavor of the results but it also allows one to understand several things.

First note that for $\eta = 0$, $H_0$ is a non-negative definite quadratic function and hence it attains its minimum on a connected set (eventually a single point). Let $\pi^* = \{\pi_{s,i}^*\}$ be a point on this set. For all $s,i$ for which $\pi_{s,i}^* > 0$, the corresponding first order condition (FOC) needs to be satisfied. Let $N_>$ be the total number of these variables. $N$ of them are fixed by the normalization conditions, which leave $N_> - N$ free variables to be fixed by the FOC. Given that $H_0$ is a quadratic function, the FOC are linear equations. The coefficients are given by the $\Omega$ vectors[12] $\rho^1, \ldots, \rho^\Omega$. Hence the number of linearly independent FOC's cannot be larger than $\Omega$. If the number of free variables $N_> - N$ is smaller than $\Omega$, the solution is unique. But when $N_> - N > \Omega$ some (linear combinations) of the free variables remains unspecified. The phase transition occurs exactly when $\Omega = N_> - N \equiv \alpha_c N$. For $N < \Omega/\alpha_c$

[12] We consider here and below vectors in the $N_> - N$ dimensional linear space of the free variables.

the equilibrium is unique, whereas when the number of agents is larger than $\Omega/\alpha_c$ the equilibrium is not unique. In the latter case ($\alpha < \alpha_c$), the FOC equations become equivalent to the set of equations $\langle A|\omega\rangle = 0\ \forall\omega$. Hence for $\alpha < \alpha_c$ the market outcome is symmetric and contains no information on the state $\omega$: We may say that the market is efficient.

Put differently, the dynamics Eq. (14) of the vector $U(t)$ spans the linear space formed by (linear combinations of) the vectors $\rho^\omega$. The $N_> - N - \Omega$ components of $U(t)$ which are orthogonal to all $\Omega$ vectors $\rho^\omega$ do not change at all in time. This means that, for $\alpha < \alpha_c$ the stationary state depends on the initial condition $U(0)$: different initial conditions select different equilibria. Figure 2 shows the behavior of $\sigma^2/N$ versus $\alpha$ for two different types of initial conditions: symmetric – when agents is initially indifferent among the strategies – and polarized – where agents initially strongly prefer one to the other strategies. The inset shows how much agents are asymptotically polarized on their initially preferred strategy, in the second case. This, when agents have only two strategies, is expressed by the quantity

$$M = \frac{1}{N}\sum_{i=1}^{N}\frac{\pi_{+,i} - \pi_{-,i}}{2}$$

where the strategy + is the initially preferred one. The curves in figure 2 are analytic results. We refer to Ref. [38,10] for a detailed account of the method.

It must be noted that this approach works only for small learning rates $\Gamma$. Ref. [9] has found the critical learning rate $\Gamma_c(\alpha)$ above which one enters into a "turbulent" dynamics characterized by large volatility. This effect is the same as the one discussed above in the case without information.

As soon as $\eta > 0$ is turned on, the situation changes completely. This is easy to understand because the FOC for the variable $\pi_{s,i}$ involves now also a term $+\eta\pi_{s,i}$. This makes them linearly independent for any value of $\alpha$. The phase transition at $\alpha_c$ then disappears. The behavior of the market is summarized in the ***phase diagram*** of fig. 3. This is discussed in detail in Ref. [42]. The lower region of the phase space is characterized by an unique solution, whereas multiple, disconnected equilibria occur in the upper region[13].

Figure 4 illustrates the behavior of the market along two cuts of the phase diagram: While for $\alpha > \alpha_c$ the transition is continuous, for $\alpha < \alpha_c$ $\sigma^2$ features a jump as the line $\eta = 0$ is crossed.

[13] This qualitative difference is captured by the mathematical approach of statistical mechanics. This analyzes simultaneously different *replicas* of the same system. When the equilibrium is unique, the solution is *replica symmetric* in the sense that it is invariant under permutation of different replicas. When there are many disconnected equilibria, different replicas may end up in different equilibria and replica symmetry can be *broken* (i.e. the solution is no more invariant under replica permutations). Replica symmetry breakdown is what happens along the line which separates the two regions in figure 3.

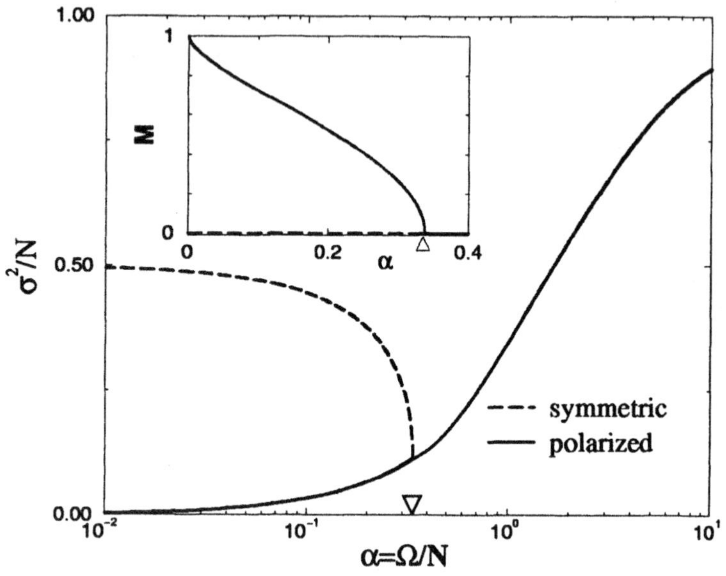


**Fig. 2.** Global efficiency $\sigma^2/N$ for $S = 2$ and $\eta = 0$, as a function of $\alpha = \Omega/N$. These curves, derived within the statistical mechanics approach, are exact in the limit $N \to \infty$. The curves refer to two different initial conditions: symmetric (dashed line) and biased (full line). Inset: polarization $M$ as a function of $\alpha$ for the two initial conditions. The triangles mark the location of the phase transition.

Also in this case we see that the assumption of price taking behavior ($\eta = 0$) is by no means an innocent one: as soon as agents depart from this behavior, the collective properties of the system change drastically. It can be shown [38] that the payoff of each individual agent increases with $\eta$ for $\eta \in [0, 1]$. Hence, not only the situation improves globally, but it also improves for every agent.

### 3.7 Market's ecology

There are different types of participants in a market, depending on their goals. Some trade to gain money from transactions with no particular interest in the asset they buy and sell. Only asset price fluctuations matter for this kind of traders – the speculators. A further type of market participants are those who use the market for exchanging goods. This is indeed the reason why market exist: for example, they enable people to exchange what they produce for what they consume or to reallocate their resources across time. This type of agents is interested in the asset itself: They will buy it or sell it irrespective of the history of recent fluctuations (as long as these were not extreme). Following Ref. [10], we shall call this type of agents *producers*.

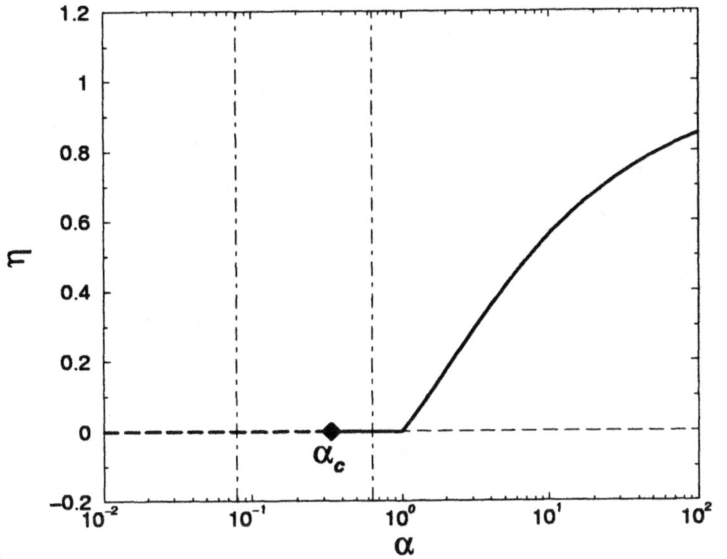


**Fig. 3.** Phase diagram of the minority game in the $(\alpha, \eta)$ plane. The equilibrium is unique below the line whereas many different equilibria are possible above. Along the dashed line, corresponding to $\eta = 0$ and $\alpha < \alpha_c$, a continuum of equilibria exists. The change in the collective behavior is continuous for $\alpha > \alpha_c$ (full line) and discontinuous along the dashed line. This can be better appreciated in the next figure where $\sigma^2/N$ is plotted along the two dot-dashed cuts of the phase diagram.

Challet *et al.* [10] have proposed a modification of the minority game which captures these features. Speculators are modeled as above: They have a fixed range of behavioral rules which they adapt in time. They process market information $\omega$ in the search of ***arbitrage opportunities***, i.e. correlations between the outcome $A(t)$ and the state $\omega$.

Producers, on the other hand, follow a trading rule which is constant in time. They are part of the financial world and their behavior is correlated with the state of the world $\omega$ which is thought to capture all relevant economic information. This type of traders play a role similar to that of ***hedgers*** in Ref. [43]. These agents inject information into the market: their trading activity is predictable, given the state of the world $\omega$:

$$a_i(t) = \tilde{\rho}_i^{\omega}.$$

Let $N_p$ be the number of producers and assume again that the trading strategies $\tilde{\rho}_i^{\omega}$ are drawn at random. Then, in a market composed of producers only, distribution of price changes is nearly Gaussian. The process associated to producers can be considered as the ***fundamentals***, i.e. the price process

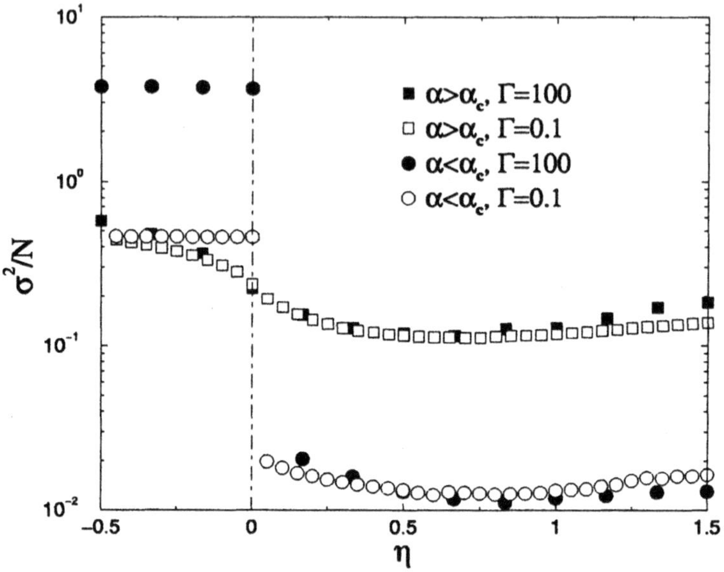


**Fig. 4.** Global efficiency as a function of $\eta$ for $N = 101$ agents, $\Omega = 64$ (i.e. $\alpha > \alpha_c$) and $\Omega = 8$ ($\alpha < \alpha_c$). Two sets of data with fast ($\Gamma = 100$) and slow ($\Gamma = 0.1$) learning are shown. Each point is the average of 100 random realizations in the stationary state.

which reflects the economic performance of the asset. Speculative trading will *color* this process and transform its statistical properties[14].

How do these species of traders interact? First note that, in a market composed of producers only, price changes would depend on $\omega$. Such an highly predictable market is very favorable for speculators who may derive considerable gains. However when more and more speculators are added to the market, its predictability decreases and speculators gains are meager and meager. This is illustrated in Fig. 5, which also shows that producers instead benefit from the presence of speculators because their losses are reduced. When the number of speculators increases beyond a critical value, which depends on the reduced number $N_p/\Omega$ of producers, the market enters into the symmetric phase where $H = 0$ and the outcome $A(t)$ is unpredictable (see fig. 5). These results show that the relation between these two species is more similar to symbiosis than to competition: Producers feed speculators by

[14] Ref [10] extends the discussion to a further type of agents, the so called *noise traders*. Noise traders totally disregard the state of the world $\omega$ or have no information at all on it. They rather follow rules of behavior which are statistically uncorrelated with $\omega$ (such as astrology) and with the behavior of other agents. The presence of these agents does not brings new qualitative features. So we limit our discussion to producers and speculators and refer the interested reader to Ref. [10].

injecting information in the market and benefit, in their turn, of the liquidity provided by speculators. We refer the interested reader to Refs. [10,14] for a more detailed discussion of market behavior.

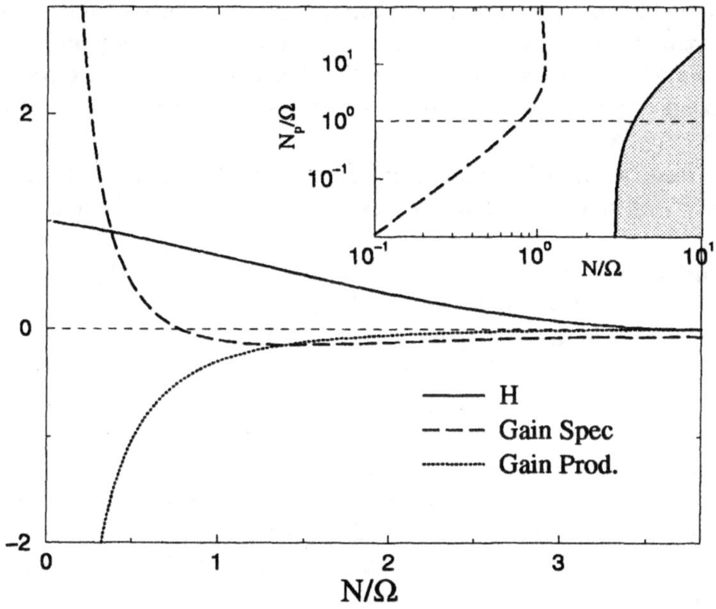


**Fig. 5.**

### 3.8 Stylized facts in toy models

Financial markets are characterized by some well established statistical regularities of high frequency data, often referred to as "stylized facts". In the reference model of price dynamics, which is the simplest one accounting for no-arbitrage hypothesis and market's efficiency, the logarithm of prices performs a random walk and hence returns are Gaussian.

The first stylized fact is that the probability distribution of returns has fat tails, much fatter than that of a Gaussian. The second is that volatility is clusterized in time. This means that market's activity has an intermittent behavior with periods of high activity separated by quiescent periods.

Can simple toy models, such as those discussed here, account for these facts? What is the theoretical explanation, if any, that these models provide? These issues have been recently addressed by several authors. We shall focus mostly on Refs. [14,15]. Quite similar results have been found in similar models based on the MG [11–13].

Ref. [14] observes that the statistics of price increments changes when markets approach efficiency. The models considered are generalized versions of the MG, where agents have different weights representing their wealth. When agents reinvest their gains, so that their wealth depends on time, the model displays quite interesting complex features. The market is again characterized by the occurrence of a phase transition, separating an efficient $H = 0$ phase from an inefficient one, when the number of speculators exceeds a critical value. It turns out that stylized facts observed in high frequency data of financial markets [6] arise just close to the phase boundary. More precisely, the distribution of price increments acquires fatter and fatter tails and volatility clustering appears [14].

This results suggests that financial markets operate close to a critical point, a conclusion hinted at by several authors [4,44,5] on a speculative basis.

Remarkably we find that both of these features crucially depend on the assumption of price taking behavior: Fat tails and volatility clustering disappear, in the MG models, if agents abandon price taking behavior.

Finally Refs. [15,11–13] extended the MG by allowing speculators not to trade. A quite rich and yet little explored phenomenology emerges in these models [15]: When the number of speculators exceeds a threshold, the market enter into a marginally efficient phase where only a fraction of the speculators is active. This fraction increases as the number of producers grows (with a square root law) so as to maintain always the market in a nearly efficient $H \simeq 0$ state. The models features fat tailed return distributions, volatility clustering and market crashes [15].

## 4 Conclusions

Toy models of financial markets exhibit an extraordinarily rich collective behavior. This can be fully characterized using techniques developed in statistical mechanics. This uncovers a non-trivial dependence of macroscopic properties on the microeconomic behavior of agents. The most remarkable example of this is probably the relation between market's volatility and price taking behavior [9,38].

Of course toy models describe a highly simplified world. In real financial markets things are much more complex: agents trade at different frequencies and with different time horizons. The may trade different shares in different markets, using a large number of instruments such as derivatives; the importance of agents varies greatly, from big investment funds and banks to small gamblers. Will the simplified picture, which results from such simple models, survive even when all these issues have been taken into account?

Some features, such as the existence of a phase transition for price taker agents, has been shown to be quite robust to a number of modifications [10,14,15]. However, much more work needs to be done on these lines of research, hopefully by economists and physicists in close collaboration.

**Acknowledgements** Most of my understanding of the material exposed in these pages has emerged thanks to the interaction with a number of collaborators – J. Berg, D. Challet, A. Chessa, A. De Martino, F. Ricci-Tersenghi, A. Rustichini, R. Zecchina and Y.-C. Zhang – and to discussions with colleagues – J.-P. Bouchaud, D. Farmer, S. Franz, J. Garrahan, D. Levine, D. Sherrington. I gratefully acknowledge all of them.

## References

1. M. Mezard, G. Parisi, M. A. Virasoro (1987) Spin glass theory and beyond (World Scientific).
2. Arthur W. B. (1994) Inductive reasoning and bounded rationality. Am. Econ. Assoc. Papers and Proc **84**, 406
3. Challet D. and Zhang Y.-C. (1997) Emergence of cooperation and organization in an evolutionary game. Physica A **246**, 407
4. Zhang Y.-C. (1998) Evolving models of financial markets. Europhys. News **29**, 51
5. J. D. Farmer, Comp. Fin., Nov-Dec (1999), pp. 26-39.
6. R. Mantegna and E. Stanley (1999) Introduction to Econophysics (Cambridge University Press)
7. Challet D. and Marsili M. (1999) Phase Transition and Symmetry Breaking in the Minority Game. Phys. Rev. **E 60**, 6271
8. Berg J., Marsili M., Rustichini A. and Zecchina R. (2001) Statistical mechanics of asset markets with private information. Quant. Finance, **1** (2), XXX
9. Marsili M. and Challet D., (2001) Trading behavior and excess volatility in toy models of financial markets. Adv Compl. Sys. **4** 5
10. Challet D., Marsili M. and Zhang Y.-C. (2000) Modeling Market Mechanisms. Physica A **276**, 284
11. Jefferies P. et al. Int. (2000) J. Th. and Appl. Fin. **3**-3
12. Lo T. S. et al. (2000) From market games to real-world markets. e-print cond-mat/0008387
13. Bouchaud J.-P., Giardina I., Mezard M. (2000) On a universal mechanism for long ranged volatility correlations. e-print cond-mat/0012156.
14. Challet D., Chessa A., Marsili M. and Zhang Y.-C. (2001) From Minority Games to real markets. Quant. Fin. **I**-168
15. Challet D. Marsili M. and Zhang Y.-C. (2001) Stylized facts of financial markets and market crashes in Minority Games. e-print cond-mat/0101326
16. Bell A. M., Sethares W. A. and Bucklew J. W. (1999) Coordination failure as a source of congestion in information networks. Working paper.
17. Casti J. L. (1996) What if... New Scientist **151**, 36
18. Casti J. L. (1996) Seeing the light at El Farol. Complexity **1**, 7
19. Shapley L. and Shubik M. (1977) Trade Using One Commodity as a Means of Payment. Journal of Political Economy, **85**, 937-968
20. Erev I. and Roth A. E., (1998) Predicting how people play games: Reinforcement learning in experimental games with unique, mixed strategy equilibria. Am. Ec. Rev. **88**, 848
21. D. Fudenberg and D. K. Levine (1998) The theory of learning in games (MIT Press)

22. Camerer C. and Ho T.-H. (1999) Experience-weighted attraction learning in normal form games. Econometrica **67**, 827
23. A. Rustichini (1999) Optimal properties of Stimulus-Response Learning Models. Games and Economic Behavior **29**, 224
24. Savit R., Manuca R., and Riolo R. (1990) Adaptative competition, market efficiency and phase transition. Phys. Rev. Lett. **82**, 2203
25. Johnson N.F. et al. (1998) Volatility and Agent Adaptability in a Self-Organizing Market. Physica A **258**, 230
26. Cavagna A. (1999) Irrelevance of Memory in the Minority Game. Phys. Rev. E **59**, R3783
27. Chiarella C. (1992) The dynamics of speculative behavior. Ann. Oper. Res. **37**, 101-123
28. Lux T. and Marchesi M. (1999) Scaling and criticality in a stochastic multi-agent model of a financial market. Nature **397**, 498
29. For a commented collection of papers and preprints on the minority game see the web site `http://www.unifr.ch/econophysics/`.
30. de Cara M. A. R., Pla O., Guinea F. (1999) Competition efficiency and collective behavior in the El Farol bar problem. Eur. Phys. J. B **10**, 187-191
31. Johnson N. F. et al. (1999) Self Organized Segregation within an Evolving Population. Phys. Rev. Lett. **82** 3360
32. Johnson N. F. et al. (2000) Evolutionary freezing in a competitive population. Physica A **283** 568
33. Burgos E., Ceva H. and Perazzo R.P.J. (2000) Quenching and Annealing in the Minority Game. e-print cond-mat/0007010
34. Reents G., Metzler R. and Kinzel W. (2000) A New Stochastic Strategy for the Minority Game. e-print cond-mat/0007351
35. Johnson N.F. et al., (1999) Crowds Effects and Volatility in a Competitive Market. Physica A **269**
36. A. Cavagna et al. (1999) A thermal model for adaptive competition in a market. Phys. Rev. Lett. **83**, 4429
37. Challet D., Marsili M. and Zecchina R. (2000) Statistical Mechanics of Systems with Heterogeneous Agents: Minority Games. Phys. Rev. Lett. **84**, 1824
38. Marsili M., Challet D. and Zecchina R. (2000) Exact solution of a modified El Farol bar problem. Physica A **280**, 522
39. Challet D. and Marsili M. (2000) Relevance of memory in the minority game. Phys. Rev. E **62**, 1862
40. Challet D. and Zhang Y.-C. (1998) On the Minority Game : analytical and numerical studies. Physica A **256**, 514
41. Challet D., Marsili M. and Zecchina R. (2000) Comment on: A thermal model for adaptive competition in a market. Phys. Rev. Lett. **85**, 5008 (see also Cavagna et al.'s reply on page 5009)
42. De Martino A. and M. Marsili (2001) Replica symmetry breaking in the minority game. J. Phys. A **XX**, XXX
43. H. Working (1953) Am. Econ. Rev. **43**, 314
44. Zhang Y.-C. (1999) Towards a theory of marginally efficient markets. Physica A **269**, 30

# Price Bubbles and the Long Run Profitability of a Trend Following Technical Trading Rule

**David Goldbaum**

Department of Economics - Rutgers University - 360 Dr. Martin Luther King Jr. Blvd. - Newark, NJ 07102 - Phone: (973) 353-5164 - Fax: (973) 353-5819 - goldbaum@andromeda.rutgers.edu

**Abstract:** The model allows individual traders to endogenously choose between fundamental information and a technical trading rule. Since the technical trading rule works as a crude signal extractor, its profitability relies on price movements reflecting changes in the intrinsic value of the security. Sustainable profits can be earned using the technical trading rule. The trading rule's popularity grows with its recent trading success, but excessive popularity diminishes the trading rule's ability to extract a reliable signal. Parameter settings that allow large fluctuations in the technical trading rule's popularity lead to unsustainable periods of positive profits coupled with long-term losses.

Keywords: Asset Markets, Bounded Rationality, Technical Analysis

JEL Classifications: G14, G11, C15

The Efficient Markets Hypothesis provides the theoretical foundation with which technical trading rules are rejected as a viable trading strategy. In an environment of efficient markets past price patterns cannot be used to forecast future returns. Technical trading rules (TTRs), which provide the user with a signal of when to buy or sell an asset based on such price patterns, should not be useful for generating excess returns. Despite academic admonitions, use of TTRs is widespread. Taylor and Allen (1992) document the popularity of the trading rules in their survey of currency traders in London. Recent efforts have also examined the profitability of TTRs and the implications for market efficiency. Those that explicitly examine TTRs include Brock, Lakonishok, and LeBaron (1992), Osler and Chang (1995), Goldbaum (1999), and Lo, Mamaysky, and Wang (2000).

Technical traders (the users of TTRs) and chartists tend to put little faith in strict efficient markets. In defense of a trend following rules, traders describe how prices are influenced by private information, but fail to fully reflect it. Prices may even be slow to react to public information. Users claim that TTRs provide an indication of when those with superior information have begun to trade and what position they take in the market. The Rendleman, Jones, and Latané (1982) finding that price movement starts before an earnings report is made public and continues to adjust afterwards lends credence to the technical trader's belief.

For this work, a model is developed and examined in which individual optimizing behavior leads to inefficiencies in an asset market that can be exploited through use of a TTR. The objective for this paper is to explore parameter settings that generate profitability for the TTR. The approach relies on heterogeneity in the agent's information and subsequent decisions. Much of the model is developed analytically, aggregating within groups of traders. The lack of a closed formed solution, though, necessitates computer simulation of the market as a final step to discern the characteristics of the model.

A number of recent papers introduce techniques relevant to this work. Brock and Hommes (1997 and 1998), Brock and LeBaron (1996), Lux (1998), and LeBaron, Arthur, and Palmer (1999) all model endogenous information selection by individual traders. These papers share the feature that traders endogenously select an information source to use based on past performance. Farmer (2000) does not have shifting populations of traders, but instead bases each trader's current access to capital on past trading success, which leads to similar results with regards to the influence of a trading strategy on the market.

Other papers related to this analysis include models examining the profitability of TTRs. Using the standard approach of Grossman and Stiglitz (1980), Grundy and McNichols (1989) and Brown and Jennings (1989) employ two rounds of trading in order to examine the information revealed in past prices. Blume, Easley, and O'Hara (1994) work with a similar model, but introduce features to make trading volume a useful statistic. As do many models examining the use of TTRs, these models employ an unknown, but stationary, intrinsic value. With stationarity, a sequence of market-generated data provides information on the unknown parameter.

A distinguishing feature of the model presented in this paper includes the basis by which traders evaluate the information options. The traders determine whether current conditions call for the acquisition of fundamental information in a forward looking manner, rather than relying on past performance. Another feature has the traders in the model employing the trading rule in a manner more reflective of how they are used in practice. A simple trading rule issues nothing more than a basic "buy" or "sell" indicator, an approach appropriate for examining the profitability of

TTR strategies. Finally, the model presented in this paper lacks stationarity in the security's intrinsic value. If the trading rule is to be useful, it must extract information about future realizations of the market generated price series or the exogenous dividend.

# 1. The Model

A more detailed development of the model used in this analysis can be found in Goldbaum (2000). In each period, each trader makes two decisions. The first decision is which information source should form the basis for the second decision. The second decision is the portfolio selection, which entails setting demand for the risky asset. In the first decision, the information options are between using fundamental information and a simple TTR. After each trader makes the selection, information is received. Each trader submits a demand function for the risky asset. The final events of the period are to set the market clearing price (a Walrasian equilibrium) and conduct transactions.

Based on the shared public information available at the beginning of each period, the trader forecasts the profits earned by each information source based on presumptions about unobservables in the current state and the evolution of the state variables in the next period.

## 1.1. The Market

A large but finite number of agents, indexed by $i = 1, ..., N$, trade a risky asset and a risk free bond. The risk free bond, with a price of one, pays $R$. The risky asset is purchased at the market price $p_t$ in period $t$ and pays a stochastic dividend $d_{t+1}$ and sells for the market price $p_{t+1}$ in period $t+1$. The stochastic dividend follows a random walk:

$$d_{t+1} = d_t + \varepsilon_{t+1}, \text{ with } \varepsilon_{t+1} \sim IIDN(0,\sigma_d^2). \tag{1}$$

## 1.2. The Fundamental Trader

Those traders deciding to employ fundamental information base their forecast of next period's dividend on individual private research. Each trader's forecast involves individual error, which is captured by modeling each trader as receiving a noisy private signal centered on the dividend being forecasted,

$$y_{i,t} = d_{t+1} + e_{i,t}, \text{ with } e_{i,t} \sim IIDN(0,\sigma_e^2) \tag{2}$$

The trader's mean squared error minimizing forecast of $d_{t+1}$ is thus,

$$\mathrm{E}[d_{t+1} | H_{i,t}] = (1 - \beta)d_t + \beta y_{i,t}, \text{ with} \tag{3}$$

$$\beta \equiv \text{cov}(d_{t+1}, y_{i,t})/\text{var}(y_{i,t}) = \frac{\sigma_d^2}{\sigma_d^2 + \sigma_e^2}.$$

The traders maximize an exponential utility function on wealth, $w_{t+1}$, conditional on the private information set $H_{i,t}$,

$$\max_{a_i, b_i}\{E[-\exp(-\gamma w_{i,t+1}) \mid H_{i,t}]\}$$
$$\text{s.t., } p_t a_{i,t}^F + b_{i,t}^F = w_{i,t} \tag{4}$$
$$w_{i,t+1} = (p_{t+1} + d_{t+1}) a_{i,t}^F + R b_{i,t}^F .$$

The private information includes the public market information, $I_t = \{d_t, p_{t-1}, ttr_{t-1}\}$, plus $\{y_{i,t}\}$. The variable *ttr* is the TTR signal defined below. The resulting demand for the risky asset is

$$a_{i,t}^F = \frac{E[p_{t+1} + d_{t+1} \mid H_{i,t}] - R p_t}{\gamma \, \text{var}(p_{t+1} + d_{t+1} \mid H_{i,t})}. \tag{5}$$

## 1.3. The Technical Trader

The TTR considered is a Simple Moving Average rule. Each period, the trader computes a moving average (MA) of the price of the risky asset over some extended period, $l$.

$$MA_t = \frac{1}{l}\sum_{s=0}^{l-1} p_{t-s} .$$

Let $ttr_t = 1$ if $p_t > \text{MA}_t$ (buy), or -1 if $p_t < \text{MA}_t$ (sell). If $p_t = \text{MA}_t$, a "hold" signal is issued with $ttr_t = 0$. The demand for the risky asset based on the TTR signal is simply

$$a_t^{TR} = fd \cdot ttr_t \tag{6}$$

where $fd$ is the exogenously set fixed magnitude of demand built into the TTR.

## 1.4. Solving for Price

The market price, $p_t$, satisfies

$$(1 - n_t)\bar{a}_t^F + n_t \bar{a}_t^{TR} = 0 \tag{7}$$

where $n_t$ represent the proportion of the trader population using the TTR and $\bar{a}^k$ representing the average demand within population $k$. Thus, *per capita* demand equals *per capita* supply, which is set exogenously to zero. As a reference price, in the absence of the technical traders ($n_t = 0$), the closed formed solution for the "fundamental" price is

$$p_t^F = b_1 d_t + b_2 d_{t+1} \tag{8}$$

with $b_1 = (1-\beta)/(R-1)$ and $b_2 = \beta/(R-1)$.

With the technical traders present in the market ($0 < n_t \leq 1$), the price solution becomes

$$p_t = b_{0t} + b_1 d_t + b_2 d_{t+1}. \tag{9}$$

The coefficients again solve to $b_1 = (1-\beta)/(R-1)$ and $b_2 = \beta/(R-1)$. The time dependent value of the intercept, $b_{0t}$, reflects the impact of the demand by the group of technical traders. The solution to $b_{0t}$ depends on how the traders forecast next period's $b_{0t+1}$. This paper is limited to examining the case in which the forecast $b_{0t+1}^e = b_{0t-1}$, the last observed price deviation from $p^F$.[1] The intercept $b_{0t}$ solves to

$$b_{0t} = \frac{n_t a_t^{TR} \sigma^F}{(1-n_t)(R-1)} \tag{10}$$

with $\sigma^F = \text{var}(p_{t+1} + d_{t+1} \mid H_{i,t}) = (R^2\sigma_e^2 + \beta^2)\sigma_d^2 / (R-1)^2$.

## 1.5. Selecting an Information Source

The realized profits for a period are defined by the holding of the risky security and its excess return over the risk free rate

$$\pi_t^k = a_t^k (p_{t+1} + d_{t+1} - Rp_t),\ k = F, TR. \tag{11}$$

The expected profit to using the fundamental information sources is a forward-looking expectation:

$$E[\pi_t^F] = E\left[\frac{E[p_{t+1} + d_{t+1} - Rp_t \mid H_t]}{\gamma\sigma^F}(p_{t+1} + d_{t+1} - Rp_t) \mid I_t\right]. \tag{12}$$

Substituting equations (8) and (9) and taking expectations, (11) solves to

$$E[\pi_t^F] = \left(\frac{a_{t-1}^{TR} n_{t-1}}{n_{t-1} - 1}\right)^2 \sigma^F.^2 \tag{13}$$

---

[1] This is the "continuation regime." This and the alternative "exploitation regime," where traders forecast $b_{0t+1} = 0$, are examined in Goldbaum (2000). Each regime is roughly self-fulfilling. Actually forecasting $b_{0,t+1}$ would require knowledge of or a forecast of $n_{t+1}$, which in the model is computationally infeasible and unlikely to be know by traders in practice. Thus, roughly, without knowledge of $n_{t+1}$, a guess, which proves to be a good guess *ex post*, is for each trader to employ the forecast appropriate for the prevailing regime.

[2] Traders forecast $n_{t+1}$. Consistent with the regime, they use $n_{t-1}$ as a point estimate.

For the TTR, the forecast of the profits is based on a weighted average of the past performance of the trading rule. This is not an unreasonable stipulation since the advice from technical traders is to look for rules that have recently performed well.[3] Forecasted TTR profits are

$$\pi_t^{TRe} = \mu\pi_{t-2}^{TR} + (1-\mu)\pi_{t-1}^{TRe}\ .^4 \tag{14}$$

As a way of introducing heterogeneity, the individual trader's final choice is modeled as a randomized discrete choice in the nature of Manski and McFadden (1981) and Anderson, de Palma, and Thisse (1992). The proportion of the population choosing each signal converges to the probability that individual traders choose that signal. Thus,

$$n_t = \frac{\exp(\rho\pi_t^{TRe})}{\exp(\rho\pi_t^{TRe}) + \exp(\rho E_t[\pi_t^F])} \tag{15}$$

is the proportion of the population using the TTR in period $t$.

The "intensity of choice" parameter, $\rho$, effectively sets the sensitivity of the population to differences in the expected profits. With $\rho \to \infty$, small differences in the expected profits lead to large population shifts toward the source with the greater expected profit for the period. With $\rho$ set to zero, differences in the expected profits are ignored.

## 2. Simulation Results

Without a computable solution, computer simulation is necessary for examining the model and the impact of the TTR on the market's behavior. The important control parameter is the intensity of choice, $\rho$. The TTR employed in this examination is a five period moving average rule. For the most part, examining different length MA rules produces results similar to those presented in this paper. The following parameters settings are kept unchanged in the examined simulations: $\gamma = 1$, $R = 1.02$, $fd = 0.0001$, and $\mu = 0.3$; also $\sigma_d^2 = 1$ and $\sigma_e^2 = 1$ which yields $\beta = 1/2$. The market is examined by running simulations based on the random dividend process. The simulations presented consist of 40,000 periods, after dropping the first 200. For comparison purposes, the same set of shocks is used in each simulation.

[3] For example, major software packages popular among traders for tracking charts and TTRs actively promote a feature which automatically informs the user which of the rules tracked in real-time have recently performed well and on which securities.

[4] The profit from the position taken in $t$-1 can only be determined once the time $t$ price is realized.

Market behavior under different values of ρ can be divided into two categories. At lower values, the popularity of the TTR is limited by the traders' low intensity of choice. As a result, the technical traders' impact on the price is also limited so that the market price remains near its intrinsic value. Price movements tend to reflect changes in intrinsic value. In this environment, it will be demonstrated that the TTR can be employed for long run profits.

At higher setting of ρ, $n_t$ takes greater swings between zero and one. The TTR can move the market price substantially away from the asset's intrinsic value. Price movements reflect changes in $n_t$ rather than movements in the intrinsic value. The TTR fails to provide profits comparable to the fundamental information.

Table 1 list summary statistics for some of the important variables of the model. The statistics are reported for ρ equal to 0, 300, and 500. Notice that for all three values of ρ, the profit earned by the TTR is positive. The mean profit for using the fundamental information is negative for ρ = 0, near zero for ρ = 300, and positive, and even exceeding the profits for the TTR for ρ = 500. To understand the TTR profit, consider an example where a shock increases $d_{t+1}$ from $d_t$. From (8), based on the fundamental traders alone, $p_t^F$ would increase, but only partially reflect the private information. The TTR functions as a crude signal extractor from the actual price, so that the increase in $d_{t+1}$ triggers a "buy" signal based on the price increase. The demand of the technical traders pushes the actual market clearing price, $p_t$, higher than $p_t^F$. For a give *fd*, there is an $n_t$ sufficiently small that the magnitude of the price distortion remains small enough to leave $p_t$ less than $d_{t+1}R/(R-1)$, the one period ahead intrinsic value of the asset. In the next period, when $d_{t+1}$ and $p_{t+1}$ are realized, the TTR earns a profit.

**Table 1**. Summary Statistics of Simulated Market

| | *n* | | | p-p(F) | | |
|---|---|---|---|---|---|---|
| ρ | 0 | 300 | 500 | 0 | 300 | 500 |
| mean | 0.500 | 0.515 | 0.506 | -0.0005 | -0.0228 | -0.0639 |
| std. | 0.000 | 0.127 | 0.197 | 9.63 | 14.19 | 24.37 |
| min. | 0.500 | 0.154 | 0.000 | -9.63 | -104.54 | -426.29 |
| max. | 0.500 | 0.916 | 0.978 | 9.63 | 87.08 | 256.77 |
| | π(*F*)x1000 | | | π(*TR*)x1000 | | |
| ρ | 0 | 300 | 500 | 0 | 300 | 500 |
| mean | -0.309 | 0.098 | 2.024 | 0.309 | 0.271 | 0.209 |
| std. | 3.984 | 6.995 | 27.287 | 3.984 | 4.174 | 4.738 |
| min. | -15.97 | -69.88 | -253.37 | -15.35 | -17.83 | -44.77 |
| max. | 15.35 | 123.76 | 1982.37 | 15.97 | 17.35 | 41.53 |

$T = 40{,}000$ observations; $\gamma = 1$; $R = 1.02$; $fd = 0.0001$; $\mu = 0.3$; $\sigma_d^2 = 1$; $\sigma_e^2 = 1$.

At about ρ = 2000, the mean realized TTR profit turns negative. For ρ = 2000, mean $\pi^{TR} = -4.80\text{x}10^{-5}$ while mean $\pi^{F} = 201.89$.

If the technical traders are making profits through signal extraction at the expense of the fundamental traders, how do the fundamental traders also make a profit? As suggested by Table 1, the answer depends on ρ and the movement in $n_t$. Figure 1 displays a number of time-series generated by the simulation that are useful for understanding the workings of the model. The figure includes the plots of the two different ρ settings, 300 and 500. An arbitrary 100 periods, 8100 to 8200, are displayed. The top frame plots $p_t$ and $p_t^F$ separately. The second frame plots $p_t$ - $p_t^F$. The third frame plots both fundamental and TTR realized profits, and the bottom frame plots $n_t$.

Each "buy" and "sell" signal by the TTR causes $p_t$ to differ from $p_t^F$. As revealed in (10), the magnitude of that difference depends on $n_t$. With $n_t$ = 0.5, the price deviation from $p_t^F$ is 9.63 for the given parameters, which is small when compared to the period to period change in price caused by the shocks to the intrinsic value.[5] In Figure 1, the noticeable deviations from $p_t^F$ occur at instances when the random shocks to the dividend cause $p_t^F$ to trend. Three increasing trends with bubbles peak at $t$ = 8111, 8122, and 8148. A decreasing trend troughs at $t$ = 8182. A trend in the intrinsic value reinforces the TTR's success by creating a string of correct and profitable signals. With each profitable signal, the subsequent popularity of the TTR is increased. The increased $n_t$ increases the price distortion, contributing further to the trend in the realized market price. The TTR's success actually contributes to its own success in the following periods, as long as the bubble is supported by the trend in the intrinsic value. When the trend stops or reverses, the realized profits of the TTR no longer sustain the TTR's popularity, which brings about a fall in $n_t$, a collapse in the price bubble, negative profits for the TTR, and positive profits for the remaining fundamental traders.

For the trader using the TTR from the beginning of the bubble through to the end, overall profits are positive since the intrinsic value after the bubble's collapse remains profitable compared to the purchase price at the beginning. The profits earned by a trader who sticks with the fundamental information depends on the size of the bubble and how small 1-$n_t$ is when the bubble collapses. The fundamental information earns negative profits throughout the price bubble's development, but

---

[5] The population standard deviation of $p_{t+1}^F$ - $p_t^F$ is 35.12.

since $n_t$ is always smaller at the beginning of the bubble than at the end, the losses at the beginning are spread out among a larger population of fundamental traders while the gains at the bubble's collapse are concentrated among a smaller population.

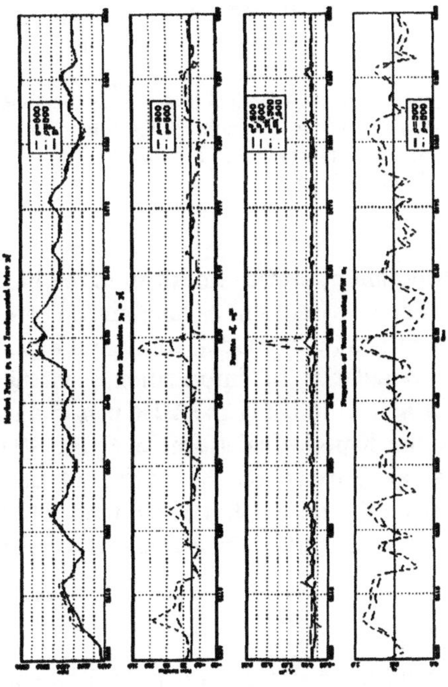

**Fig. 1.** Time-Series from Market Simulation. This figure plots a sample time series produced by the simulation with the dividend following a random walk. Parameter settings have $\rho = 300$ and 500. Top Frame: market price $p_t$, and fundamental information price $p_t^F$. Second Frame: price deviation $p_t - p_t^F$. Third Frame: profits $\pi_t^F$ and $\pi_t^{TR}$. Bottom Frame: population proportion using the TTR $n_t$.

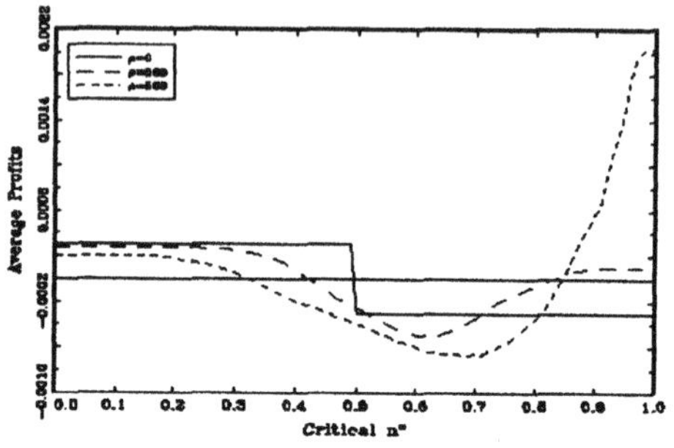


**Fig. 2.** Profits Based on a Critical $n^c$. The hypothetical traders uses the TTR if $n_t \geq n^c$ and uses the fundamental information if $n_t < n^c$. The far left trader relies solely on the TTR. The far right trader relies solely on the fundamental information. $T = 40{,}000$. Vertical axis: average profits. Horizontal axis: the setting for the critical value.

Figure 2 captures the distribution of profit. Recall from the model that each trader selects his information source based on a randomized process, but to help understand how profits are earned in the model, consider a hypothetical trader who, following the swings in popularity of information, switches from using the fundamental information if $n_t < n^c$ to using the TTR if $n_t \geq n^c$. Figure 2 plots the average profits earned by the hypothetical trader over the 40,000 periods of the simulation on the vertical axis, based on the setting of the critical $n^c$ on the horizontal axis. The trader on the far left has a low threshold for using the TTR. The far right trader uses the fundamental information regardless of the popularity of the TTR.

As demonstrated by the $\rho = 0$ example, under low $\rho$ the TTR earns a positive profit by free riding on the information revealed in the price, the population shifts are insufficient to generate profits for the fundamental trades when the bubble collapses. The greater $\rho$, the larger the bubble and the smaller the population of fundamental traders who share in the profit derived from its collapse.

With both the dedicated fundamental information and the dedicated technical traders earning long run profits, the losses are being earned by the traders who switch from using the fundamental information when it is popular at the beginning of a price bubble (earning negative profits) to using the TTR when it is popular at the end of the bubble (earning negative profits again when the bubble collapses).

# 3. Concluding Remarks

Because fundamental information gathered through individual research provides the trader with a noisy signal of a security's intrinsic value, the trader optimally tempers his forecast by incorporating publicly available information. As a result, market prices reflect only a portion of the news that generates the individual signals. This model demonstrates how a trend following TTR is a tool that can exploit this situation by signaling periods of transition.

When the "intensity of choice" is low, price movements continue to reflect changes in the security's intrinsic value. This is the best environment for the TTR to earn long run profits. At higher $\rho$ values, price movements tend to reflect changes in $n_t$ rather than the intrinsic value. The ability of the TTR to extract useful information from the price is diminished. The profits from the fundamental information far exceed those to using the TTR. At still higher $\rho$, the TTR fails to earn long run positive profits.

Price bubbles develop because of the technical traders and at the expense of the TTR's long run profitability. As a price bubble grows, it increases the concurrent profit earned by the TTR, contributing to additional growth in the popularity of the TTR and further development of the price bubble, a symbiotic relationship. The bubble's collapse produces negative profits for the TTR. The larger $\rho$, the greater the magnitude of the price bubbles, the more profitable is the fundamental information in the long run, and the lower the long run profits earned by the TTR.

The short run success of the TTR hinges on a market price that changes in response to private fundamental information, but only partially reflecting the value of that information. To be competitive with the long run profits of the fundamental information, the TTR depends on a market price that tracks the fundamental price fairly closely, without swings in $n_t$ that cause price bubbles to develop. The TTR will not earn a positive long run profit if the price movements fail to reflect changes in the intrinsic value, but instead reflect changes in $n_t$.

# References

Anderson, S.P., de Palma, A., Thisse, J.F.: Discrete Choice Theory of Product Differentiation. Cambridge, MA: MIT Publishers 1992

Blume, L., Easley, D., O'Hara, M.: Market Statistics and Technical Analysis: The Role of Volume. The Journal of Finance 49(1), 153-181 (1994)

Brock, W.A., Hommes, C.H.: Rational Routes to Randomness. Econometrica 65(5), 1059-1095 (1997)

Brock, W.A., Hommes, C.H.: Heterogeneous Beliefs and Routes to Chaos in a Simple Asset Pricing Model. Journal of Economic Dynamics and Control 22, 1235-1274 (1998)

Brock, W.A., Lakonishok, J., LeBaron, B.D.: Simple Technical Trading Rules and the Stochastic Properties of Stock Returns. The Journal of Finance 47(5), 1731-1764 (1992)

Brock, W.A., LeBaron, B.D.: A Dynamic Structural Model for Stock Return Volatility and Trading Volume. The Review of Economics and Statistics 78(1), 94-110 (1996)

Brown, D.P., Jennings, R.H.: On Technical Analysis. The Review of Financial Studies 2(4), 527-551 (1989)

Farmer, J.D.: Market Force, Ecology, and Evolution. Working paper 98-12-117. Santa Fe Institute 2000

Goldbaum, D.H.: A Nonparametric Examination of Market Information: Application to Technical Trading Rules. Journal of Empirical Finance 6, 59-85 (1999)

Goldbaum, D.H.: Technical Analysis, Price Trends, and Bubbles. Working paper. Rutgers University 2000

Grossman, S.J., Stiglitz, J.E.: On the Impossibility of Informationally Efficient Markets. The American Economic Review 70(3), 393-408 (1980)

Grundy, B.D., McNichols, M.: Trade and the Revelation of Information through Prices and Direct Disclosure. The Review of Financial Studies 2(4), 495-526 (1989)

LeBaron, B.D., Arthur, W.B., Palmer, R.: Time series Properties of an Artificial Stock Market. Journal of Economic Dynamics and Control 23, 1487-1516 (1999)

Lo, A.W., Mamaysky, H., Wang, J.: Foundations of Technical Analysis: Computational Algorithms, Statistical Inference, and Empirical Implementation. Journal of Finance 55(4), 1705-65 (2000)

Lux, T.: The Socio-economic Dynamics of Speculative Markets: Interacting Agents, Chaos, and the Fat Tails of Return Distributions. Journal of Economic Behavior & Organization 33(2), 143-65 (1998)

Manski, C.F., McFadden, D.: Structural Analysis of Discrete Data with Econometric Applications. Cambridge, MA: MIT Press 1981

Osler, C., Chang, K. Head and Shoulders: Not Just a Flaky Pattern. Staff Report No. 4. Federal Reserve Bank of New York 1995

Rendleman, R., Jones, C., Latané, H.: Empirical Anomalies Based on Unexpected Earnings and the Importance of Risk Adjustments. Journal of Financial Economics 10(3), 269-287 (1982)

Taylor, M.P., Allen, H. The Use of Technical Analysis in the Foreign Exchange Market. Journal of International Money and Finance 11, 304-314 (1992)

# Firms' Financial Heterogeneity and Business Cycles

**Marco Gallegati**

Dipartimento di Economia, Università di Ancona, 60100 Ancona, Italy

**Abstract.** In this paper I try to generalise the Greenwald-Stiglitz's framework with equity rationing introducing firms' financial heterogeneity as suggested by the hierarchy of finance model. After distinguishing among firms' alternative financing regimes it is possible to establish both a short run non-neutrality proposition according to which monetary policy may affect real output through changes of the expected future price level, and a long run non-neutrality proposition based on the Fisher effect. Moreover, changes in the degree of financial fragility of the system, 'proxied' by changes in the proportion of liquidity constrained firms, may determine shifts of the dynamic behaviour of the system.

**JEL Classification**: E32, E44, G32

*Keywords: Financing Hierarchy, Firms' Heterogeneity, Business Cycles*

# 1. Introduction

The Greenwald-Stiglitz's theoretical framework (from now on G-S) has been applied both to the analysis of business fluctuations and to the determination of the values of employment, output and prices, in opposition, respectively, to the real business cycles theory and the "vertical aggregate supply" set-up of the New Classical Macroeconomics. The New Keynesian macroeconomics models specify in imperfections in goods (Mankiw, 1985, Akerlof and Yellen, 1985, Mankiw and Romer, 1991)[1] and capital markets (Greenwald and Stiglitz, 1993) explanations for money non-neutrality in the short run. In particular, in G-S's framework the *non-neutrality proposition* according to which monetary policy affects real output is limited to the case of a 'price shock'. In this way, money non-neutrality is limited, as in the Lucas' 'price surprise' model, to an unanticipated change in monetary policy. As regards the analysis of business fluctuations, the NCM provides a description of the business cycle based on the Slutsky-Frisch "impulse-

[1] With imperfections in goods market we refer to imperfect competition and real or nominal rigidities.

propagation" approach where serially uncorrelated shocks (the impulses) affect output through a propagation mechanism and the fluctuations can be interpreted as deviations from the *steady-state* equilibrium due to exogenous stochastic shocks. The alternative approach to the traditional one where fluctuations around the *steady-state* persist until the system is perturbed by exogenous shocks is represented by the endogenous approach to the business cycle. According to the endogenous approach the economic activity is naturally characterised by oscillations which represent the *modus operandi* of a market economy. Business cycle is self-sustaining and doesn't need repeated exogenous shocks in order to have the typical oscillatory behaviour.

In G-S's framework the macroeconomic implications of informational imperfections in financial markets result in the assumption that firms, being rationed in the equity market, finance their production using a mix of retentions and debt. In this way the firms of the G-S model are, as regards their marginal source of funds, all identical. On the contrary, the hierarchy of finance model suggests that the interaction between firms' financing and investment (production) decisions may give place to different financing regimes, where the firm is *liquidity constrained* when internal funds are low relative to firms' investment opportunities, and *demand constrained* when internal funds are high relative to firms' investment opportunities. So while *liquidity constrained firms* will finance their own investment projects using retentions and issuing new debt, the investments projects of *demand constrained firms* will be entirely financed through retained earnings.

The hypothesis of asymmetric information, which has a leading role inside the New Keynesian Economics framework, means that agents differ each other, so that the representative agent hypothesis becomes inconsistent inside this approach. Moreover, with heterogeneous agents the possibility of extending microeconomic behaviour to macroeconomic relationships is denied by the stringency of the conditions necessary to obtain a consistent aggregation of heterogeneous agents (Martel, 1996). If agents are heterogeneous and interact each other, the propagation mechanism, and then aggregate behaviour, will depend on the distribution of the agents and it is likely to be affected by changes in the distributions of the agents. Once we abandon the representative agent hypothesis, business fluctuations and cycles may emerge as the result of a change in the distribution of the agents rather than as the consequence of exogenous shocks to which each individuals react, as in the standard real business cycle literature (Lucas, 1975, Long e Plosser, 1983).

On the basis of the above considerations I try to generalise the G-S's framework with equity rationing introducing firms' financial heterogeneity as suggested by the hierarchy of finance model, as firms' heterogeneity may attenuate the conflict between theory and aggregate data which emerge in models adopting the representative agent hypothesis. (Forni and Lippi, 1997). The paper is organised as follows: in section 2 I derive the individual firm supply function modifying the G-S framework to take into account firms' heterogeneity through the distinction between *liquidity* e *demand constrained* firms. In section 3 I examine the properties of the aggregate supply curve, while in section 4 it is analysed the relationship between financial fragility and dynamic behaviour of the system. Section 5 concludes the paper.

## 2. The Greenwald-Stiglitz's framework with firms' financial heterogeneity

The theoretical framework developed by Greenwald-Stiglitz (1988, 1990, 1993) represents one example of a New Keynesian macromodel laying on the assumption of capital market informational imperfections.[2] Even if the G-S's framework takes into account the problem of microfounding macroeconomics relationships it does not consider the microeconomic implications of the hierarchy of finance model where a financing hierarchy between internal and external sources of funds may emerge as a result of asymmetric information between lenders and borrowers (Myers, 1984). The aim of this paper is then to generalise G-S's framework taking into account firms' financial heterogeneity, where heterogeneity is determined on the basis of the marginal source of funds used by each single firm in financing its own production decisions.

- *The structure of the economy*

Following G-S I consider an economy where each firm produces an homogeneous good in a standard competitive market using a decreasing return to scale technology with labour as the only input, that is

$$y_t^i = f\left(n_t^i\right). \tag{1}$$

Thus, the inverse function of (1) will be

$$n_t^i = \phi\left(y_t^i\right) \tag{2}$$

with $\phi = f^{-1}$. The economy is sequential as there is a one-period lag between the production decisions and their realisations. At the beginning of each period the firm chooses its own level of production (and the necessary amount of inputs) which will be available for sale only at the end of the period. The time structure of production and sale is such that firms must hire and pay inputs (workers) before being able to sell output.[3] With informational imperfections restricted to the equity

---

[2] Alternative approaches about macroeconomic implications of asymmetric information in capital markets are Bernanke-Gertler (1989, 1990) and Kiyotaki-Moore (1997).

[3] If labour is the only input, then production costs will coincide with the wage bill.

market, firms finance their wage bill by means of internal funds (net worth or equity base) and by borrowing for the difference between the wage bill and internally generated funds. But contrary to G-S's framework production costs are not, by assumption, higher than firm's equity base in our model. In this way, we may have both firms whose internal funds are lower than their production costs (*liquidity constrained* firms), and firms whose internal funds are higher than their production costs (*demand constrained* firms).[4]

Moreover, the time discrepancy between the production and the sale stages is at the origin of firm's uncertainty about the selling price of its own good $P^i_{t+1}$, which will be a random variable with finite variance and expected value equal to the overall price level $P_{t+1}$, i.e.

$$E\left(P^i_{t+1}\right) = P_{t+1} \tag{3}$$

As a consequence, the relative price $u^i_{t+1}$, given by

$$u^i_{t+1} = P^i_{t+1} / P_{t+1} \tag{4}$$

will be a random variable with finite variance and expected value equal to unity, independent and identically distributed with probability density function g($u$) and cumulated distribution function G($u$). Finally, the time structure of transactions and the following price uncertainty expose the firm to a bankruptcy risk[5] in the event that the relative price $u^i_{t+1}$ be lower than the value at which the firm is just solvent $\bar{u}^i_{t+1}$. Thus, the bankrupt probability of each individual firm will depend on the probability than the relative price be lower than its critical threshold value, that is

$$\text{Bankrupt Probability} = \Pr\left(u^i_{t+1} < \bar{u}^i_{t+1}\right).$$

[4] In G-S's framework firm's production costs are by assumption higher than internal funds, and this difference, given firm's inability to raise new equity, have to be covered by borrowing. In terms of the hierarchy of finance model it means that G-S's model considers only those firms whose investment opportunities are high relative to their own internally generated funds (*liquidity constrained* firms), and omits completely those firms whose investment opportunities are low relative to their own internal funds (*demand constrained* firms). In this way the base G-S's model may be seen as a particular case of the more general one where firms may be both *liquidity* or *demand constrained.*

[5] Bankruptcy risk is important for production and employment decisions as going bankrupt is costly. Other than indirect costs as the opportunity and managers' reputation costs, a firm going bankrupt faces direct costs, given by the legal and administrative costs associated to the bankruptcy procedure.

- *Firm's behaviour*

Let's now analyse the firm's decisional problem. Firm's objective function consists of maximising expected real net profits, that is expected real profits, $E_t(\pi^i_{t+1})$, net of expected real bankruptcy costs, $C^i_t$

$$\max_{y^i_t} V = E_t\left(\pi^i_{t+1}\right) - C^i_t \ . \tag{5}$$

Due to the time discrepancy between production and sale firm's profit refers to period $t+1$, as it is only in $t+1$ that the firm knows the selling price $P^i_{t+1}$ of the good produced in period $t$. The expected profit function will be different for *liquidity* and *demand constrained* firms as different are the ways firms finance production. In particular if the $i^{th}$-firm was *liquidity constrained* then its profits in nominal terms would be, as in *G-S*,[6]

$$_L\Pi^i_{t+1} = P^i_{t+1} y^i_t - (1+I_t)\left(P_t w_t \phi(y^i_t) - A^i_t\right) \tag{6}$$

so that real profits would be

$$_L\pi^i_{t+1} = u^i_{t+1} y^i_t - \frac{P_t}{P_{t+1}}(1+I_t)\left(w_t \phi(y^i_t) - a^i_t\right) \tag{7}$$

and real expected profits

$$E\left(_L\pi^i_{t+1}\right) = y^i_t - h^e_t R_t \left[w_t \phi(y^i_t) - a^i_t\right] \tag{8}$$

where $h^e_t = P_t/P^e_{t+1}$ e $R_t = (1+I_t)$ is the capitalisation factor. The *liquidity constrained* firm goes bankrupt when it makes negative profits, that is when the relative price $u^i_{t+1}$ is lower than its critical threshold value $_L\bar{u}^i_{t+1}$:

$$u^i_{t+1} < \frac{h^e_t R_t \left(w_t \phi(y^i_t) - a^i_t\right)}{y^i_t} = {_L\bar{u}^i_{t+1}} \tag{9}$$

In this way the probability of bankruptcy may be expressed as

$$Pr(u^i_{t+1} < {_L\bar{u}^i_{t+1}}) = G\left(_L\bar{u}^i_{t+1}\right) = G\left(\frac{h^e_t R_t \left(w_t \phi(y^i_t) - a^i_t\right)}{y^i_t}\right) \ . \tag{10}$$

---

[6] The notation employed is the following: capital letters refer to variables in nominal terms, small letters refer to variables in real terms. Y is output, W wage, P price level, N employment, I nominal interest rate on loans, and A equity base or net worth.

Let $C_t^i$ be the cost of bankruptcy for a *liquidity constrained* firm and $Pr(u^i_{t+1} < \bar{u}^i_{t+1})$ its bankrupt probability, real expected bankruptcy costs will be

$$C_t^i = c_t^i \left[ Pr\left( u_{t+1}^i < {}_L\bar{u}_{t+1}^i \right) \right] \tag{11}$$

and then using equation (10) we obtain the final expression for $C_t^i$ as

$$C_t^i = c_t^i \left[ G\left( \frac{h_t^e w_t R_t \phi(y_t^i) - h_t^e R_t a_t^i}{y_t^i} \right) \right] \tag{12}$$

If, otherwise, the $i^{th}$-firm was *demand constrained* its profits in nominal terms would be

$${}_D\Pi_{t+1}^i = P_{t+1}^i y_t^i - P_t w_t \phi(y_t^i) \tag{13}$$

so that real term profits would be

$${}_D\pi_{t+1}^i = u_{t+1}^i y_t^i - \frac{P_t}{P_{t+1}} w_t \phi(y_t^i) \tag{14}$$

and real expected profits

$$E\left( {}_D\pi_{t+1}^i \right) = y_t^i - h_t^e w_t \phi(y_t^i)\ . \tag{15}$$

Firm's optimal choice of the $i^{th}$-firm will take place without *a priori* knowledge about its own belonging to the *liquidity* or *demand constrained* status. Let $p_1$ be the *ex ante* probability for a firm of being *liquidity constrained* and $(1 - p_1)$ the *ex ante* probability of being *demand constrained.* The objective function V may be expressed as weighted average of the profits in the two states with weights represented by the probability for the firm of being in one of the two states, that is[7]

$$\pi_{t+1}^i = p_1 ({}_L\pi_{t+1}^i) + (1 - p_1)({}_D\pi_{t+1}^i)\ . \tag{16}$$

Firm's objective function V may then be expressed as:

$$V = \left[ p_1 E_t ({}_L\pi_{t+1}^i) + (1 - p_1) E_t ({}_D\pi_{t+1}^i) \right] - c_t^i\ p_1 \left[ G({}_L\bar{u}_{t+1}^i) \right] \tag{17}$$

where the first expression in square brackets represents the expected profit, and the latter the expected bankrupt probability. Using equations (8) and (15) relative to the

[7] Equation (16), for $p_1$=1, corresponds to the profit equation of the G-S model where all firms are *liquidity constrained*

expected profits of both type of firms and equation (10) relative to the bankrupt probability of *liquidity constrained* firms, the expression for V becomes

$$V = y_t^i - p_1 h_t^e R_t \left(w_t \phi(y_t^i) - a_t^i\right) - (1-p_1)\left(h_t^e w_t \phi(y_t^i)\right) - c_t^i \left[ p_1 G\left( \frac{h_t^e w_t R_t \phi(y_t^i) - h_t^e R_t a_t^i}{y_t^i} \right)\right] \tag{18}$$

Rearranging the terms relative to the expected profits we have

$$V = y_t^i - h_t^e w_t \left(p_1 R_t + (1-p_1)\right)\phi(y_t^i) + p_1 h_t^e R_t a_t^i - c_t^i \left[ \frac{h_t^e w_t p_1 R_t \phi(y_t^i) - p_1 h_t^e R_t a_t^i}{y_t^i} \right] \tag{19}$$

and letting $Z_t = h_t^e w_t p_1 R_t$, and $S_t = h_t^e w_t (1-p_1)$ we may write the final expression for V as

$$V = y_t^i - (Z_t + S_t)\phi\left(y_t^i\right) + p_1 h_t^e R_t a_t^i - c_t^i \frac{Z_t \phi\left(y_t^i\right) - p_1 h_t^e R_t a_t^i}{y_t^i}. \tag{20}$$

- *Profit maximisation*

The first order condition for the firm's maximisation problem is:

$$1 - (Z_t + S_t)\phi'(y_t^i) = \rho_t^i \tag{21}$$

where $\rho_t^i$ is the marginal expected bankruptcy cost of firm *i* at time *t*, i.e.

$$\rho_t^i = c \frac{Z_t \phi'\left(y_t^i\right) y_t^i - Z_t \phi\left(y_t^i\right) + p_1 h_t^e R_t a_t^i}{\left(y_t^i\right)^2} \tag{22}$$

Equation (21) shows that the price, normalised at unity, will be equal to the sum of the marginal production cost and the marginal bankruptcy cost. Solving equation (21) we obtain:

$$y_t^i = \arg\max V = y\left(Z_t, S_t, p_1, R_t, a_t^i, F_{\bar{u}}\right) \tag{23}$$

where $Z_t = Z(h_t^e w_t, p_1, R_t)$ and $S_t = S(h_t^e w_t, p_1)$. Now, being $h_t^e w_t = \frac{W_t}{P_{t+1}^e} = w_t^e$, we may write the individual supply function as:

$$y_t^i = y(w_t^e, p_1, R_t, a_t^i, F_{\bar{u}}) \tag{24}$$

where $y_{w/p^e} < 0, y_R < 0, y_{p_l} < 0, y_a > 0, y_{F_{\bar{u}}} < 0$. The level of production of the $i^{th}$-firm depends negatively on the expected real wage, the nominal interest rate, the probability of being *liquidity constrained* [8] and the relative price uncertainty, and positively on the real equity base level. All these variables affect firm's production level indirectly through the expected marginal bankruptcy cost, and directly, with the exception of the degree of uncertainty of the relative price, through the marginal cost of production. The labour demand curve, given (2) and (24) will be then a function of the expected real wage as follows:

$$n_t^i = \phi(y_t^i) = \phi(y^i(w_t^e, R_t, a_t^i, p_l, F_{\bar{u}})) . \quad (25)$$

Figure 1 show the implications of equation (21) for the determination of the $i^{th}$-firm's equilibrium output and employment. Firm's output equilibrium level occurs at the point where price (equal to one in the model) equals marginal costs, with marginal costs given by the sum of marginal costs of production and marginal bankruptcy costs.

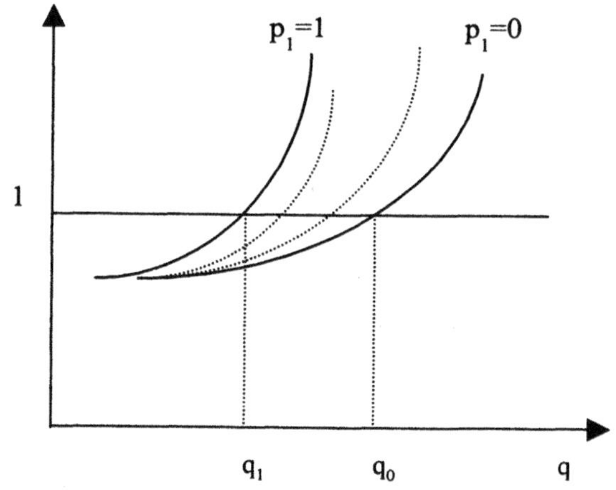


**Figure 1** - Determination of the $i^{th}$-firm's equilibrium output and employment as a function of the probability of being *liquidity constrained*

[8] The marginal cost of production is greater when $p_l=1$ than when $p_l=0$. Moreover, the marginal bankruptcy cost is positive if $p_l=1$, and zero if $p_l=0$. This imply that when the probability of being *liquidity constrained* rises, firm's marginal cost of production rises too, and as a consequence, the optimal levels of production and employment fall.

As there are as marginal costs curves as probabilities of being *liquidity constrained*, there will be many possible output equilibrium and employment values. In particular, the range of possible output equilibrium values goes from $q_1$ to $q_0$, corresponding to a probability for the firm of being *liquidity constrained* going from one, $p_1$=1, to zero, $p_1$=0, respectively. Again, the output equilibrium level $q_1$ corresponding to a probability of being *liquidity constrained* equal to one parallels the output equilibrium level in G-S. Moreover, given that for increasing values of the probability of being *liquidity constrained* the marginal cost curve shifts towards the left, we may conclude that a higher value of the probability of being *liquidity constrained* will lead to a reduction of the equilibrium output and employment levels.

Finally, once determined from equation (21) firm's optimal production (employment) level, we may establish, *ex-post*, the *liquidity* or *demand constrained*, status of the $i^{th}$-firm taking $a_t^i$ as predetermined. If, for example,

$$w_t\phi(y_t^i) = w_t\phi\left(y(w_t^e, p_1, R_t, a_t^i, F_{\bar{u}}\right) > a_t^i \tag{26}$$

we have a *liquidity constrained* firm, otherwise a *demand constrained* one.

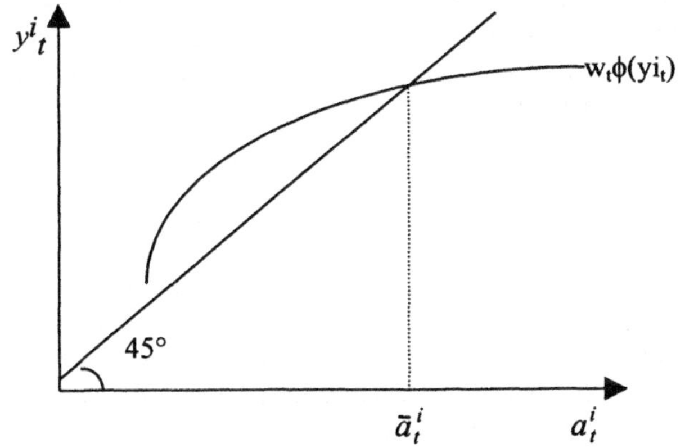


*liquidity constrained* *demand constrained*

**Figure 2** - Firm's status and net worth

Moreover, as the left member of (26) is an increasing function of $a_t^i$, $i^{th}$-firm's belonging to one of the two regimes will strictly depend on the relative slope of the concave function with respect to that of the 45° degree line. In particular, if the situation was the one represented in figure 2 the firms with a net worth lower than $\bar{a}_t^i$ would be *liquidity constrained*, while those with a net worth higher than $\bar{a}_t^i$ would be *demand constrained*. Thus, if firms' net worth was distributed according to a repartition function of the type $H(a_t^i)$, the proportion of liquidity constrained over all the firms would be

$$\theta_t = H(\bar{a}_t^i). \tag{27}$$

# 4. Aggregate supply

The aggregate supply function may be obtained by summing the supply functions of individual firms. Even assuming that all firms have an identical production function $f$ and face the same price uncertainty (the same distribution probability *G(.)*), they will differ with respect to their equity base level. Moreover, in order to get the aggregate supply curve we need to take into account both the distribution of the individual equity base and the distribution of the probability of being *liquidity constrained.*

As the binomial distribution deals with populations characterised by a dichotic alternative, firm's belonging to the *liquidity* or *demand constrained* "regime" may be described by a binomial random variable X which will assume the value 1 (*liquidity constrained*) with probability $p_1$, $Pr(X=1)= p_1$, and the value 0 (*demand constrained*) with probability $1-p_1$, $Pr(X=0)= 1-p_1$. A random variable X has a probability distribution given by

$$P(X=x)=p_1^x(1-p_1)^{1-x},$$

where $x = 0,1$ is the observed value of X, with mean and variance equal to $p_1$ and $p_1(1-p_1)$, respectively. Suppose there are $n$ firms in the economy, we will have $n$ random variables $X_1$, $X_2$,...., $X_n$, each one characterised by a probability distribution analogous to that of the random variable X. Let Y be random variable given by the sum of the $n$ random variables $X_1$, $X_2$,...., $X_n$, that is $Y = X_1 + X_2 + .... + X_n$, and suppose that $\theta$ represents the number of *liquidity constrained* firms, then the probability distribution of the random variable Y will be

$$P(Y\theta)=\binom{n}{\theta}p^{\theta}(1-p)^{n-\theta}$$

while the mean and the variance of this distribution will be, respectively $np_1$ and $np_1(1-p_1)$. Approximating (24) by taking a Taylor series expansion around the average level of firms' equity base, we obtain the following aggregate supply function::

$$y_t = y\left(w_t^e, R_t, \bar{a}_t, np_1; \sigma, np_1(1-p_1)\right) \tag{28}$$

where $\bar{a}_t$ is the "average" equity base level and $\sigma^2$ the variance of firm equity levels.

The specific form of the aggregate supply function will depend on the labour market equations. Aggregate labour demand, obtained by aggregation of the individual firms' labour demand, may be expressed, according to (25), as follows:

$$n_t^d = \Phi(y_t) = \Phi\left(y\left(w_t^e, R_t, \bar{a}_t, np_1; \sigma, np_1(1-p_1)\right)\right) \tag{29}$$

As regards the aggregate supply of labour, if we assume that workers choose rationally between leisure and consumption in a competitive labour market, we may obtain, if leisure and consumption are normal goods and an internal solution exists, the following labour supply function:[9]

$$n_t^s = n\left(w_t, \frac{1}{h_t^e}\right) \tag{30}$$

where labour supply depends positively on current real wage and negatively on the expected inflation rate,[10] i.e. $n_w > 0$ e $n_{1/h^e} < 0$. A labour supply function like (30) include as particular cases the extreme versions of the labour supply function of *NKE* and *NCM*. If we assume, as in *G-S*, that labour supply depends on the current real wage only,[11] i.e. $n_{1/h^e} = 0$, an increase of the expected future price level give. rise, through a decrease of the expected real wage, to a fall in firms' marginal production costs. The consequent increase in firms' labour demand will give rise, for a given labour supply, to a higher level of production. A similar result, even if lower in size, may be obtained when the labour supply function is characterised by a sufficiently low intertemporal substitution effect such that labour supply reacts more to changes in current real wage than to changes in expected inflation.[12] Finally, if we assume as in *NCM* models, that only the expected real wage affects labour supply, we may obtain the well known result of full neutrality of the aggregate supply function to changes in the expected future price level.[13] Thus, if

---

[9] The model has been elaborated by Lucas-Rapping (1969) and Sargent (1979).

[10] Being $1/h_t^e = P_{t+1}^e/P_t = 1 + \pi_t^e$, $1/h_t^e$ may be indicated for short as the expected inflation rate.

[11] In this case workers and firms will have an asymmetric behaviour towards future price, as workers supply labour looking at current real wage, while firms demand labour looking at the expected real wage.

[12] See Tamborini (1996).

[13] The log-linear specification of the labour supply function stresses the necessary condition for the full neutrality result of the *NCM* to hold. Given the following log-linear specification of equation (30)

$$n_t^s = \alpha_0 + \alpha_1(w_t - p_t) - \alpha_2(p_{t+1}^e - p_t)$$

we exclude the case of equality in magnitude of the parameters relative to the variables $w_t$ and $1/h_t^e$ in (30), the short run aggregate supply curve will be the one represented in figure 3. This aggregate supply curve is upward sloped as output in the short run increases with the expected future price level. Moreover, it shifts up as a result of a rise in current real wage, nominal interest rate on loans and in the probability of being *liquidity constrained*, and shifts down when the "average" real equity base rises.

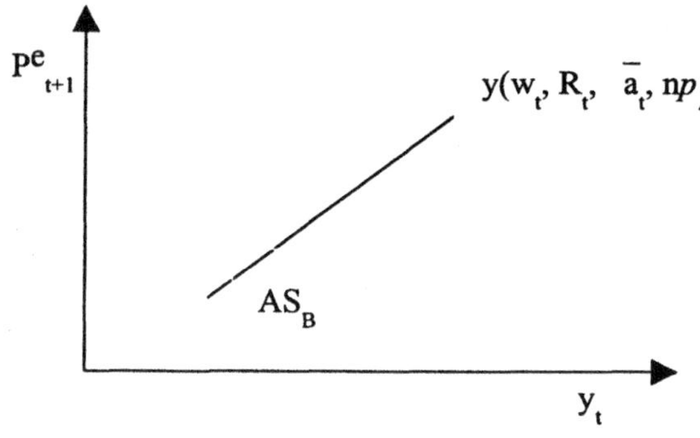


**Figure 3** - Short run AS curve

In conclusion, real output, in the short run, can be positively affected not only by "price shocks", as in *G-S* and in *NCM* models, but even by changes of the expected future price level. In this way we may establish a short run *non-neutrality policy effectiveness proposition* according to which monetary policy affects real output not only when monetary policy changes are not anticipated, but even when they are perfectly anticipated.[14] As a consequence, the *non-neutrality proposition* emerging from the generalised *G-S* model is inconsistent with the policy ineffectiveness proposition of the *NCM* and of the same *G-S*'s framework.

The dependence of the aggregate supply function on the nominal, rather than the real, interest rate let us establish even a long run *non-neutrality policy effectiveness proposition* based on the so-called *Fisher effect*. Given the defintion of the real interest rate as

$$r_t = R_t - \pi_{t+1}^e$$

---

it is only when $\alpha_1 = \alpha_2$, i.e. when the coefficients of current real wage and expected inflation rate are equal in magnitude, that labour supply depends on the expected real wage only.

[14] The result of money non-neutrality even when agents are not caught by surprise is obtained even in Delli Gatti and Mauro Gallegati (1997).

and assuming, as in Greenwald and Stiglitz (1993, p.97), both the constancy of the real interest rate and its equality to the individual's rate of time discount, δ,

$$r_t = \delta$$

we have that, being inflation expectations realised in the long run, *i.e.* $\pi^e_{t+1} = \pi_{t+1}$, changes of the inflation rate will be reflected in a one to one change of the nominal interest rate. In this way if we assume that the inflation rate is mainly determined by the money supply growth rate it is possible to establish a long run *non-neutrality policy effectiveness proposition,* in the sense that a higher money growth rate, determining a higher rate of inflation, will give rise to a higher level of the nominal interest rate and then, given (28), a reduction of the aggregate production level.

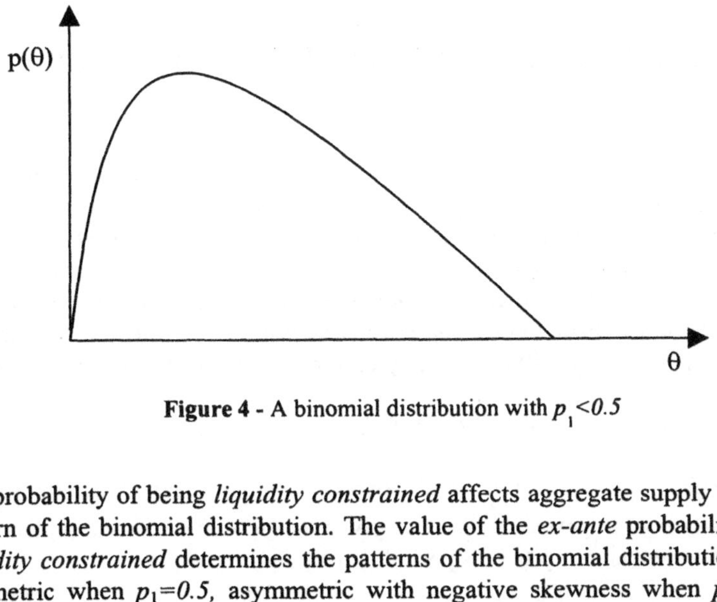


**Figure 4** - A binomial distribution with $p_1 < 0.5$

The probability of being *liquidity constrained* affects aggregate supply through the pattern of the binomial distribution. The value of the *ex-ante* probability of being *liquidity constrained* determines the patterns of the binomial distribution which is symmetric when $p_1 = 0.5$, asymmetric with negative skewness when $p_1 > 0.5$, and asymmetric with positive skewness when $p_1 < 0.5$.[15] Figure 4 shows an example of an asymmetric binomial distribution with positive skewness, $p_1 < 0.5$, where we put on the *x-axis* the values of $\theta$ and on the *y-axis* the probabilities associated with each value of θ.

The dependence of the aggregate supply curve from the pattern of the binomial distribution means that, as the distribution may have different shapes (symmetric, asymmetric with positive skewness, asymmetric with negative skewness), there will be nor a "unique" *natural level of output* nor a "unique" *natural level of the*

[15] When *n* increases the degree of asymmetry of the binomial distribution tends to fall so that the normal distribution approximates the binomial distribution.

*unemployment rate*. Indeed, given $w_t^e$, $R_t$ and $\bar{a}_t$, there will be as many output natural levels as many patterns of the binomial distribution probability. In this way the concepts of a *natural level of output* and of a *natural level of the unemployment rate* lose their meaning as there will be as many natural levels of output and unemployment rate as many values of the *ex-ante* probability of being *liquidity constrained*. This, in its turn, being determined by the interaction between firm's production decisions and firm's financial conditions, may be thought as a function of the entrepreneurs' state of confidence about the expected flow of future profits. These short-run expectations[16], being referred to the expected profits stemming from the use of the existing stock of capital, will determine output and employment current levels.

## 4. Financial fragility and dynamic behaviour of the system

If the dynamics of the equity base differs according to firm's financing regime, the dynamics of the aggregate equity base of the generalised G-S model will be represented by a weighted average of the dynamics of the equity base of the *liquidity* and *demand constrained firms*. Let's analyse the dynamics of the aggregate real equity base after introducing the distinction between *liquidity* and *demand constrained* firms. For a *liquidity constrained* firm the value of the real equity base in period *t+1* will be given by current profits net of dividends payment,

$$ {}_L a_{t+1}^i = u_{t+1}^i y_t^i - h_t R_t \left( w_t \phi\left(y_t^i\right) - a_t^i \right) - d_{t+1}^i = {}_L \pi_{t+1}^i - d_{t+1}^i \tag{31} $$

where $h_t = P_t/P_{t+1}$ and $R_t = (1+I_t)$, while for a *demand constrained* firm the value of the real equity base in period *t+1* will be given by the sum of current profits net of dividends payment and of previous period real equity base,

$$ {}_D a_{t+1}^i = u_{t+1}^i y_t^i - h_t w_t \phi\left(y_t^i\right) - d_{t+1}^i + a_t^i - w_t \phi\left(y_t^i\right) = {}_D \pi_{t+1}^i - d_{t+1}^i + {}_D a_t^i - w_t \phi\left(y_t^i\right). \tag{32} $$

Aggregate real equity base will be the weighted sum of the real equity base of *liquidity* and *demand constrained* firms, with weights given by the *ex-post* proportion of *liquidity* and *demand constrained* over all firms, endogenously determined by (27):

$$ a_{t+1} = \theta_t \, {}_L a_{t+1}^i + \left(1 - \theta_t\right) {}_D a_{t+1}^i \tag{33} $$

---

16 In his *General Theory* (1936) Keynes distinguishes between short run and long run expectations and specifiy in the latest, as linked to investments' future profitability, the determinants of investment.

Substituting (31) and (32) into (33), taking expected values and assuming that the dividends are an increasing function of the previous period equity base (Greenwald and Stiglitz, 1993) yields:

$$a_{t+1} = \theta_t \{ y_t - [h_t^e R_t (w_t \phi(y_t) - a_t)] - d(a_t) \} + (1-\theta_t)[y_t - (1+h_t^e) w_t \phi(y_t) - d(a_t) + a_t] \quad (34)$$

After rearranging, the equation for the aggregate real equity base can be written as follows:

$$a_{t+1} = y_t - (1 - \theta_t + \theta_t R_t h_t^e - \theta_t h_t^e + h_t^e) w_t \phi(y_t) + (1 - \theta_t + \theta_t R_t h_t^e) a_t - d(a_t) \ . \quad (35)$$

Equation (35), corresponds to G-S's law of motion of aggregate real equity base when the following hypothesis holds:

- all firms are *ex-post liquidity constrained*, i.e. $\theta_t$=1, and
- lenders are perfectly informed and risk neutral,i.e. $1 + r_t = E[R_t h_t^e]$.

Equation (35) is a first-order non-linear difference equation whose characteristics may be analysed using the phase diagram. The phase diagram of equation (35) is a quadratic map[17] which may give rise, depending on the slope of the quadratic map at the point of intersection with the 45° line (the *steady state* value of the real equity base), to different dynamics behaviour of the system. In particular, if the slope at the *steady state* value is between zero and one $0<\partial a_{t+1}/\partial a_t<1$, the system converges uniformly towards the *steady state* equilibrium. If the slope at the *steady state* value is between minus one and zero, $-1<\partial a_{t+1}/\partial a_t<0$, the system converges to the *steady state* equilibrium by means of dampening oscillations. And, when the negative slope of the quadratic map at the *steady state* value of the real equity base is, in absolute value, greater than one, $\partial a_{t+1}/\partial a_t<-1$, the system may assume different dynamics behaviour, i.e. limit cycles, chaotic dynamics, divergent dynamics with exploding oscillations (Day, 1994). In particular, for increasing, and greater than one, values of the slope of the phase diagram at the *steady state* value the long run dynamic behaviour of the system may pass, through the "bifurcation" process, from limit cycles of period 2, 4, 8, .......2n, to chaotic dynamics, in which case the periodicity of the orbits becomes from fixed to random (Baumol e Benhabib, 1988). Finally, for values of $\partial a_{t+1}/\partial a_t$ outside the lower extreme of the interval for which a solution to (35)[18] exists, the system will diverge from the *steady state* by means of explosive oscillations.

---

[17] See Delli Gatti and Mauro Gallegati, 1996.

[18] The existence of a solution for a first order non-linear difference equation of the type

$$y_{t+1} = f(y_t) = \Gamma y_t (1-y_t)$$

(obtainable from [27] through a change of variable, see Delli Gatti and Mauro Gallegati, 1996) requires the following condition: $0<\Gamma\leq 4$ (Brock *et al.*, 1991). In this case for $\Gamma>4$ no solution exists.

After all, when $\partial a_{t+1}/\partial a_t<0$, that is when the slope of the phase diagram at the steady state equilibrium is negative the system will be characterised by an oscillatory behaviour, not monotonic, represented by :

- fluctuations around the *steady state*, that is oscillations generated by exogenous stochastic shocks whose persistence is a necessary condition to have an irregular behaviour of the economy,[19] if the slope in, in absolute value, smaller than one;
- limit cycles (that is deterministic cycles characterised by periodic orbits) if the slope is greater than one in absolute value, and by endogenous chaotic dynamics[20] and/or divergent dynamics with explosive oscillations for increasing values of the slope.

The slope of the phase diagram of equation (35), represented by the derivative $\partial a_{t+1}/\partial a_t$ and equal to

$$\frac{\partial a_{t+1}}{\partial a_t}=\frac{\partial y}{\partial n}\frac{\partial n}{\partial a_t}-\left(1-\theta_t+\theta_t R_t h_t^e-\theta_t h_t^e+h_t^e\right)w_t\frac{\partial n}{\partial a_t}+1-\theta_t+\theta_t h_t^e R_t-\frac{\partial d}{\partial a_t} \tag{36}$$

can assume many different values which may change the dynamic behaviour of the system. In particular, as (36) contains the *ex-post* value of the proportion of *liquidity constrained* firms, and interpreting this parameter as *proxy* of the degree of financial fragility of the system we may stress the role of the financial fragility/solidity of the system for the dynamic behaviour of the system.

Now assume that the degree of financial fragility of the system, in each period, is proportional to the same period total debt, and that total debt is a positive function of the proportion of *liquidity constrained* firms, that is

$$D_t=d(\theta_t)=d\left(H\left(\bar{a}_t^i\right)\right) \tag{37}$$

with $d'>0$. As the proportion of *liquidity constrained* firms is between zero and one, $0<\theta_t<1$, we may classify as "financially fragile" a system whose proportion of *liquidity constrained* firms is close to one, and "financially sound" a system whose proportion of *liquidity constrained* firms is close to zero. Equation (36) shows the slope of the phase diagram in the general case, that is when the proportion of *liquidity constrained* at time $t$ is between zero and one, $0<\theta_t<1$. Looking at what happens to the slope of the phase diagram at the extreme values of the interval (0,1) we may derive the way a change in the financial fragility of the system affects the

---

[19] In this case fluctuations are explained with the Slutsky-Frisch's approach according to which non correlated exogenous shocks (the impulses) affects the system through a propagation mechanism given by a distributed lag relationship.

[20] See Ardeni *et al.* (1996).

dynamic behaviour of the system. When firms are all *liquidity constrained*, i.e. $\theta_t = 1$, (36) becomes

$$\theta_t = 1: \ \frac{\partial a_{t+1}}{\partial a_t} = \frac{\partial y}{\partial n}\frac{\partial n}{\partial a_t} - R_t h_t^e w_t \frac{\partial n}{\partial a_t} + h_t^e R_t - \frac{\partial d}{\partial a_t} \tag{38}$$

and coincides coincide with the slope of the phase diagram in *G-S*. While, on the contrary, when all firms are *demand constrained*, i.e $\theta_t = 0$, (36) becomes

$$\theta_t = 0: \ \frac{\partial a_{t+1}}{\partial a_t} = \frac{\partial y}{\partial n}\frac{\partial n}{\partial a_t} - \left(1 + h_t^e\right) w_t \frac{\partial n}{\partial a_t} + 1 - \frac{\partial d}{\partial a_t} \tag{39}$$

Using equations (38) and (39), we may verify how the degree of financial fragility affects the dynamic behaviour of the system. Indeed, comparing the slopes of the phase diagram in the two extreme cases, $\theta_t=0$ and $\theta_t=1$, we may note that the phase diagram when the system is financially fragile has a negative slope greater than that of a financially sound system.[21] This implies that the phase diagram will be characterised by an increasing negative slope as the proportion of *liquidity constrained* firms increases. Thus a change in agents' distribution, given in this model by a shift in the proportion of *liquidity* and *demand constrained* firms, may determine, through a change in the financial fragility of the system, a change of the dynamic behaviour of the system. For example, we may think to a system initially characterised by convergence – monotonic or by means of dampening oscillations - towards the *steady state* which, following a change in the proportion of *liquidity constrained* firms, may be characterised by limit cycles and endogenous chaotic dynamics. Alternatively, a system characterised by chaotic dynamics may, following an increase of the financial fragility of the system induced by an increase of the proportion of *liquidity constrained* firms, may display divergence – monotonic or by means of explosive oscillations - from the *steady state*.

---

[21] The slope of the phase diagram of a "financially fragile" system os greater than that of a "financially sound" system when the following inequality holds:

$$R_t h_t^e \left(1 - w_t \frac{\partial n}{\partial a_t}\right) > 1 - \left(1 + h_t^e\right) w_t \frac{\partial n}{\partial a_t} \tag{40}$$

The solution of (36), given by

$$\left(1 - R_t h_t^e + h_t^e\right)\left(w_t \frac{\partial n}{\partial a_t}\right) > 1 - R_t h_t^e,$$

is always true, being the term $w_t \left(\partial n / \partial a_t\right)$ always positive.

# 5. Conclusive remarks

Introducing sources of funds' heterogeneity in G-S model let us obtain some interesting results concerning the NEK *vs* NMC debate on the "aggregate supply curve" and *non-neutrality policy effectiveness proposition* on one side, and the relationship between financial conditions and business fluctuations on the other side. First of all, it is possible to establish both a short run and a long run *non-neutrality policy effectiveness proposition.* In the short run, with the exception of the *NCM*'s extreme version of the labour supply function in which only the expected real wage affects workers' labour supply, monetary policy may affect real activity because of a short run aggregate supply curve positively sloped with respect to the firms' expected future price level. In this way even a perfectly anticipated monetary policy change may have real effects, in opposition to *NCM* and the same *G-S* models where only not anticipated monetary policy changes may affect real output. Thus, the *non-neutrality proposition* emerging from the generalised *G-S* model is inconsistent with the policy ineffectiveness proposition of the *NCM.* Monetary policy ability to affect real activity even in the long run depends on the so-called *Fisher effect,* as the aggregate supply curve is negatively related to the nominal interest rate. Assuming, indeed, that in the long run expected and actual inflation will coincide, changes of the inflation rate will be reflected in one to one changes of the nominal interest rate and then, given the long run relationship between money growth rate and inflation rate, monetary policy may affect, through its effects on the nominal interest rate on loans, the level of economic activity even in the long run.

Finally, distinguishing between *liquidity* e *demand constrained* firms stresses the role of financial conditions in business fluctuations if we interpret the proportion of *liquidity constrained* firms as a *proxy* of the degree of financial fragility of the system. Thus, a change of the above proportion, being interpretable as a change of the financial fragility of the system may give rise to a change of the dynamic behaviour of the system as the law of motion of real aggregate equity base is a first-order non-linear difference equation which can generate a wide range of different types of dynamics: convergence (monotonic or by means of dampening oscillations), limit cycles, chaotic dynamics, divergence (convergence (monotonic or by means of explosive oscillations). As the dynamic behaviour of the system is, in its turn, dependent on the degree of financial fragility of the system, changes of the degree of financial fragility/solidity of the system may determine the shift from a type of dynamic behaviour (damped oscillations) to another (limit cycles) and vice versa. In this way fluctuations and cycles may emerge as the consequence of a change in the distribution of the agents (i.e. the proportion of *liquidity constrained* firms) rather than as the result of the reaction of identical agents to exogenous shocks, as in the standard real business cycle literature.

# References


Akerlof G.A. e J.L. Yellen (1985), "A Near-rational Model of Business Cycle, with Wage and Price Inertia", *Quarterly Journal of Economics*, 100, 823-38.

Ardeni P., A. Boitani, D. Delli Gatti e Mauro Gallegati (1996), "La Nuova Economia Keynesiana (Microfondazioni)", in M. Messori (eds.) *La Nuova Economia Keynesiana*, Società editrice il Mulino, Bologna.

Baumol e Benhabib (1989), "Chaos: Significance, Mechanism, and Economic Applications", *Journal of Economic Perspectives*, 3, 77-105.

Bernanke B. e M. Gertler (1989), "Agency Costs, Net Worth, and Business Fluctuations", *The American Economic Review*, 79, 14-31.

Bernanke B. e M. Gertler (1990), "Financial Fragility and Economic Performance", *Quarterly Journal of Economics*, 105, 87-114.

Brock W., D. Hsieh e B. Le Baron (1991), *Nonlinear Dynamics, Chaos, and Instability*, MIT Press, Cambridge.

Day R. (1994), *Complex Economic Dynamics*, MIT Press, Cambridge.

Delli Gatti D. e Mauro Gallegati (1996), "Financial Market Imperfections and Irregular Growth Cycles", *Scottish Journal of Political Economy*, 146-58.

Delli Gatti D. e Mauro Gallegati (1997), "Financial Constraints, Aggregate Supply, and the Monetary Trasmission Mechanism", *The Manchester School*, 65, 101-26.

Forni M. e M. Lippi (1997), *Aggregation and the Microfoundations of Dynamic Macroeconmics*, Clarendon Press, Oxford.

Greenwald B. e J.E. Stiglitz (1988), "Imperfect Information, Finance Constraints and Business Fluctuations", in M. Kohn and S.C. Tsiang (eds.) *Finance Constraints, Expectations and Macroeconomics*, Oxford University Press, Oxford.

Greenwald B. e J.E. Stiglitz (1990), "Macroeconomic Models with Equity and Credit Rationing", in R.G. Hubbard (eds.) *Asymmetric Information, Corporate Finance and Investment*, The University of Chicago Press, Chicago.

Greenwald B. e J.E. Stiglitz (1993), "Financial Market Imperfections and the Business Cycles", *Quarterly Journal of Economics*, 108, 77-114.

Keynes J.M. (1936), *The General Theory of Employment, Interest and Money*, MacMillan, London.

Kiyotaki N. e J. Moore (1997), "Credit Cycles", *Journal of Political Economy*, 105, 211-48.

Long J.B. e C.I. Plosser (1983), "Real Business Cycles", *Journal of Political Economy*, 91, 39-69.

Lucas R.E. Jr. (1975), "An Equilibrium Model of the Business Cycle", *Journal of Political Economy*, 83, 1113-44.

Lucas R.E. and L. Rapping (1969), "Real Wages, Employment and Inflation", *Journal of Political Economy*, 77, 721-54.

Mankiw N.G. (1985), "Small Menu Costs and large Business Cycles: a macroeconomic Model of Monopoly", *Quarterly Journal of Economics*, 100, 529-39.

Mankiw N.G. e D. Romer (eds.) (1991), *New Keynesian Economics,* MIT Press, Cambridge.

Martel R.J. (1996), "Heterogeneity, Aggregation, and a meaningful Macroeconomics", in D. Colander (eds.) *Beyond Microfoundations: Post Walrasian Macroeconomics*, Cambridge University Press, Cambridge.

Myers S.C., (1984), "The Capital Structure Puzzle", *The Journal of Finance*, 39, 575-92.

Sargent T.J. (1979), *Macroeconomic Theory*, Academic Press, New York.

Tamborini R., (1996), "An investigation into the New Keynesian macroeconomics of imperfect capital markets", in M. Messori (eds.), *Liquidity constraints and market failures. The microeconomic foundations of New Keynesian economics,* Aldershot, Elgar

# Part IV.

# Coalitions

# Stable Coalition Structures with Fixed Decision Scheme*

Guillaume Haeringer[1]

CODE, Departament d'Economia i d'Historia Economica, Universitat Autònoma de Barcelona, 08193 Bellaterra (Barcelona), Spain

**Abstract.** This paper studies the stability of coalition structures when coalitions must choose an alternative in a one-dimensional space, using a fixed decision scheme. This decision scheme depends on the preferred alternative of all agents in the coalition. We assume that the number of agent is finite. Hence, when an agent, or a group of agent, leaves a coalition and enter in another coalition, this will affect the outcome of the decision scheme in the former and the new coalitions. We provide in this paper sufficient conditions for the existence of stable coalition structures, for two stability concepts, namely $\mathcal{C}$-stability and Tiebout-stability. The model presented here can be understood as a local public good economy with horizontal differentiation.

## 1 Introduction

The aim of this paper is to study the stability of coalition structures, in non-cooperative and cooperative frameworks. We advocate that coalitions result from a contest between two opposite forces. On the one hand, individuals might be attracted by big communities, clubs or networks, e.g., when the power of coalitions increases with its size — as suggested by Demange [4] — or when there are increasing returns to scale. On the other hand, individuals may feel better off in small structures, and seek to join small groups rather than big ones. For instance, in small groups agents are usually more likely to

* I wish to thank Salvador Barberà, Marcus Berliant, Carmen Beviá, Pierre Dehez, Carmen Herrero, Michel Le Breton, Jean-François Mertens, Thomas Palfrey, Ulrich Schmidt, Gisèle Umbhauer, Myrna Wooders, seminar audiences at CORE, 1999 "Ecole de Printemps" of CORE-GREQAM in Aix-en-Provence, 2000 Public Economic Theory Conference in Warwick, and 2000 World Congress of the Econometric Society in Seattle for helpful comments and criticisms. I am particularly grateful to Hideo Konishi and Jordi Massó for comments and suggestions that greatly improved the exposition of this work. Part of this work was done while I was visiting the Institute of Mathematical Economics at Bielefeld University under the European research programme "Game-theoretic approaches to cooperation and exchange of information, with economic applications." Financial support from the European Commission, contract number HPMF-CT-1999-00284, is gratefully acknowledged. All remaining errors are mine.

be part of some kind of decision making process than in bigger ones. If one of these two forces has no or little influence, trivial equilibria may be reached.[1] By contrast, when both forces have comparable effects, the outcome of the coalition formation process becomes harder to predict, and finding stable coalition structures is more intricate.

We propose here a coalition formation model in a similar vein to local public good economies, as defined by Tiebout [18],[2] with the following characteristics. Agents in a **finite** population have to choose the location of a public good/facility. They may form jurisdictions (that we call coalitions), and each coalition finances its own public good. We suppose that an agent's utility in a coalition increases with the number of agents in the coalition to which she belongs (which refers to the tax payed by agents being decreasing with respect to the number of agents consuming the public good), and decreases with the distance between the agent's location and that of the public good produced by her coalition (which refers to a transportation costs). A major assumption that we make is the existence of **a decision scheme**, which is referred to a collective choice within each coalition. That is, the choice of the location of the public good in a coalition is made using a decision scheme, which depends on the positions of the agents living in the coalition. This latter can be either the result of a vote, or the decision of a social planner. Our purpose is to study the impact of this assumption on the existence of stable coalition structures, i.e. states in which no player, or no group of players, moves from one coalition to another or create a new coalition, and to identify which decision schemes ensure the existence of stable coalition structures.

A major consequence of a fixed decision scheme is that when an agent joins a coalition, she will affect the location of the public good of her former and her new coalition. This in turn may affect the payoffs of those individuals who already belong to these coalition. Put differently, in models where no decision scheme is assumed, like in Greenberg and Weber [10], or in Konishi, Le Breton and Weber [15], the decision of the coalition can be affected by a migrant. With a decision scheme and a finite number of agents, a migrant is **not** always welcome. It is worth noticing that this impact of the fixed decision scheme assumption is not shared with models with a continuum of agents. In these economies, each individual has a zero weight in the population. Hence, when an agent enters a coalition, she has no influence on the outcome of the decision scheme.

Another consequence of the decision scheme assumption is that the local public good economy model can be understood as a two-stage game. The first stage is devoted to the formation of coalitions, and the second stage is devoted to the choice of the public good. Thus, agents can know their payoffs in all the coalition they may join, which enables them to make a ranking

[1] In this paper, "equilibrium" means a state such that no player migrates from a coalition to another one.

[2] See Wooders [20] for a survey.

of coalitions in terms of preferences. This allows us to define an associated **hedonic game**, that has proved useful for the study of coalition formation. Hedonic games are games in which the utility of a player depends on the identity of other members of her coalition. These games were first defined by Drèze and Greenberg [5], and recently developed by Banerjee, Konishi and Sönmez [2], and Bogomolnaia and Jackson [3].

In this paper, we focus on a specific decision scheme, that we call the mean of the extremes. corresponds to the mid-point between the most "marginalistic" agents of a coalition, i.e., the pair of agents with the greatest distance between their respective location. We call this decision scheme the ***mean of the extremes***. This latter is attractive for two reasons. First, it easily enables us to prove the existence of stable coalition structures for the two main stability concepts used in the literature, namely the $\mathcal{C}$-stability (introduced by Guesnerie and Oddou [11]), and the Tiebout equilibrium. A second interesting feature is about its welfare properties. We show that a necessary assumption for the existence of stable coalition structures (under both stability concepts) is a symmetry assumption. This assumption merely says that agents must face the same tradeoff between the distance and the coalition size. That is, that the functional form of utility functions must be identical up to a linear transformation. This implies that the mean of the extremes is the only Rawlsian decision scheme. Since the agents that are more likely to disrupt a coalition are those being at the border of the coalition, it seems natural that a decision scheme aimed at ensuring the stability of coalitions must particularly take care of these agents.

The paper is organized as follows. The next section presents the model and the main assumptions. In section 3 we study the case in which multilateral moves are allowed, but without the free entry assumption. These equilibria are called $\mathcal{C}$-stable coalition structures, and correspond to the core of the associated NTU-game. In section 4, we study equilibria with free mobility, that we call Tiebout equilibria.

## 2 The Model

We denote by $N$ the set of agents (that we call players), and assume that it is finite, $N = \{1, \ldots, n\}$. A **coalition** $S$ is any subset of $N$, and $|S|$ denotes its size. A **coalition structure** $\mathrm{B} = \{S_1, \ldots, S_k\}$ is any partition of $N$ in nonempty coalitions: $S_h \cap S_{h'} = \varnothing$, $\forall\ h \neq h'$, and $\bigcup_{h=1}^{k} S_h = N$. For each coalition structure B we designate by $S(i, \mathrm{B})$ the coalition $i$ belongs to. When no confusion is possible, we write $S(i)$ instead of $S(i, \mathrm{B})$. Throughout the paper, $\subset$ and $\subseteq$ will respectively denote the strong and the weak set inclusion.

Let $\Omega$ be any subset of the real line. Positions of players are given by $p = \{p_1, \ldots, pn\} \in \Omega^n$. We say that a coalition $S$ is **connected** if $i, k \in S$

implies that $j \in S$ for all $j$ such that $p_i \leq p_j \leq p_k$. A coalition structure B is **connected** if all coalitions $S \in$ B are connected.

**Definition 1.** The distribution of players $(p_1, \ldots, p_n)$ is **monotone** if either,

$$
\begin{aligned}
& |p_i - p_{i-1}| \geq |p_{i+1} - p_i|, \ \forall\, i = \{2, \ldots, n-1\}\,, \\
\text{or} \quad & |p_i - p_{i-1}| \leq |p_{i+1} - p_i|, \ \forall\, i = \{2, \ldots, n-1\}\,.
\end{aligned}
$$

We say that the distribution is **strictly monotone** if it is monotone and a strict inequality holds for at least one $i = \{2, \ldots, n-1\}$. The distribution is **uniform** if the equality holds for all $i = \{2, \ldots, n-1\}$.

**Definition 2.** A **decision scheme** $F$ is an application that maps for each coalition $S$ the vector of players' positions, $(p_i)_{i \in S}$, to a point in $\Omega$,

$$
F : \bigcup_{S \in 2^N \setminus \{\varnothing\}} \Omega^{|S|} \to \Omega\,. \tag{1}
$$

To simplify the notation we write $F(S)$ instead of $F((p_i)_{i \in S})$. We assume that $F(S)$ does not depend on players outside $S$. We impose some properties on decision schemes.

**Assumption 1 (Pareto optimality)** *For any coalition $S$, $F(S)$ is higher or equal to the lowest position in $S$, and lower or equal to the highest position in $S$, that is, $F(S) \in [\min_{i \in S} p_i, \max_{i \in S} p_i]$.*

Pareto optimality has a simple interpretation. Any change of the value of $F(S)$ makes at least one player in $S$ worse off.

**Assumption 2 (Sensitivity)** *Let any $S \subseteq N$ and $i \notin S$. Then $|F(S \cup \{i\}) - p_i| \leq |F(S) - p_i|$.*

Sensitivity merely says that if a player enters a coalition, then the outcome of the decision scheme of the coalition with this new member cannot move in the opposite direction of that of this player.

**Assumption 3 (Invariance)** *Let $g : \mathbb{R} \to \mathbb{R}$ be any concave of convex function and $\mu \in \mathbb{R}$. Define $p'_i := g(p_i) + \mu$. Then for all $S \subseteq N$, $F((p'_j)_{j \in S}) = g(F((p_j)_{j \in S})) + \mu$.*

This assumption merely says that the relative positions to the outcome of the decision scheme should not depend on the scale chosen for $\Omega$, nor its origin.

Let F be the set of all decision schemes that satisfy Pareto Optimality, Sensitivity, and Invariance.

Several decision schemes satisfy these properties. The most common one is the median voter, defined by the player $i$ such that the sets $\{j \in S : p_j \leq p_i\}$ and $\{j \in S : p_j \geq p_i\}$ have the same cardinality.

Another decision scheme that satisfies theses properties is the **mean of the extremes.**[3] It consists of taking the point at the mid-distance between the two players with the lowest and highest positions. Let $p_{\min(S)} = \{p_i : p_i \leq p_j, \forall j \in S\}$, and $p_{\max(S)} = \{p_i : p_i \geq p_j, \forall j \in S\}$. The mean of the extremes is defined as

$$F^{ext}(S) = \frac{p_{\min(S)} + p_{\max(S)}}{2} . \tag{2}$$

We define the utility of player $i$ as a mapping $u_i : \mathbb{N} \times \mathbb{R}_+ \to \mathbb{R}$. That is, the utility of player $i$ when in coalition $S$ is given by $u_i(|S|, \delta_i(S))$, where, $\delta_i(S) = |F(S) - p_i|$ when $i \in S$, and $u_i$ is **non-decreasing** with respect to its first argument and **non-increasing** with respect to its second argument. Moreover, we also suppose that $u_i$ is concave with respect to coalition size and distance:

**Assumption 4 (Concavity)**

*(i)* ***with respect to the coalition size:*** *For all $s \in \mathbb{N}$ and $\delta \in \mathbb{R}_+$,*

$$u_i(s+1, \delta) - u_i(s, \delta) \geq u_i(s+2, \delta) - u_i(s+1, \delta) .$$

*(ii)* ***with respect to the distance:*** *For all $s \in \mathbb{N}$ and $\delta, \delta' \in \mathbb{R}_+$, such that $\delta \leq \delta'$,*

$$u_i(s, \delta) - u_i(s, \delta + \varepsilon) \leq u_i(s, \delta') - u_i(s, \delta' + \varepsilon), \ \forall \varepsilon > 0 .$$

Concavity with respect to the coalition size assumes decreasing returns to scale in the production of the public good, and concavity with respect to the distance assumes convex transportation costs. Many models of local public good economies assume that there are some congestion effects. As we shall show in section 4, adding congestion effects would not change the results.

Finally, we suppose that players' utilities functional form are equal, up to a linear transformation.

**Assumption 5 (Symmetry)** *For all $x, x' \in \mathbb{N}$, $y, y' \in \mathbb{R}$,*

$$u_i(x, y) \geq u_i(x', y') \Rightarrow u_j(x, y) \geq u_j(x', y'), \ \forall i, j \in N .$$

[3] Notice that [1,13,19] all made the assumption of a uniform distribution of players. Thus, all their results still hold when their decision scheme is replaced by the mean of the extremes. [1, section VI, p. 1044] weakens the uniform distribution assumption, but they only presented how the model would change — such as the outcome of the decision scheme.

For the sake of simplicity, we shall use a different assumption in the proofs (results are still valid with assumption 5): There exists a $u : \mathbb{N}\times\mathbb{R}_+ \to \mathbb{R}$ such that $u_i(x,y) = u(x,y)$, $\forall\, x \in \mathbb{N}$, $y \in \mathbb{R}^+$, $\forall\, i \in N$. As we shall show, assumption 5 is crucial for the existence of a Tiebout equilibrium (see example 4) or a $\mathcal{C}$-stable coalition structure (see example 1). The modified version of assumption 5 implies a nice property on the mean of the extremes. This decision scheme is the unique Rawlsian decision scheme. Indeed, in any coalition $S$, the player with the lowest utility is player $i^*$ such that $\delta_i(S) \geq \delta_j(S)$, $\forall\, j \in S$. Clearly, if the outcome of the decision scheme is not the mid-point between the two extreme players, then one can increase the utility of player $i^*$ by moving the outcome to $F^{ext}(S)$.

**The coalition formation game**

Our model can be presented as a **hedonic** coalition formation game, which was first introduced by refined Drèze and Greenberg [5]. Pure hedonic games are those in which no re-distributional issue has to be solved within coalitions. In these games, a player's payoff only depends on the composition of members of the coalition she belongs to. Our model can also be seen as a pure hedonic game, since the decision scheme $F$ is fixed for all coalitions and known by all players. Hence, we can deduce for each player $i \in N$, a preorder $\succcurlyeq_i$ over $\mathrm{N}_i$, the set of coalitions to which $i$ can belong, $\mathrm{N}_i = \{S \subseteq N : S \ni i\}$. The preorder is defined as follows.

$$u(|S|, \delta_i(S)) \geq u(|T|, \delta_i(T)) \Leftrightarrow S \succcurlyeq_i T,\ \forall S,\ T \in \mathrm{N}_i\ .$$

The strict preference relation and the indifference relation are respectively denoted $\succ_i$ and $\sim_i$. We say that a coalition $S \in \mathrm{N}_i$ is **individually rational** for player $i$ if $S \succcurlyeq_i \{i\}$, which means that $i$'s utility level in $S$ is higher or equal to the utility she would receive if she was alone.

## 3 $\mathcal{C}$-stable coalition structures

The second equilibrium concept we use is that of $\mathcal{C}$-stable coalition structures, which was first introduced by Guesnerie and Oddou [11] for TU-games, and later adapted by Greenberg and Weber [9] for NTU-games. As in the Tiebout equilibrium, a player is free to leave her coalition, but now, a player can enter a coalition only if **all** its members are better off when accepting this new entrant.

**Definition 3.** A coalition structure $\mathrm{B}$ is **blocked** by a coalition $S \notin \mathrm{B}$ if,

$$S \succ_i S(\mathrm{B}, i), \quad \forall\, i \in S\ .$$

**Definition 4.** A coalition structure $B = \{S_1, \ldots, S_k\}$ is **$C$-stable** if there is no coalition $S \notin B$ that blocks B.

We now introduce the main result of this section.

**Proposition 1.** *If $F$ is the mean of the extremes then there always exists a $C$-stable coalition structure.*

The assumption that players have the same utility function plays a crucial role. The next example shows that even if positions are uniformly distributed and $F$ is the mean of the extremes, a $C$-stable coalition structure may not exist.

*Example 1.* Consider 7 players, $a, b, c, d, e, f$ and $g$, whose positions on $\Omega$ are $p_a = 0$, $p_b = 1$, $p_c = 3$, $p_d = 5$, $p_e = 7$, $p_f = 8.9$, $p_g = 10.9$. Utilities are given by,

$$u_a(|S|, \delta_a(S)) = \begin{cases} |S| - 10^{-10}\delta_a(S) & \text{if } \delta_d(S) \leq 3\,, \\ |S| - 100\delta_a(S) & \text{if } \delta_d(S) > 3\,, \end{cases}$$

$$u_b(|S|, \delta_b(S)) = |S| - 10^{-10}\delta_b(S)\,,$$

$$u_c(|S|, \delta_c(S)) = \begin{cases} |S| - \delta_c(S) & \text{if } \delta_d(S) \leq 1\,, \\ |S| - 100\delta_c(S) & \text{if } \delta_d(S) > 1\,, \end{cases}$$

$$u_d(|S|, \delta_d(S)) = \begin{cases} 2|S| - 0.9\delta_d(S) & \text{if } \delta_d(S) \leq 2\,, \\ 2|S| - \frac{3}{2}\delta_d(S) & \text{if } 2 < \delta_d(S)\,, \end{cases}$$

$$u_e(|S|, \delta_e(S)) = 2|S| - 100\delta_e(S)\,,$$

$$u_f(|S|, \delta_f(S)) = 2|S| - 10^{-10}\delta_f(S)\,,$$

$$u_g(|S|, \delta_g(S)) = 2|S| - 10^{-10}\delta_g(S)\,.$$

We then get the following preferences over coalitions (like in example 2, only the coalitions that are individually rational are presented),

$\{abcd\} \succ_a \{ab\} \succ_a \{a\}$,
$\{abcd\} \succ_b \{bcd\} \succ_b \{bfg\} \succ_b \{ab\} \succ_b \{bd\} \succ_b \{bf\} \succ_b \{bg\} \succ_b \{b\}$,
$\{abcd\} \succ_c \{bcd\} \succ_c \{c\}$,
$\{def\} \sim_d \{bcd\} \succ_d \{abcd\} \succ_d \{bd\} \sim_d \{df\} \sim_d \{d\}$,
$\{def\} \succ_e \{e\}$,
$\{bfg\} \succ_f \{def\} \succ_f \{fg\} \succ_f \{df\} \sim_f \{bf\} \succ_f \{f\}$,
$\{bfg\} \succ_g \{fg\} \succ_g \{bg\} \succ_f \{g\}$.

Observe that if $e$ is alone, the only possible coalition structures must contain either $\{abcd\}$ or $\{bcd\}$. Indeed, any coalition structure containing $\{e\}$ and neither $\{abcd\}$ nor $\{bcd\}$ is blocked by one of these two coalitions. But any coalition structure including these coalitions is blocked by $\{def\}$, and any coalition structure with the coalition $\{def\}$ is blocked by $\{bfg\}$. In this case, $e$ remains alone, and we then have a cycle. Thus there is no $C$-stable coalition structure.

## 4 Tiebout equilibrium

The first equilibrium concept we study is that of **Tiebout equilibrium**. This refers to the case where only individual decisions are allowed, but all players are free to enter and exit any existing coalition. We call it an equilibrium with free mobility. This concept was introduced by Tiebout [18], who described it as a situation in which players "vote with their feet."

**Definition 5.** A partition $\mathrm{B} = \{S_1, \dots, S_k\}$ is a **Tiebout equilibrium**, or a **Tiebout-stable** coalition structure, if for all $i \in N$, we have,

$$S(i) \succcurlyeq_i T \cup \{i\}, \ \forall\, T \in \mathrm{B} \cup \{\varnothing\}\,.$$

Adding $\{\varnothing\}$ to B guarantees that a Tiebout equilibrium is always individually rational.

In order to prove the existence of a Tiebout equilibrium, consider the uniform distribution with $F = F^{ext}$. Using Lemma 1 in the Appendix, we can restrict ourselves to connected coalitions.

Consider a uniform distribution and an player $i$ who is at the border of her coalition. That is, consider coalitions, $\{p_i, p_{i+1}, \dots, p_{i+k}\}$, with $k = 0, \dots, n-i$ or coalitions, $\{p_k, p_{k+1}, \dots, p_i\}$, with $k = 1, \dots, i$ Let $\mathrm{S}_i$ be the set of coalitions that satisfy one of these conditions. It is easy to see that if we restrict to coalitions in $S \in \mathrm{S}_i$, player $i$'s preferences are single peaked with respect to the size of the coalition. That is, there exists $h^*$ such that

$$|S| \leq |T| \leq h^* \Rightarrow T \succcurlyeq_i S\,, \text{ and } h^* \leq |T| \leq |S| \Rightarrow T \succcurlyeq_i S\,. \tag{3}$$

Had we made the assumption of congestion effects the results would not have changed. Indeed, congestion effects would merely change the slope of the utility functions (for players at the border of the coalition), making it more "flat", and lower the critical size $h^*$. We now state our existence result.

**Proposition 2.** *Assume that the distribution of players' positions on $\Omega$ is monotone and $F$ is the mean of the extremes, then a Tiebout-equilibrium always exists.*

The following examples show that a Tiebout equilibrium may not exist in our model when the distribution of players is not monotone (Example 2), and when the concavity (Example 3), and the symmetry assumptions (Example 4) are not satisfied.

*Example 2.* Let the set of players be $N = \{a, b, c, d, e, f\}$ with the following non-monotone positions on the real line, $p_a = 0$, $p_b = 9$, $p_c = 19$, $p_d = 27$, $p_e = 34$, and $p_f = 40$. Consider the following utility function,

$$u(|S|, \delta(S)) = \begin{cases} |S| & \text{if } \delta(S) \leq 4\,, \\ |S| - \beta\delta(S) & \text{if } 4 < \delta(S) < \frac{19}{2}\,, \\ |S| - 10000\delta(S) & \text{if } \delta(S) \geq \frac{19}{2}\,. \end{cases}$$

with $\frac{2}{13} < \beta < \frac{2}{11}$.

Given the possible values that $\beta$ can take, it is easily checked that this utility function is concave with respect to the size of the coalition and the distance. First, observe that $b$ is closer to $a$ than $c$. Hence, $b$ prefers $\{ab\}$ to $\{bc\}$. Moreover, the distances between $b$ and $d$, and $c$ and $d$ are respectively greater that the distances between $d$ and $f$ and $d$ and $e$. Thus, when $F$ is the mean of the extremes, $d$ prefers $\{def\}$ to $\{bcd\}$. Notice also that $\{abc\}$ and all coalitions with four players or more are not individually rational. We then obtain the following preference,[4]

$\{ab\} \succ_a \{a\}$ ,
$\{bcd\} \succ_b \{ab\} \succ_b \{bc\} \succ_b \{b\}$ ,
$\{bcd\} \succ_c \{cd\} \succ_c \{bc\} \succ_c \{cde\} \succ_c \{c\}$ ,
$\{cde\} \succ_d \{de\} \succ_d \{cd\} \succ_d \{def\} \succ_d \{bcd\} \succ_d \{d\}$ ,
$\{def\} \succ_e \{ef\} \succ_e \{ed\} \succ_e \{cde\} \succ_e \{e\}$ ,
$\{ef\} \succ_f \{def\} \succ_f \{f\}$ .

In this example, no Tiebout equilibrium exists. We can observe that as soon as $c$ is alone, then $d$ joins $\{c\}$. But $b$ prefers $\{bcd\}$ to coalitions $\{ab\}$ and $\{b\}$. As soon as $b$ joins $\{cd\}$, then $d$ joins $\{ef\}$, which yields $b$ to leave $\{bc\}$ to join $\{a\}$. To conclude, observe that if $\{cde\}$ is formed, then $e$ leaves this coalition to join $\{f\}$.

*Example 3.* Consider five players, $a$, $b$, $c$, $d$ and $e$, whose positions on the line are $p_a = 0$, $p_b = 2 - \varepsilon$, $p_c = 3$, $p_d = 4$ and $p_e = 5$. This distribution is monotone, so we know by proposition 2 that a Tiebout equilibrium exists. Consider the following utility function,[5] with $\varepsilon < 0.1$.

$$u_i(|S|, \delta_i(S)) = \begin{cases} |S| & \text{if } \delta_i(S) \leq \frac{1}{2} + \frac{\varepsilon}{2} , \\ |S| - 2.1\delta_i(S) & \text{if } -\frac{1}{2} + \frac{\varepsilon}{2} < \delta_i(S) \leq |\frac{3}{4} , \\ |S| - 2.1 & \text{if } \frac{3}{4} \leq \delta_i(S) \leq 1 , \\ |S| - 2.1 - \frac{1}{10}\delta_i(S) & \text{if } 1 < |\delta_i(S) < \frac{3}{2} + \frac{\varepsilon}{2} , \\ |S| - 5 - 100\delta_i(S) & \text{if } -\frac{3}{2} + \frac{\varepsilon}{2} \leq \delta_i(S) . \end{cases}$$

We obtain the following preference profile (again, only individually rational coalitions are represented).

$\{abc\} \succ_a \{a\}$ ,
$\{abc\} \succ_b \{bc\} \succ_b \{bcd\} \succ_b \{b\}$ ,

[4] We do not present coalitions that are not individually rational. For instance, $\{abc\}$ is not individually rational for $a$ and $c$, but it is for $b$. Hence, $\{abc\}$ does not appear in $b$'s preferences.

[5] As the size of the coalition and the distance to the outcome of the decision scheme plays somewhat similar roles, an example in the same vein of example 3 where concavity with respect to the coalition size does not hold and non-existence of Tiebout equilibria can be found.

$\{bc\} \sim_c \{cd\} \succ_c \{cde\} \succ_c \{abc\} \succ_c \{c\}$ ,
$\{cde\} \succ_d \{cd\} \sim_d \{de\} \succ_d \{bcd\} \succ_d \{d\}$ ,
$\{de\} \succ_e \{cde\} \succ_e \{e\}$ .

We leave to the reader to check that there does not exist any Tiebout-stable coalition structure.

*Example 4.* Let $N = \{a, b\}$ be the set of players and $p_a = 0$, and $p_b = 4$ be their positions on $\Omega$. If $F$ is the mean of the extremes, then $F(\{a, b\}) = 2$. Let the utility functions be, $u_a(|S|, \delta_a(S)) = 3|S| - \delta_a(S)$, $u_b(|S|, \delta_b(S)) = 3|S| - \delta_b(S)^2$. We then get $u_a(1,0) = u_b(1,0) = 3$, and $u_a(2, \delta_a(\{ab\})) = 4$ and $u_b(2, \delta_b(\{ab\})) = 2$. Since $\{ab\} \succ_a \{a\}$ and $\{b\} \succ_b \{ab\}$ no Tiebout equilibrium exists.

An interesting property of our model is that a $C$-stable coalition structure is not necessarily Tiebout-stable.

*Example 5.* Let the set of players be $N = \{a, b, c\}$ with the following positions on the real line $p_a = 0$, $p_b = 3$ and $p_c = 5$, and the following utility function:

$$u(|S|, \delta_(S)) = \begin{cases} |S| & \text{if } \delta(S) \leq 1 \,, \\ |S| - 0.6\delta(S) & \text{if } 1 < \delta(S) < 1.1 \,, \\ |S| - 0.65\delta(S) & \text{if } 1.1 \leq \delta(S) \,. \end{cases}$$

For any decision scheme between the mean of the extremes and the median voter, it is easy to see that the only Tiebout stable coalition structure is $\{abc\}$, and that the only $C$-stable coalition structure is $\{\{a\}, \{bc\}\}$.

## 5 Conclusion

In this paper, we studied the stability of coalition structures when a fixed decision scheme is imposed on coalitions when determining the location of the public good. We showed that if the distribution of players is monotone, then a Tiebout equilibrium always exists under the the Rawlsian decision scheme. This latter proved useful when constructing stable coalition structure. Indeed, under the symmetry assumption, the two extreme players have the same utility level. It can be noticed that these two players are usually the firsts to deviate (particularly in the Tiebout case). Hence, it does not seem surprising that if one wants to ensure the stability of coalition structures, the decision scheme should particularly take care of extreme players. In [12], we show that all other decision schemes that ensure the existence of Tiebout equilibria also need the symmetry assumption. This is to be contrasted with $C$-stability, which does not need any assumption on the distribution of agents' positions. A last remark is about the contrast between our results and that of previous paper, especially Greenberg and Weber's paper [10]. Their game has positive externalities, which implied that the set of $\mathcal{C}$-stable coalition structures is also a Tiebout equilibrium. In our game, players' moves may induce negative externalities due to the fixed decision scheme, which can be somehow reinterpreted as congestion effects.

# Appendix

In [12], we prove the following result.

**Lemma 1.** *Let $F$ be a decision scheme satisfying Pareto Optimality and Sensitivity. If* B *is a Tiebout equilibrium, then* B *is connected.*

*Proof of Proposition* 2. Construct the following coalition structure. If the distribution is monotone, w.l.o.g. say that $|p_2-p_1| \geq |p_3-p_2|$, then form coalition $S_1$ with player 1 and player 2, add player 3, 4, and so on, until player $i-1$ such that $u_i(|S_1|+1, \delta_i(S_1) \cup \{i\}) < u_i(1,0)$. Then, form a second coalition, $S_2$, with player $i, i+1$, and so on, until player $j-1$ such that $u_j(|S_2|+1, \delta_j(S_2) \cup \{j\}) < u_j(1,0)$. Repeat this procedure until the last player is reached, yielding the coalition structure $\mathrm{B} = \{S_1, \ldots, S_k\}$. Observe that the only players that would wish to move are those in $S_{k-1}$, who move to $S_k$. Indeed, by construction, no player in $S_{h+1}$ wants to move to $S_h$, $h = 1, \ldots, k-1$, that is, for any $i \in S_{h+1}$, we have $u_i(|S_h|+1, \delta_i(S_h \cup \{i\})) < u_i(1,0)$. Because the distribution is monotone, $\delta_j(S_h \cup \{j\}) \geq \delta_i(S_h \cup \{i\})$, which implies that $u_j(|S_h|+1, \delta_j(S_h \cup \{j\})) < u_j(1,0)$, for any $j \in S_{h-1}$. Clearly, the former inequality is true if $F$ is the mean of the extremes. Consider now $S_k$ and $S_{k-1}$. If no player moves from the latter to the former, we are done. Suppose then that there exists $i \in S_{k-1}$ who moves to $S_k$. Consequently, some player in $S_{k-2}$ can decide to move to $S_{k-1}$. We claim that $i$ will not decide to move back to $S_{k-1}$. Consider player $i+1$. By construction, we know that $u_{i+1}(|S_{k-1}|+1, \delta_{i+1}(S_{k-1} \cup \{i+1\})) < u_{i+1}(1,0)$. If $i$ enters $S_{k-1} \cup \{j\} \backslash \{i\}$, then, again because the distribution is monotone, $\delta_i(S_{k-1} \cup \{j\}) < \delta_{i+1}(S_{k-1} \cup \{i+1\})$. Thus $u_i(|S_{k-1}|+1, \delta_i(S_{k-1} \cup \{j\})) < u_i(1,0)$. After $j$ moves, another player in $S_{k-1}, S_{k-2}$ or $S_{k-3}$ may move to $S_k, S_{k-1}$ or $S_{k-2}$ respectively, but repeating the argument of before, it can be shown that no player, consequently to such moves, will want to move to a coalition with a lower index.

*Proof of Proposition* 1. Let $N = \{1, \ldots, n\}$ be the set of players. The proof proceeds by construction.
*Step 1* For each player $i$, let $S_i^* \in \mathrm{S}_i$ be the coalition such that $u_i(|S_i^*|, \delta_i(S_i^*)) \geq u_i(|T|, \delta_i(T))$, $\forall\, T \in \mathrm{S}_i$. Observe that $S^*$ is connected, and that if $i$ is the

player with the lowest (resp. highest) position in $S^*$, then $u_i(|S_i^*|, \delta_i(S_i^*)) = u_j(|S_i^*|, \delta_j(S_i^*))$, where $j$ is the player with the highest (resp. lowest) position in $S^*$. Construct now the coalition $S_1$ such that $S_1 = S_i^*$ for some $i \in N$ and $u_i(|S_i^*|, \delta_i(S_i^*)) \geq u_j(|S_j^*|, \delta_j(S_j^*))$, for all $j \in N$. We are then left with the set of player $N \backslash S_i^*$.
*Step k* Let $N_{k-1} = \{S_1, \ldots, S_{k-1}\}$ be the set of coalitions already formed. Define the set

$$\hat{\mathrm{S}}_{k,i} = \{S \subseteq N \backslash N_{k-1} : i \text{ has the lowest or highest position in } S\} .$$

For each $S \in \hat{\mathrm{S}}_{k,i}$, consider the coalitions $T_1^S, \ldots, T_l^S$ in $N_{k-1}$ that are surrounded by $S$. That is, for all $1 \leq h \leq l$, $\exists\, i, k \in S$ such that $p_i < p_j < p_k$, for all $j \in T_h^S$. Call a coalition $T_h^S$ *candidate for* $S$ if all players in $T_h^S$ are better off if they join $S$: $u_i(|T_h^S| + |S|, \delta_i(T_h^S \cup S)) > u_i(|T_h^S|, \delta_i(T_h^S))$, for all $i \in T_h^S$. Let $T_1^{c,S}, \ldots, T_{l'}^{c,S}$ be the collection of candidate coalitions for $S$, and let $\mathrm{T}(S, k-1, i) = \{T_1^{c,S}\} \cup \cdots \cup \{T_{l'}^{c,S}\}$. Define now the set

$$\mathrm{S}_{k,i} = \{S \subseteq \mathrm{T}(S, k-1, i) \cup N \backslash N_{k-1}\} .$$

That is, $\mathrm{S}_{k,i}$ is such that if $S$ contains some players in $T_h^{c,S}$, then $S$ contains all players in $T_h^{c,S}$.

Let $S_i^{k,*}$ be the coalition such that $u_i(|S_i^*|, \delta_i(S_i^*)) \geq u_i(|T|, \delta_i(T))$, $\forall\, T \in \mathrm{S}_{k-1,i}$, Define the coalition $S_k \equiv S_i^{k-1,*}$, where $i$ is the player such that

$$u_i(|S_i^{k,*}|, \delta_i(S_i^{k,*})) \geq u_j(|S_j^{k,*}|, \delta_j(S_j^{k,*})),\ \forall j \in N \backslash N_{k-1} .$$

Continue this procedure until all players have be assigned to some coalition (it may be that there are one-player coalitions), and call B the coalition structure obtained. We claim that B is $\mathcal{C}$-stable.

Clearly, by construction, no collection of coalitions can coalesce to form a coalition, say $T$. Indeed, consider the player, say $i$, with the lowest position in $T$. Observe that $i$ was also the player with the lowest position in $S(i, \mathrm{B})$, we have $S(i, \mathrm{B}) = S_i^{k,*}$, where $k$ is the step when $S(i, \mathrm{B})$ was formed. This contradicts the fact that $i$ blocks with $T$, i.e., $u_i(|T|, \delta_i(T)) > u_i(|S(i, \mathrm{B})|, \delta_i(S(i, \mathrm{B})))$. Thus, the only possibility is that if some coalition $T$ blocks, it disrupts another coalition $S \in \mathrm{B}$. Two cases occur, whether a disrupted coalition is surrounded by $T$ or at the border of $T$.

Consider the first case. Let $S$ be the disrupted coalition, with $i, j, k \in S$, $p_i \leq p_j \leq p_k$, $i, k \in T$ and $j \notin T$. Assume that $i$ and $k$ are respectively the players in $S$ with the lowest and highest position. Because $S$ is surrounded by $T$, there exists some player $h \in T$ such that $p_k \leq p_h$. Hence, for any player in $l \in S$, we have $F(T \cup \{l\}) = F(T)$. Observe that $i$ and $k$ have the same utility level in $S$. Hence, if $i$ blocks with $T$, so does $k$. Let $\Delta_k = \delta_k(T) - \delta_k(S)$. (Define analogously $\Delta_i$ and $\Delta_j$.) Clearly, we have either $\Delta_i \geq \Delta_j$ and $\delta_i(T) \geq \delta_j(T)$ or $\Delta_k \geq \Delta_j$ and $\delta_k(T) \geq \delta_j(T)$. Assume that the first pair of inequalities holds. Because $i$ blocks with $T$, we have $u_i(|S|, \delta_i(S)) > u_i(|T|, \delta_i(T))$. Hence, by the concavity assumption, it must be also the case that $u_j(|S|, \delta_j(S)) > u_i(|T| + 1, \delta_j(T))$. In other words, $j$ also blocks with $T$.

Consider now the second case, when a disrupted coalition is at the border of $T$. Let $i \in S$ be the player at the border of $T$. W.l.o.g. suppose that $i$ has the lowest position in $T$. Consider now $j$, such that $p_i \leq p_j$ and $j$ is at the border of $S$. Suppose first that $i \neq j$. Clearly, $u_i(|S|, \delta_i(S)) > u_j(|S|, \delta_j(S))$. But then this contradicts the fact that $j$ is such that $u_j(|S_j^{k,*}|, \delta_j(S_j^{k,*})) \geq u_i(|S_i^{k,*}|, \delta_i(S_i^{k,*}))$, where $k$ is the step when $S$ is formed. If $i = j$, then $S$ being formed instead of $T$ implies that $u_i(|S|, \delta_i(S)) \geq u_i(|T|, \delta_i(T))$, a contradiction with the fact that $i$ blocks with $T$, which completes the proof of the claim.

## References


1. Alesina, A., Spolaore, E. (1997) On the Number and Size of Nations. Quarterly Journal of Economics. **112**, 1027–1056.
2. Banerjee, S., Konishi H., Sönmez, T. (1998) Core in Simple Coalition Formation Game. Social Choice and Welfare, forthcoming.
3. Bogomolnaia, A., Jackson, M.O. (1998) The Stability of Hedonic Coalition Structures. Mimeo, CalTech.
4. Demange, G. (1994) Intermediate Preferences and Stable Coalition Structures. Journal of Mathematical Economics. **23**, 45–58.
5. Drèze, J., Greenberg, J. (1980) Hedonic Coalitions: Optimality and Stability. Econometrica. **48**, 987–1003.
6. Eppel, D., Filimon, R., Romer, T. (1984) Equilibrium Amon Local Jurisdictions: Towards and Integrated Treatment of Voting and Residential Choice. Journal of Public Economics. **23**, 281–308.
7. Eppel, D., Filimon, R., Romer, T. (1993) Existence of Voting and Housing Equilibrium in a System of Communities with Property Taxes. Regional Science and Urban Economics. **23**, 585–610.
8. Greenberg, J., Shitovitz, B. (1988) Consistent Voting Rules for Competitive Local Public Goods Economies. Journal of Economic Theory. **46**, 223–236.
9. Greenberg, J., Weber, S. (1986) Strong Tiebout Equilibrium under Restricted Preferences Domain. Journal of Economic Theory. **38**, 101–117.
10. Greenberg, J., Weber, S. (1993) Stable Coalition Structures with a Unidimensional Set of Alternatives. Journal of Economic Theory. **60**, 62–82.
11. Guesnerie, R., Oddou, C. (1979) On Economic Games that are not Necessarily Superadditive. Economics Letters. **3**, 301–306.
12. Haeringer, G. (2000) Stable Coalition Structures with Fixed Decision Scheme. Universitat Autónoma de Barcelona, working paper ♯ 471.00.
13. Jehiel, P., Scotchmer, S. (1997) Constitutional Rules of Exclusion in Jurisdiction Formation. University of California, Berkeley, working paper ♯ 231.
14. Konishi, H. (1996) Voting with Ballots and Feet: Existence of Equilibrium in a Local Public Good Economy. Journal of Economic Theory. **68**, 480–509.
15. Konishi, H., Le Breton, M., Weber, S. (1998) Equilibrium in a Finite Local Public Good Economy. Journal of Economic Theory. **79**, 224–244.

16. Le Breton, M., Weber, S. (2000) The Art of Making Everybody Happy: How to Prevent a Secession. Mimeo.
17. Rose-Ackerman, S. (1979) Market Models of Local Government: Exit, Voting, and the Land Market. Journal of Urban Economics. **6**, 319–337.
18. Tiebout, C. (1956) A Pure Theory of Local Public Expenditures. Journal of Political Economy. **65**, 319–337.
19. Westhoff, F. (1977) Existence of Equilibria in Economies with a Local Public Good. Journal of Economic Theory. **14**, 84–112.
20. Wooders, M. H. (1999) Multijurisdictional Economies, the Tiebout Hypothesis, and Sorting. Proceedings of the National Academy of Sciences. **96**, 10585–10587.

# Coalition Formation with Heterogeneous Agents

Davide Fiaschi[1] and Pier Mario Pacini[1]

Dipartimento di Scienze Economiche, University of Pisa

**Abstract.** The paper analyzes a game of coalition formation in which agents with limited computational abilities possess heterogeneous endowments and seek to coalesce into groups to produce and divide an output. The basic game is modelled as a two stage game; in the first stage every agent sends other agents various messages consisting in the proposal of a coalition. In the second stage, knowing the coalitions, agents choose the actions to perform. We show that there exists at least a Strong Nash equilibrium characterized by the consecutive property, i.e. the richest agents form a coalition among themselves and so do the poorest agents. The numerical simulations show that agents play the SNE with high probability; moreover, the more unequal the initial distribution of resources, the lower the average utility in the economy. The transition paths highlight the fact that the rich agents are the first to coalesce, so that the inequality of individual utilities increases in the early periods and then decreases when poor agents also start coalescing. In contrast, average utility is generally increasing, so that there exists a non linear relationship between inequality and average utility.

## 1 Introduction

The formation of coalitions is a common aspect of many economic activities; for example scientific research groups, consumers' associations, production partnerships, consumption and production cooperatives are all instances of voluntary agreements among independent parties that coalesce to obtain the same goal.

Coalition formation has already been observed and studied in a number of works from different points of view: among these, Guesnerie and Oddou (1988) deals with the problem of coalition formation for the provision of public goods, Farrell and Scotchmer (1988) discusses formation of cooperatives of workers with different working capacity, while Bennet (1985) analyzes coalition formation in more general settings (see Greenberg (1994) for an excellent review).[1] In these works the formation of cooperating groups is affected by a moral hazard problem whenever private actions cannot be monitored. A common finding is that the cooperative agreement is self-sustained if the coalition

[1] A large part of coalition formation literature is interested in games with externality, see for example Yi (1997) and Thoron (1998).

size is below a certain threshold, whereas it would not be incentive compatible in larger coalitions. These conclusions are in agreement with Olson (1983) (see also Ecchia and Mariotti (1998) and Yi (1997)).

In this paper we analyze a game of coalition formation in which agents have heterogeneous endowments and try to coalesce into groups to produce and divide an output; the formation of a coalition does not affect the utilities of agents not belonging to such coalition (i.e. this is a game without externalities). There is an incentive to aggregation due to an increasing returns to scale technology, but imperfect information on other agents' actions does not allow for the formation of a unique coalition in which all agents participate. Agents are not free to join an existing coalition without the consensus of its members (i.e. we consider an exclusive membership game, see Block (1996))[2]. The basic game is modelled as a two stage game; in the first stage every agent sends the other agents messages consisting in the proposal of a coalition; this allows individuals to identify which other individuals are willing to coalesce with them and therefore to coordinate their strategies. In the second stage, knowing the coalitions resulting from the coordination achieved in the first stage, agents choose which actions to perform; the resulting output is then divided among the members of the coalitions that have been formed.

In this framework we show that there exists at least a Strong Nash equilibrium (SNE) characterized by the consecutive property, i.e. the richest agents form a coalition among themselves, and so do the poorest agents. One of the main aims of the paper is to verify whether, in a dynamic context in which the game is repeatedly played, this is the equilibrium towards which agents converge or whether there are other stable structures emerging from social interaction. In addition, we examine how the outcomes of social interaction depend on the basic characteristics of the economy, i.e. the returns to scale of the available technology and the distribution of the endowments.

In the repeated game, three aspects are crucial: (i) the type of expectations an agent forms concerning other players' messages when s/he has to decide which message to send; (ii) how an agent chooses his/her action within a coalition and (iii) how the coalition output is divided among the members of a coalition (the second and third aspects are clearly related). We assume, respectively, that (i) an agent expects that his/her own message will be matched by (and hence a coalition will form among) those agents that would obtain a higher utility in the proposed coalition than in the current situation; (ii) agents within a coalition play the Pareto superior Nash equilibrium strategy and (iii) coalitional production is equally shared among the members of a coalition.

[2] From an analytical point of view, the closest paper to ours is Hart and Kurz (1983); in particular we consider a type of game the Authors of the cited paper call game $\Delta$.

We assume that agents elaborate their messages by genetic algorithms (GAs)[3], so that the probability of announcing a message is directly proportional to the expected utility of playing that message. The use of GAs in elaborating agents' messages is motivated by important contributions from the theory of cognitive processes (see Holland et al. (1986)); in this approach, agents, when called upon to make a choice in a complex environment, do not engage in explicit optimization, but rather operate on a limited set of rules (or mental schemes) which they continuously modify by reacting to the effects of their own behaviour.

The numerical simulations show that agents play the SNE with high probability; moreover, in all other cases they do not play an alternative SNE. As regards the properties of this equilibrium, given a certain amount of resources, the more unequal the initial distribution of resources, the lower the average utility in the economy. Moreover the inequality of individual utilities is generally higher than the inequality of initial endowments and this difference is mainly increasing in returns to scale. The transition paths highlight the fact that rich agents are the first to coalesce, so that the inequality of individual utilities increases in the first periods and then decreases when poor agents also start coalescing. In contrast, average utility is generally increasing, as is the average size of coalitions, so that there exists a non linear relationship between inequality and average utility.

The paper is organized as follows: the structure of economy is described in Section 2; Section 3 describes how agents are modelled and the sequential structure of the economy in the simulated environment. Finally Section 4 reports on the computational implementation and results. The conclusions close the paper.

## 2 The structure of the economy

We assume that the dynamics leading to the formation of coalitional structures is a process taking place in a sequence of periods in discrete time $t = 0, 1 \ldots$. In every period there is a finite constant number of agents, indexed by $i = 1, \ldots, I$; they form a population $\mathfrak{I}$. Given the population $\mathfrak{I}$, a subset $S \subseteq \mathfrak{I}$, $S \neq \emptyset$, is a *coalition* and $|S|$ is the cardinality of $S$, i.e. the number of agents in $S$; clearly $|\mathfrak{I}| = I$. A coalitional structure $\sigma = \{S_1, S_2, ..., S_K\}$ is a partition of $\mathfrak{I}$ in non empty coalitions. When coalitions and coalitional structures refer to a given period, they are indexed by $t$, i.e. $S^t$ and $\sigma^t$ respectively.

In any period $t$ every agent $i \in \mathfrak{I}$ receives an amount of resources $\omega_i^t$; this is the endowment s/he owns and s/he can decide whether to supply it to the production activity of the group $S^t$ to which s/he happens to belong in that

[3] See Fiaschi et al. (2001) for a short introduction to GAs in a similar context. GAs have been increasingly applied in economics: on the applications to game theory, see Birchenhall (1996).

same period. If s/he contributes his/her endowment, $i$ is said to ***cooperate*** in $S^t$, otherwise s/he is said to ***defect*** in $S^t$. For the sake of simplicity, we assume that the latter decision is dichotomic, in the sense that $i$ can choose whether to cooperate ($l_i^t = 1$) or defect ($l_i^t = 0$), without the possibility of (probabilistically) mixing the two actions. The structure of a game is one of imperfect information, i.e. agent $i$ does not know the current generation's actions $l_{-i}^t = \{l_h^t\}_{h \neq i}$, or even those of the members of his/her own coalition, and there is no available technology that permits the monitoring of such choices.

In each period $t$, an agent has to take a twofold decision: (i) which coalition to announce and, given the participation in a certain coalition (which may be different from the one proposed by the agent him/herself), (ii) which action to take, i.e. to cooperate or defect within the coalition which s/he belongs to. In this respect this is a two-stage game, where in the first stage an agent signals which coalition s/he wishes to take part in, while in the second stage, the agent decides which action to perform (i.e. whether to collaborate or defect) in the coalition that has been established.

In the following we define the basic framework of economy, namely preferences, distribution of endowments, available technology and the distributive rule; these allow us to determine the agent's behavior of agent, given a certain coalition, and thus to characterize the second phase of game. The first phase, concerning the process of coalition formation, is described in Section 2.2.

*Preferences* In the rest of the paper it is assumed that participation in a coalition $S$ gives $i$ an unconditional right to receive an amount of output denoted by $y_i$ (this amount is determined by Assumption 2 below); the amount $y_i$ is completely consumed and produces utility for the receiving agent. Total utility depends not only on the share of the coalitional output received, but also on the action performed ($l_i$): thus while cooperation means contribution of the individual endowment to the coalitional production activity, defection gives $i$ the possibility of enjoying an outside opportunity which is assumed to produce a constant utility $\mu > 0$ that is constant over time and equal across agents. The utility function of every agent is assumed to be additive with respect to the above arguments, i.e.

$$U_i(y_i, l_i) = y_i + \mu \cdot (1 - l_i) . \tag{1}$$

Given this utility function and the assumption that the action $l_i$ is neither observable nor verifiable by the other members of the coalition to which $i$ belongs, a ***moral hazard*** problem arises, since agents may have incentives not to cooperate and to take advantage of a free ride on others' efforts depending on the amount of resources they can contribute and on the value of their outside opportunity. In any period, any agent is faced with the same problem, but there is no concern for the future effects of a decision, so that all choices are made without taking into account the consequences on future evolution of the social interaction.

*Individual endowments* In any period $t$ an agent $i$ receives the same amount of endowments $\omega_i$ and the vector $\Omega = (\omega_1, \ldots, \omega_I)$ represents the distribution of resources in the population. The distribution of endowments satisfies the following:

**Assumption 1** *1. $\Omega \in [\underline{\omega}, \bar{\omega}]^I$ (finite support);*
*2. the endowments in $\Omega$ are depleted completely within one period $t$;*
*3. $\Omega$ and point 2 are common knowledge.*

In other words the amount of available resources is finite and they must be used within a single period and there is no purpose in carrying them over time (no accumulation); moreover any agent knows the resources owned by any other agents and the fact that those resources will be completely used within that period. Without any loss of generality we assume that agents are indexed in direct order with respect to the amount of endowments they own, i.e. $\omega_1 \geq \omega_2 \geq \ldots \geq \omega_I$.

*Production* Given a coalition $S$, the contributions of cooperators are pooled into the coalitional production activity; the total product $Y_S$ of a coalition $S$ is assumed to be given by:

$$Y_S = A \cdot \left( \sum_{i \in S} l_i \cdot \omega_i \right)^\alpha, \; A, \alpha > 0, \tag{2}$$

where $A$ is a scale parameter and $\alpha$ measures the output elasticity to the total coalitional input $\sum_{i \in S} l_i \cdot \omega_i$ supplied by cooperators in $S$. These technological parameters are independent of time, i.e. we assume that there is no technical progress.

By this formulation the production activity of a coalition has no external effect on the production of other coalitions simultaneously present in a given period, and in this respect, the present approach is differentiated from Yi (1997) and Thoron (1998), which are mainly concerned with the case in which coalitions can affect each other through production activity.

*Distribution and equilibria within a coalition* Once an agent adheres to a coalition, s/he is entitled to receive a share of the coalitional product, independently of his/her decision in favor of cooperation or defection (which is unknown to his/her partners). The rule governing distribution of the coalitional output among the members of a coalition is assumed to be of the *equal-sharing* type, i.e.

**Assumption 2** *For whatever $S \subset \mathfrak{I}$, any agent $i \in S$ receives an amount $y_i$ of the coalitional output $Y_S$ given by*

$$y_i = \frac{Y_S}{|S|}.$$

We assume that such distributive rule is part of the institutional setting of the economy and known to all players[4].

## 2.1 Solving the second stage of the game

In view of the previous distributive rule, agents receive different incentives to cooperation within the same coalition $S$ depending on the values of the technological parameters $A$ and $\alpha$. As an example take the coalition $S = \{1,2\}$ and $\omega_1 = 0.6$, $\omega_2 = 0.9$; if $A = 1$, the agents in $S$ face the following strategic situations corresponding to $\alpha = 1$ and $\alpha = 2$ respectively

| | $l_2 = 1$ | | $l_2 = 0$ | |
|---|---|---|---|---|
| $l_1 = 1$ | 0.75 | 0.75 | 0.3 | 0.8 |
| $l_1 = 0$ | 0.95 | 0.45 | 0.5 | 0.5 |

| | $l_2 = 1$ | | $l_2 = 0$ | |
|---|---|---|---|---|
| $l_1 = 1$ | 1.13 | 1.13 | 0.18 | 0.68 |
| $l_1 = 0$ | 0.91 | 0.41 | 0.5 | 0.5 |

In the first situation there is a Nash equilibrium prescribing defection to both agents, while the second situation admits both full defection and full cooperation as Nash equilibria[5]. In the rest of the paper we make the following:

**Assumption 3** *In any coalition $S \subseteq \Im$ there is a unique Pareto dominating Nash equilibrium and this is what agents play when $S$ forms.*

This assumption is certainly very strong since it removes any coordination problem and implies that if cooperation is incentive compatible for all, then everybody cooperates[6]. Furthermore, it rules out the formation of deviating coalitions which use incentive incompatible strategies to escape from a given coalitional structure. However, Assumption 3 has a notable advantage, since it makes it possible to define the ***coalition value*** $\pi$, namely an operator that associates to each coalition $S$ a vector of individual utilities in $\Re^{|S|}$; formally any component of $\pi$ is given by

$$\pi_i(S,\omega) = \frac{A \cdot \left(\sum_{h \in S} l_h^* \cdot \omega_h\right)^\alpha}{|S|} + \mu \cdot (1 - l_i^*), \forall i \in S, \tag{3}$$

[4] This is indeed a rather strong assumption; the fact that agents participating in a coalition share out equally what they produce is justifiable as the outcome of a (Nash) bargaining intra-coalitional process only if we assume agents to have the same physical characteristics. Outside the framework of this case different distributive rules would be called for, and a thorough analysis would also require distribution to be an endogenous result deriving from individual bids. We omit such further but substantial problems and retain the assumption that such a rule is commonly accepted.

[5] However notice that, in the first situation, the profile of payoffs is such that agent 2 could improve his/her situation by deviating from the coalition and forming a singleton group; this is much the same situation as is faced in the present context and which determines the formation of coalitional structures different from the grand coalition.

[6] In other words, in this case we consider full cooperation within the coalition as a focal point.

specifying that the members of $S$ receive the utility levels $\pi_i(S,\omega)$ corresponding to the Pareto superior Nash equilibrium in $S$. Since our game is without externalities, the C-value can be used to define the ***coalition structure value*** (CS-value) $\pi^\sigma$ (see Hart and Kurz (1983)) specifying for every $i$ the utility s/he receives in the coalition structure $\sigma$ as

$$\pi_i^\sigma(\sigma,\omega) = \pi_i(S,\omega), \forall i \in S \text{ and } S \in \sigma. \tag{4}$$

## 2.2 The first stage of the game: messages and coalitional formation

Since the choice of actions in the second stage of the game is solved by means of Assumption 3, the problem comes back to the first stage in which agents have to decide which coalition to announce. In this stage we assume that every agent chooses and revises signals $\theta_i$ (*messages* hereafter) indicating the agents with whom s/he wishes to coalesce, in other words a proposal of coalition; formally a message $\theta_i$ is a subset of $\Im$ including $i$ himself and $\Theta_i = \{\theta_i \subseteq \Im, i \in \theta_i\}$ is their set. An $I$-tuple $\Theta = (\theta_1, \theta_2, ..., \theta_I)$ is a configuration of messages, one for each agent, containing individual declarations of coalitions.

Once messages have been sent and the profile $\Theta$ is formed, coalitions begin to take shape and a coalitional structure is determined. We assume that the formation of coalitions follows the rule detailed in Hart and Kurz (1983) so that all agents announcing the same message form a coalition, even though the coalition formed may not include all the agents indicated in the messages. It can be formalized as a function $R^\Delta : \prod_{i\in\Im} \Theta_i \to \Sigma_\Im$ (where $\Sigma_\Im$ is the set of all possible partitions of $\Im$) such that

$$R^\Delta(\Theta) = \{S \subset \Im | i,j \in S \Leftrightarrow \theta_i = \theta_j\}.$$

If this rule governs the formation of coalitional structure, it is clear that messages indicate the largest set of agents a subject is willing to be associated with, while the coalition formed is the largest set of agents that show compatible messages. As an example suppose that the set of messages is $\Theta = (\{1,4\}, \{1,2,3,4\}, \{1,2,3\}, \{1,2,3,4\})$; then $R^\Delta(\Theta) = \{\{1\}, \{2,4\}, \{3\}\}$ is the resulting coalition structure. Notice that this construction allows for the possibility that an agent (or a group of agents) may abandon a coalition without entailing the disruption of the previously existing group.

## 2.3 The constituent game

According to the previous description, the constituent game that is going to be played in every period can be summarized as $\Gamma^{R^\Delta}\left(\Im, \{\Theta_i\}_{i\in\Im}, \pi^\sigma\right)$ where $\Theta_i$ is the set of messages available to agent $i$ to maximize his/her expected payoff, given that the rule $R^\Delta$ determines the formation of coalitions and that, once the coalition structure is formed, the payoff to each agent is given by $\pi^{\sigma}$. Thus under Assumption 3, $\{\Theta_i\}_{i \in \Im}$ is the pure strategies space of our game, since actions in the second stage are already determined. As can be easily checked this is nothing other than what Hart and Kurz (1983) termed the game $\Delta$.

The set of Nash equilibria of this game may be very large; Hart and Kurz (1983) report a possible refinement, based on the concept of strong Nash equilibrium (SNE), first proposed by Aumann (1967)[7]:

**Definition 1.** The coalition structure $\sigma$ is a SNE if there exists no $S \subset \Im$ and coalition structure $\sigma' = \{S \cup \sigma_{\Im \setminus S}\}$ such that $\pi_i^{\sigma'}(\sigma', \omega) > \pi_i^{\sigma}(\sigma, \omega)$ for all $i \in S$.

From Hart and Kurz (1983) we know that in general a SNE fails to exist; however for our particular game there is at least a SNE in pure strategies A simple way to characterize such an equilibrium is described in the Appendix where the following proposition is also proved:

**Proposition 1.** *Suppose $\alpha > 1$ in (2); then there exists at least a SNE for $\Gamma^{R^{\Delta}}$ and it has the consecutive property.*

In addition to existence of a SNE, Proposition 1 states that coalitions belonging to this equilibrium are formed by agents with similar endowments, in other words in equilibrium there cannot exist a coalition where the richest and the poorest agents form a group together without agents who have an average endowment. We denote this fact as consecutive property which corresponds to the formation of a hierarchy among coalitions in terms of their average endowments. We will verify this property in numerical simulations (see Section 4.1).

The existence of a SNE is relevant because it represents a possible resting point of the dynamic process by means of which coalition structures form. However the convergence of such a process crucially depends on many factors, the most important of which is how agents decide their messages. This is the focus of the next Section.

# 3 The repeated game and the structure of the simulation

In the previous paragraph we characterized the main features of the strategic situation that agents face when they embark on formation of a coalitional

[7] In the literature there is another refinement proposed by Bernheim et al. (1987) called coalition-proof Nash equilibrium; however we will ignore the latter because the concept of SNE is closer to our approach. It is worth pointing out that in our game the SNE coalition structure corresponds to the core stable coalition structure (see Greenberg (1994) for a definiton of the latter).

structure. In the following we assume that the game $\Gamma^{R^{\Delta}}$ is repeatedly played by the agents in a sequence of periods, in each of which agents are called upon to revise their choices, knowing the results of their previous interactions. More precisely, at the beginning of a period $t$ every agent observes the set of messages $\Theta^{t-1}$, the corresponding coalition structure $\sigma^{t-1}$ and the profile of payoff $\pi_i^{\sigma^{t-1}}(\sigma^{t-1},\omega)$ realized in the previous period. Given this information, (a subset of) agents revise their strategies $\theta$, signalling their willingness to aggregate with (sub)sets of other individuals; at the end of period $t$ a new configuration of messages $\Theta^t$ arises, and a new coalitional structure $\sigma^t$ is formed according to the rule $R^{\Delta}$. Actions in any coalition that has been formed are determined according to Assumption 3 and a new payoff profile $\pi^{\sigma}(\sigma^t,\omega)$ is generated and distributed. The same situation is repeated in the subsequent period $t+1$ and so on.

The sequence $\{\sigma^t\}_{t\geq 0}$ represents the time evolution of coalitional structures, while the sequence of the set of messages $\{\Theta^t\}_{t\geq 0}$ represents the continuous adaptation of the system caused by the decentralized and uncoordinated quest by every agent to maximize his/her utility.

### 3.1 Artificial agents

In simulating this dynamic process, we model our agents as artificial entities that use computational routines to make their decisions. More precisely an artificial agent is an object that, at any point in time,

1. *forms expectations* on the agents that could accept the proposed coalition (given that s/he knows that any other one will choose his/her own action according to Assumption 3) and hence *computes* the (expected) utility of a message
2. given these expectations, *chooses* the best message to send.

Any of these abilities is described by means of an algorithmic procedure that is discussed shortly in the following; in general notice that the complexity and depth of such procedures directly measure the degree of sophistication of the artificial agent.

**Expected utility of a message** As already noted, the process of coalition formation is driven by agents' messages and a message is a coordination device by means of which a player signals his/her willingness to form the most profitable coalition for him/herself, ***knowing*** which agent in the present situation (the configuration of period $t$) will accept his/her proposal. Therefore we assume that the expected utility of a message $\theta_i \in \Theta_i$ sent in period $t$ is given by the value of $\pi_i\left(S\left(\theta_i|\Theta^{t-1}\right),\omega\right)$ where $S\left(\theta_i|\Theta^{t-1}\right)$ is such that (1) $S\left(\theta_i|\Theta^{t-1}\right) \subseteq \theta_i$ and (2) $\pi_j\left(S\left(\theta_i|\Theta^{t-1}\right),\omega\right) > \pi_j^{\sigma}\left(\sigma^{t-1},\omega\right)$ $\forall j \in S\left(\theta_i|\Theta^{t-1}\right)$. In words, the value of a message $\theta_i$ is the utility that the agent $i$ would obtain in the (sub)coalition of $\theta_i$ formed by those agents that have an incentive to adhere to the message, because in so doing they improve their payoff with respect to what they have obtained so far. Here it is clear that we assume that agents do not behave myopically, because they do not attribute to $\theta_i$ the value of the coalition that could be formed through the rule $R^{\Delta}$ simply by inserting the message $\theta_i$ into the previous configuration of signals $\Theta_i$; rather, they seek to work out the most likely reaction of others to the prospected message $\theta_i$. Nonetheless, the behaviour assumed is still adaptive, because possible partners' acceptance or refusal of a message is estimated on the basis of current payoffs, without any strategic reasoning concerning possible "reactions to reactions".

**Choice of messages** Once the expected utility of a message is computed, the agent's task in period $t$ is to select the message $\theta_i^t$ such that

$$\theta_i^t = \arg\max_{\theta_i \in \Theta_i} \pi_i \left( S \left( \theta_i | \Theta^{t-1} \right), \omega \right). \tag{5}$$

Solving such an optimization problem requires extremely high computational abilities: in the case of $I = 50$ an agent needs to scan something around 1 millions of billions of alternatives and choose the one with the highest expected utility. Here, however, we assume that agents have computational limitations; they are sufficiently sophisticated to adopt maximizing procedures, but such procedures may fail to find the global optimum of the problem (bounded rationality). In fact the individual decision process is driven by GAs, a well known iterative procedure to search for maxima in complex environments, based on the parallel exploration of the space of possible solutions with the objective of maximizing a *fitness* index (see Holland et al. (1986)). In our case the fitness index is the expected utility of a given message as described in § 3.1 and the procedure works in such a manner that the higher is this fitness index, the higher is the probability that the corresponding agent will play the message (see Fiaschi et al. (2001) for more details).

## 4 Numerical simulations

This Section presents the numerical simulation of the dynamic process described in § 3. We discuss the basic characteristics of the resting points of the simulations and the main features of the convergence paths. We point out that a candidate for convergence is the SNE described in Proposition 1. This analysis is performed under several different assumptions concerning (1) the parameters of available technology, (2) the distribution of endowments and (3) the cardinality of the population of players. Each of the previous characteristics contributes to determining what we call an *environment* of the simulation, i.e. a triple $E = [\Im, \Omega, \alpha]$ specifying the population of agents participating in the game $\Gamma^{R^{\Delta}}$, the distribution of their resources and the strength of the parameter measuring returns to scale (for the sake of simplicity we assume that the technological parameter $A$ and the preference parameter $\mu$ are constant and equal to 1 and 0.5 respectively). In particular we examine the following cases:

- three different population sizes, i.e. $|\Im| \in \{20, 50, 100\}$.
- In the case of $|\Im| = 20$ we have run simulations for 20 different distributions of the initial endowments, ranging from an absolutely concentrated to a flat distribution; in the case $|\Im| = 50$ and 100, we have tested results respectively for 10 and 5 distributions of endowments again ranging from more concentrated to flatter distributions (for every population the total amount of resources is kept constant, independently of how they are distributed - see Figure 1).
- the technological parameter $\alpha$ is assumed to belong to the set $\{1.25, 1.5, 1.75, 2\}$.

Therefore the total number of environments is 140 and, for each of them, we run 5 simulations, changing the seed of the random numbers in order to eliminate possible effects due to the stochastic component of GAs.

The number of periods in each simulation is chosen in order to ensure that the final coalition structure is stable, i.e. shows no substantial change for a sufficiently long period (the exact number of periods varies according to the number of agents); it is worth pointing out that we never observed cycles in agents' strategies.

To give an idea of the kind of distributions of endowments we have examined, we report in the following Figures a sample of the distributions used in the simulations:

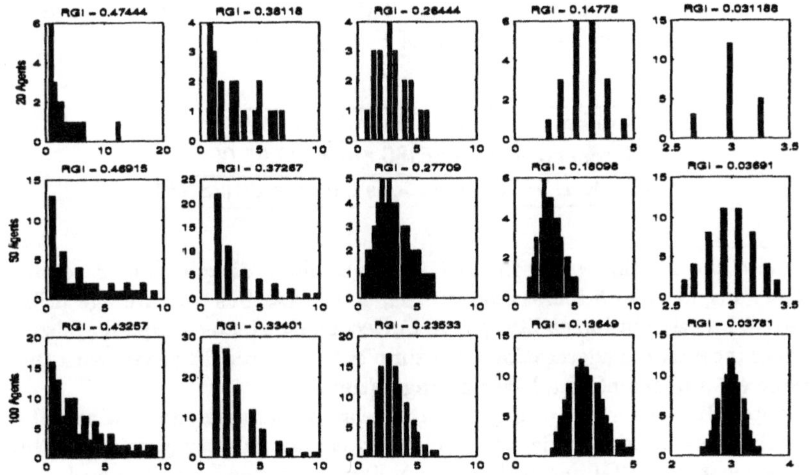


Figure 1: distributions of endowments.

In each Figure, the level of individual endowment is shown on the $x$-axis and the number of agents who have such endowment on the $y$-axis. The resources Gini index of the given distribution (RGI) is specified above each Figure: this index measures the degree of inequality (0 indicates perfect equality and 1 maximum inequality) of the endowment distributions. Figures belonging to the same row have the same number of agents (i.e. 20, 50 or 100).

### 4.1 Numerical results

The presentation of the results can be divided into two main parts; in the first we consider the stable coalition structures emerging from simulations, while in the second we seek to highlight the most relevant aspects of dynamical paths converging to such stable coalition structures.

**Stable coalition structures** Analysis of the stable coalition structures deriving from repeated interaction is accomplished in the following order: firstly we check if simulations have converged to the SNE indicated in Proposition 1; if not, we verify whether the final coalition structure is an alternative SNE. Secondly we analyze the properties of all final coalition structures emerging from simulations; naturally, this amounts to studying the properties of SNE if simulations have converged to the latter. We characterize these coalition structures in terms of (i) average utilities (AU), which can be taken as an index of efficiency[8], and of (ii) the Gini index of the distributions of utilities (UGI), a measure of inequality.[9]

*Convergence to SNE* In the following Table we show the percentage of simulations that converged to the SNE of Proposition 1:

| Number of agents | 20 | 50 | 100 |
|---|---|---|---|
| % converged to SNE | 86.80 | 66.00 | 65.00 |

Table 1: % of simulations converged to SNE

[8] Ex ante the relation between AUI and the initial distribution of resources is ambiguous; thus with increasing returns, ceteris paribus, the average utility is a positive function of inequality, but in our economy greater inequality also means a lower incentive to aggregation (it is simpler to coalesce for agents with the same resources) and therefore a lower average utility.

[9] Ex ante, given a certain technology, the relation between the RGI and the UGI is ambiguous because it depends on which coalitional structure will form. The relation between UGI and technology is likewise ambiguous: thus an increase in returns to scale causes, ceteris paribus, an increase in inequality; however higher returns to scale incentivate aggregation and therefore they produce a decrease in inequality.

The high percentage of convergence to SNE of Proposition 1 seems to suggest that the latter is the only SNE for our game; further support for this finding is obtained by observing that the stable coalition structures not converged to this SNE are not SNE. We notice that the probability of converging to the SNE is negatively correlated to the number of agents; this result is mainly caused by a lack of coordination among agents' actions due to the larger size of the economy. Finally, the lack of convergence to a SNE for many simulations means a slight difference in the coalition structure, attributable above all to the fact that such coalitions contain the poorest agents (in fact for the latter the incentives to coalesce are weaker), so that the percentage of convergence shown in Table 1 is underestimated (for further details see the next Section).

***Efficiency of stable coalition structures*** In the present game the grand coalition cannot be considered the most efficient coalition structure, because it fails to consider that incentives to aggregation and cooperation depend on the individual endowments. In this respect we consider as a term of comparison the average utility corresponding to the SNE of Proposition 1 (EAU), following the intuition that this allocation is the most efficient, given the individual incentive constraints.[10]

The following Figures show the relationships between AU, EAU for all the distributions (indicated on $x$-axis by their RGI):

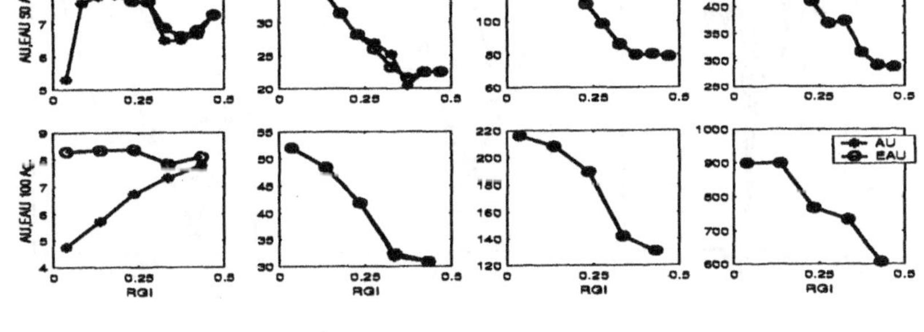


Figure 2: efficiency.

[10] In a limited number of simulations we will observe that the AU is greater than the EAU. In such cases the individual incentive constraints are not respected; in particular a restricted coalition would have increased the utilities of some agents.

A common pattern seems to emerge: the higher the RGI (i.e. the more unequal is the initial distribution of resources), the lower is AU as well as EAU. It is worth noticing that a similar result is reported in the endogenous growth literature (see Galor and Zeira (1993)). Here the result is due to the combined effect of the equal sharing rule and the possibility of defecting, implying that the greater the gap between the endowments of two agents, the lower is the incentive to coalesce. Agents find it easier to play the SNE of Proposition 1 when returns to scale are high; this is because the higher the increasing returns, the bigger the gains obtainable by coalescing. Finally, the departures from SNE in terms of AU are not very substantial, except for the simulations when $\alpha = 1.25$ and the number of agents is equal to 100; this means that many simulations not considered converged to SNE are very close to such equilibrium. As already observed, this implies that the values reported in Table 1 are underestimated.

*Equity of stable coalition structures* In this Section we analyze the relationship between UGI and the RGI. This relationship is not univocally determined ex ante; in fact aggregation causes inequality to decrease (denote it aggregation effect (AE)), while increasing returns tend to play an opposite effect (denote it technological effect (TE)). The following Figure shows the difference between UGI and RGI for all the distributions (reported on the $x$-axis by means of their RGI) for alternative technologies:

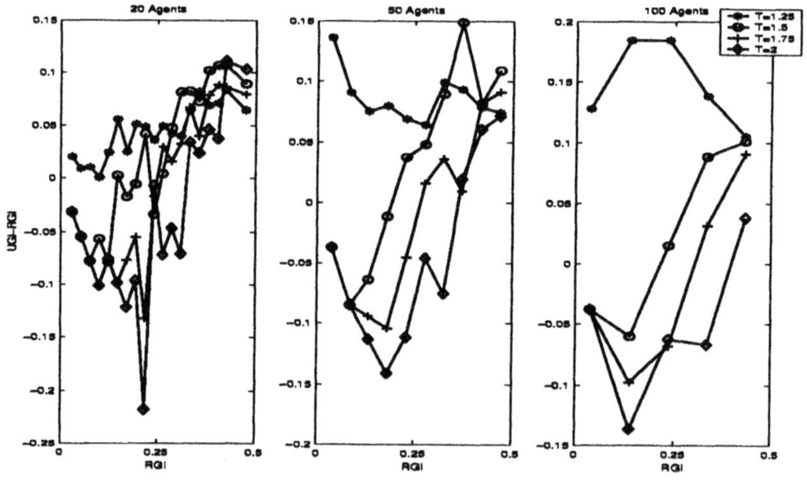


Figure 3: equity.

For very unequal distributions UGI is generally greater than RGI, i.e. the TE is greater than the AE. This is because high inequality disincentivates aggregation in an economy in which the coalitional output is equally shared.

An increase in returns to scale generally implies a decrease in UGI, which means that the AE is greater than the TE (in Figure 3 this is particularly evident in the simulations where the number of agents is equal to 100).

**Dynamic paths** In this Section we describe the characteristics of the process leading to stable coalition structures, with reference to some indexes. The first index is a measure of coordination among agents (SCI); it shows how the willingness of an agent to coalesce with another agent is matched by the signal from the latter (and viceversa)[11]. A second index is the average endowments of coalitions that have formed (AEFC), indicating which agents start coalescing first (the richer or the poorer). Moreover we consider the AU, the UGI and an index of concentration of coalition structure (CSC) of all periods[12]. All simulations show a similar behavior; an example is presented in the following Figure, relative to the case where the number of agents is equal to 100 and $\alpha = 1.5$:

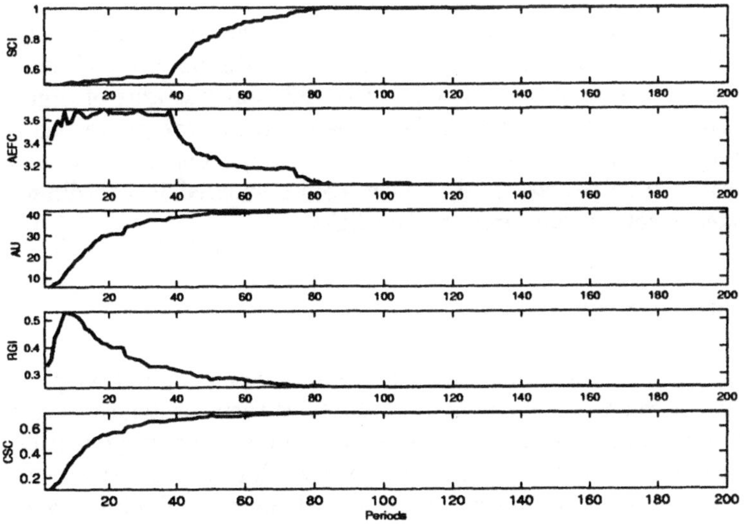


Figure 4: dynamical behavior.

We observe that AEFC is firstly increasing, corresponding to an initial phase where only the richest agents form coalitions, then generally decreasing,

[11] It is computed on the set of signals and it is normalized to 1, where 1 means full coordination. In particular $SCI^t = 1 - \frac{\sum_j \sum_i (\Theta^t_{i,j} - \Theta^t_{j,i})^2}{(I^2 - I)}$,where $\Theta^t_{i,j}$ is equal to 1 if agent $i$ signals to agent $j$ his/her willingness to coalesce or equal to 0 if s/he does not.

[12] Given a coalition structure $\sigma^t = \{S_1, S_2, ..., S_K\}$, then $CSC^t = \frac{\sqrt{|S_1|^2 + ... + |S_K|^2}}{I}$.

when poor agents also start coalescing: it eventually converges to the average endowment when a stable coalition structure is reached. The high value of SCI in the first periods signals a relevant mismatch. This is due to the willingness of poor agents to coalesce with rich agents; thus the latter, not yet belonging to a coalition, are currently receiving a low utility (in particular lower than the utility they would receive if they were in coalition with poor agents). Once the rich agents have coalesced among their own group, the poor agents turn their attention to agents with similar endowments and this results in the formation of coalitions among poor agents as well.

This pattern is reflected by the behavior of the RGI, which firstly increases, when only rich agents coalesce, and then decreases, when poor agents also start coalescing. The increase in AU reflects the ongoing process of formation of coalitions, making it possible to engage in greater and greater exploitation of the increasing returns to scale. It should be noted that UGI and AU have a non linear relationship, similar to the well-known one found by Kuznetz between inequality and income.

Finally, the monotonic increase in the CSC shows that the aggregation process among agents is continuous, and there are no sudden jumps in the form of the coalition structure, and therefore no disruption of already formed coalitions in favour of new, qualitatively different patterns of organization. Indeed the aggregation proceeds from the richest to the poorer agents without switching back, so that concentration (independently of its magnitude) increases as time passes, in that new agents succeed in coordinating with other agents not yet allocated in a coalition.

## 5 Conclusions

In this paper we studied the repeated interaction of a population of artificially intelligent agents who, period by period, revise their strategies by responding only to individual incentives and searching for a social organization where, in the absence of any external coordination device, individual payoffs are maximized, given other agents' behaviour. Four main findings emerge from the simulations. Firstly, agents play SNE strategies with high probability, although the latter is negatively related to the size of economy, since the larger the population, the greater the coordination problems. Secondly, given a production technology, a greater inequality in the distribution of initial resources corresponds to a lower average utility (a proxy for efficiency) in the SNE. The explanation for this phenomenon lies in the role played by the equal sharing rule, which discourages rich agents from coalescing with poor agents when the gap in individual endowments is high. On the other hand, the profile of individual utilities in the SNE generally amplifies the inequality present in the initial distribution of endowments; this is mainly due to the increasing returns to scale of the production technology, when it is not outweighted by the equalization effect of aggregation among agents.

Finally, the transition paths highlight the fact that the aggregation process proceeds from the richest to the poorest agents, with the former aggregating first and the others following later. This entails that the inequality in the profile of individual utilities first increases and then decreases, although both average utility as well as the concentration of the coalition structure are always increasing.

These results confirm that a decentralized and unmastered process in which there are no externalities among coalitions can converge to a coordinated outcome such as that prescribed by the SNE of the constituent game, with the equity and efficiency of the outcomes linked to the interrelation between available technology and the fairness of the initial distribution of endowments. This should be regarded as a preliminary step toward the construction of a truly dynamic model of social interaction in which the outcomes of the equilibrium configuration reached contribute, either entirely or in part, to determination of the individual endowments in the subsequent period, thus allowing the study of the endogenous evolution of a society of a society organized into groups of cooperation. The agenda of future research within the present framework will focus precisely on this aspect.

## References


1. Aumann, R. (1967): A survey of Cooperative Games Without Side Payments. In: M. Shubik (Eds.): Essay in Mathematical Economics. Princeton University Press, 3-27.
2. Bennett, E. (1985): Endogenous vs. Exogenous Coalition Formation. Economie Appliquee 38(3/4), 611-35.
3. Bernheim, B. D., B. Peleg and M. D. Whinston (1987): Coalition-Proof Nash Equilibria: Concepts. Journal of Economic Theory 42(1), 1-12.
4. Birchenhall, C.R. (1996): Evolutionary Games and Genetic Algorithms. In: Gillì M. (Eds.): Computational Economic Systems. Kluwer Academic Publishers, 3-23.
5. Block, F. (1996): Non-cooperative models of coalition formation in games with spillovers. In: C. Carraro and D. Siniscalco (Eds.): New directions in the economic theory of the environment. Cambridge: Cambridge University Press, 311-52.
6. Ecchia, G. and M. Mariotti (1998): Coalition formation in international environmental agreements and the role of institutions. European Economic Review 42, 573-582.
7. Farrell J. and Scotchmer S. (1988): Partnerships. Quarterly Journal of Economics 103, 279-297.
8. Fiaschi, D., N. Garrido and P. M. Pacini (2001): Production Partnerships Formation with Heterogeneous Agents: a Simulation with SWARM. In: F. Luna and A. Perrone (Eds.): Agent-Based Methods in Economics and Finance: Simulations in Swarm. Kluwer Academic Publishers.
9. Galor, 0. and J. Zeira (1993): Income Distribution and Macroeconomics. Review of Economie Studies 60, 35-52.

10. Greenberg, J. (1994): Coalitional Structures. In: R. J. Aumann and S. Hart (Eds.): Handbook of Game Theory. II.. Amsterdam, North-Holland, 1305-1337.
11. Guesnerie, R. and C. Oddou (1988): Increasing Returns to Size and Their Limits. Scandinavian Journal of Economics 90, 259-73.
12. Hart, S. and M. Kurz (1983): Endogenous Formation of Coalitions. Econometrica 52, 1047-1064.
13. Holland, J.H., K.J. Holyoak, R.N. and P. Thagard (1986): Induction: Processes of Inference, Learning and Discovery. MIT Press, Cambridge Mass.
14. Olson, M. (1983): The Logic of Collective Action. Harvard University Press, Cambridege Mass..
15. Thoron, S. (1998): Formation of a Coalition-Proof Stable Cartel. Canadian Journal of Economics 31(1), 63-76.
16. Yi S.-S. (1997): Stable Coalition Structures with Externality. Games And Economic Behavior 20, 201-237.

# A Existence and characterization of SNE

*Proof (Proof of Proposition 1).* For the sake of simplicity we take $A$ in the production function equal to 1.

Take a configuration of strategies $\Theta^*$ that satisfies the following requirements:

- $R^{\Delta}(\Theta^*) = \sigma^* = \{S_1^*, S_2^*, \ldots, S_n^*\}$,
- $\theta_i^* = S_k^*$, $i \in S_k^*$ and
- let $i_k^1$ be the first (richest) member of $S_k^*$ and let $S_0^* = \emptyset$; $S_k^*$ is the coalition of minimal cardinality in the set $\arg \max_{S \subseteq \Im / \{S_0^* \cup \ldots \cup S_{k-1}^*\}} \pi_{i_k^1}(S, \omega)$,

  $k = 1, ..., n$.

Firstly we derive some properties of $\sigma^*$.

**Fact 1**: if in a coalition $S_k^*$ there is full defection, then $|S_k^*| = 1$.

Suppose the contrary $|S_k^*| > 1$; then since to defect is the dominant strategy $\frac{\left(\omega_{i_k^1}\right)^\alpha}{|S_k^*|} < \mu$; moreover, since $S_k^*$ maximizes $i_k^1$ payoff, then $\left(\omega_{i_k^1}\right)^\alpha < \mu$, otherwise $i_k^1$ would have not adhered to $S_k^*$; but then the minimal coalition maximizing $i_k^1$'s payoff is $\{i_k^1\}$ itself and not $S_k^*$.

**Fact 2**: if $|S_k^*| \geq 2$ then all members of $S_k^*$ are cooperators, $k = 1, ..., n$.

Suppose the contrary that in $S_k^*$ there are some defectors $D \subset S_k^*$ and some cooperators $C \subset S_k^*$ such that $D \cup C = S_k^*$. Firstly we demonstrate that if $C \neq \emptyset$ then $i_k^1 \in C$; suppose not: then $\frac{(|C| \cdot \bar{\omega}_C)^\alpha}{|S_k^*|} + \mu > \frac{\left(|C| \cdot \bar{\omega}_C + \omega_{i_k^1}\right)^\alpha}{|S_k^*|}$ and $\frac{(|C| \cdot \bar{\omega}_C)^\alpha}{|S_k^*|} > \frac{\left(|C| \cdot \bar{\omega}_C - \omega_{i_C^1}\right)^\alpha}{|S_k^*|} + \mu$ where $\omega_{i_C^1}$ is the richest member of $C$ and $\bar{\omega}_C$ is the average endowment of the members of $C$. Combining the two equations, we get $\frac{(|C| \cdot \bar{\omega}_C)^\alpha}{|S_k^*|} + \frac{(|C| \cdot \bar{\omega}_C)^\alpha}{|S_k^*|} - \frac{\left(|C| \cdot \bar{\omega}_C - \omega_{i_C^1}\right)^\alpha}{|S_k^*|} > \frac{\left(|C| \cdot \bar{\omega}_C + \omega_{i_k^1}\right)^\alpha}{|S_k^*|}$, i.e. $(|C| \cdot \bar{\omega}_C)^\alpha > \frac{\left(|C| \cdot \bar{\omega}_C + \omega_{i_k^1}\right)^\alpha + \left(|C| \cdot \bar{\omega}_C - \omega_{i_C^1}\right)^\alpha}{2}$ which is impossible, given the assumption that $\alpha \geq 1$ and $\omega_{i_k^1} \geq \omega_{i_C^1}$. Now we demonstrate that $D = \emptyset$; suppose not: then $\forall j \in D$ $\frac{(|C| \cdot \bar{\omega}_C)^\alpha}{|S_k^*|} + \mu > \frac{(\omega_i + |C| \cdot \bar{\omega}_C)^\alpha}{|S_k^*|}$ while $\forall i \in C$ $\frac{(|C| \cdot \bar{\omega}_C)^\alpha}{|S_k^*|} > \frac{(|C| \cdot \bar{\omega}_C - \omega_i)^\alpha}{|S_k^*|} + \mu$; take the coalition $S_C \subset S_k^* : i \in S_C \Leftrightarrow i \in C$, then $\forall i \in S_C$ $\frac{\left(|S_C| \cdot \bar{\omega}_{S_C}\right)^\alpha}{|S_C|} > \frac{\left(|S_C| \cdot \bar{\omega}_{S_C} - \omega_i\right)^\alpha}{|S_C|} + \mu$ since $|S_C| \cdot \bar{\omega}_{S_C} = |C| \cdot \bar{\omega}_C$ and $|S_C| < |S_k^*|$. Moreover $\frac{\left(|S_C| \cdot \bar{\omega}_{S_C}\right)^\alpha}{|S_C|} > \frac{(|C| \cdot \bar{\omega}_C)^\alpha}{|S_k^*|}$, but then $S_k^*$ is not maximizing the $i_k^1$'s payoff (remember that $i_k^1 \in C$).

**Fact 3**: $\sigma^*$ has the consecutive property, i.e. if agent $i$ and agent $j$ belongs to some $S_k^*$, then any agent $h$ such that $i < h < j$ has to belong to $S_k^*$.

Suppose not and assume that a coalition $S_k^*$ is the union of two non-consecutive set of agents $C^1$ and $C^2$, where $i_k^1 \in C^1$ (therefore $\omega_i \geq \omega_j, \forall i \in C^1, \forall j \in C^2$); let $\bar{C}$ the set of agents such that $C^1 \cup \bar{C} \cup C^2$ has the consecutive property and take the consecutive coalition $T \subset C^1 \cup \bar{C} \cup C^2$ and such that $|T| = |S_k^*|$. Full cooperation is the Nash equilibrium of $T$ since if $\frac{\left(|S_k^*| \cdot \bar{\omega}_{S_k^*}\right)^\alpha}{|S_k^*|} > \frac{\left(|S_k^*| \cdot \bar{\omega}_{S_k^*} - \omega_i\right)^\alpha}{|S_k^*|} + \mu$ for all $i \in S_k^*$ then $\frac{(|T| \cdot \bar{\omega}_T)^\alpha}{|T|} > \frac{(|T| \cdot \bar{\omega}_T - \omega_i)^\alpha}{|T|} + \mu$ for all $i \in T$, since $\bar{\omega}_T \geq \bar{\omega}_{S_k^*}$ (in fact $\omega_h \geq \omega_j, \forall h \in T, \forall j \in C^2$). But, by the last point, it is also true that $\frac{(|T| \cdot \bar{\omega}_T)^\alpha}{|T|} > \frac{\left(|S_k^*| \cdot \bar{\omega}_{S_k^*}\right)^\alpha}{|S_k^*|}$ so that $S_k^*$ did not maximize $i_k^1$'s payoff.

Now we show that $\Theta^*$ is a SNE for $\Gamma^{R^\Delta}$. Suppose not; then there exists a coalition $T \subseteq \Im$ such that $\pi_i(T, \omega) > \pi_i^\sigma(\sigma^*, \omega)$ for all $i \in T$. Let $i_T^1$ be the first (richest) agent in $T$ and let $S_k^*$ the coalition in $\sigma^*$ such that $i_T^1 \in S_k^*$; by Fact 2 $i_T^1$ is a cooperator. Now consider the coalition $G = i_k^1 \cup T$, where $i_k^1$ is the richest agent in coalition $S_k^*$; any cooperator in $T$ (call their group $T_C$) will maintain cooperation in $G$ since if $\frac{(|T_C| \cdot \bar{\omega}_C)^\alpha}{|T|} > \frac{(|T_C| \cdot \bar{\omega}_C - \omega_i)^\alpha}{|T|} + \mu \Rightarrow \frac{\left(|T_C| \cdot \bar{\omega}_C + \omega_{i_k^1}\right)^\alpha}{|T+1|} > \frac{\left(|T_C| \cdot \bar{\omega}_C + \omega_{i_k^1} - \omega_i\right)^\alpha}{|T+1|} + \mu$, given that $\omega_{i_k^1} \geq \bar{\omega}_C$. Hence $\pi_{i_T^1}(G, \omega) > \pi_{i_T^1}(T, \omega)$. Since $i_k^1$ is certainly a cooperator he will get as much as $i_T^1$ so that $\pi_{i_k^1}(G, \omega) = \pi_{i_T^1}(G, \omega) > \pi_{i_T^1}(T, \omega) > \pi_{i_T^1}^\sigma(\sigma^*, \omega) = \pi_{i_k^1}^\sigma(\sigma^*, \omega)$ since both $i_k^1$ and $i_T^1$ were cooperators in $S_k^*$, by Fact 2. But then $S_k^*$ would not maximize $i_k^1$'s payoff, contradicting the way it was constructed.

# Location Games with Externalities

Jacques Durieu[1], Philippe Solal[1], and Sylvie Thoron[2]

[1] University of Saint-Etienne, CREUSET, 6 rue Basse des rives, 42100 Saint-Etienne, France.E-mail: durieu@univ-st-etienne.fr and solal@univ-st-etienne.fr

[2] University of ToulonGREQAM, Centre de la vieille Charit, 2 rue de la Charit, 13002 Marseille, France. E-mail: thoron@ehess.cnrs-mrs.fr

**Abstract.** We propose a two step game of coalition or city formation. In a first step, each player chooses the location in which he wants to be. The payoff function, determined in the second step by a game between the different locations reflects two effects: a public effect such that payoffs decrease with the number of non-empty locations; a private effect such that payoffs to the inhabitants of a particular location decrease with the size of the population at that location. We analyse the consequences for the set of stable profiles of an increase in the relative weight of the public effect in the payoff function. We show that the number of stable profiles increases with the public effect but that the newly added profiles are not always more concentrated.

## 1 Introduction

Economies are characterized by groups or coalitions of individuals. Firms, and cartels of firms are the simplest example. Such collective arrangements are made because it is in the interest of their members to do so. The usual explanation given for the appearance of groups is that there are increasing returns to scale. The latter may arise in many ways the most frequently cited being the simple benefits of the division of labour. Externalities, in general, play an important role and this is one reason why cities form and provide public goods (P. Pecorino, 1999), and nations sign environmental agreements (M.H. Babiker, 2001). In these particular examples, the payoffs to a coalition present a common feature that has been very little analysed in the literature: there are externalities between coalitions in the same structure or profile in the sense that the payoff to one coalition depends on the actions of non members. In this paper, we focus on positive externalities in the sense that the formation of a coalition is beneficial to non-members. For example, a city, if it is large enough, may provide public facilities for which the inhabitants are prepared to pay taxes. However these facilities may benefit many people other than the inhabitants. Indeed it may be difficult to prevent outsiders from profiting from them. Thus the cooperative efforts of the inhabitants of the city may yield widespread positive externalities. However, when we speak about a positive externality from cooperation, we have to keep in mind that, from the point of view of the members, cooperation has also a cost. The inhabitants of a city pay taxes in order to pay for the public facilities which also benefit those who do not pay taxes in the city and who may choose to live elsewhere and pay lower taxes and to come to the city from time to time to enjoy its facilities. In other words, in this kind of situation, there is a problem of free riding. Free riding arises as a result of two effects: a public effect which is positive for everybody and a private effect which is negative for those initiating the cooperation. Therefore, it is clear that players have two opposite incentives: the incentive to form coalitions, in our example, to live in cities and to benefit from the resultant positive effect to which they have contributed, versus the incentive to live elsewhere, in the hope that the inhabitants of the cities will support the cost of cooperation. Given these two conflicting incentives for individuals a natural question which arises is "which coalitions, in our example, cities will actually form?" To model this situation, we use a location game in which players can migrate from one location to another without any restriction on the part of their ex or new partners. For each location profile, payoffs with positive externalities are determined by a game between the locations of the profile.Thus the payoff to individuals in one location will depend on their own collective choice of activities such as the production of public goods and on the collective choice of the individuals in other locations which also effect them. In the payoff function of this game, the relative weight of the public and the private effects are represented by one parameter. Once this payoff function is given we can analyse the set of stable location profiles which are those in which no player has an incentive to migrate to another location. It is clear that the nature of the game between locations influence the outcome. The objective is to analyse the evolution of the set of stable location profiles when the relative weight of the public effect increases. The obvious intuition is that, as the relative weight of the positive externality or public effect increases individuals will be more likely to wish to form large conglomerations. We refer to location profiles in which there are a few hightly populated locations or cities as concentrated profiles. The question corresponding to the intuition is then: "Does an increase in the relative weight of the public effect leads to an increase in the concentration of the stable location profiles?" The answer to this question is not simple as one might hope. Although it is true that as the relative weight of the public effect increases more concentrated profiles become stable it is also true that very dispersed coalitions also become stable. Thus, when there are few large cities, the alternative for an individual is to move to another large city or to live alone. In the latter case she loses the benefit of the public good produced by her own contribution, does not now produce any significant local benefit herself, but does still retain the benefits of the public facilities paid for by the other inhabitants in the city that she previously lived in. However, when there are many small towns, individuals gain some of the direct effects of cooperation and would not gain much by moving to another town , thereby slightly increasing the size of that town, and paying higher taxes without a comparable increase in the facilities provided by that town. Perhaps most interestingly, intermediate profiles are the last to become stable, that is the arrangement of a number of middle-sized towns needs a very high public effect to survive, individuals preferring either to be in very large towns with the associated facilities or in rather small towns with low contributions. The paper is organized as follows. Section 1 introduc a normal-form game with player set $\mathcal{N} = \{1, 2, \ldots, i, \ldots, n\}$ for some positive integer $n$ and with$\mathcal{S} = \times_{i \in \mathcal{N}} S_i$ as the set of purestrategy combinations $s = (s_1, s_2, \ldots, s_i, \ldots, s_n)$, where eachpure strategy set$S_i = S$ consists of the same $mn$ distinct locations $s_i = k$, $k = 1, 2, \ldots, m$ available to each player $i \in \mathcal{N}$.The mapping $\pi : \mathcal{S} \rightarrow I\!R^n$ gives the combinedpayoff-function whichassigns a payoff vector $\pi(s) = (\pi_1(s), \pi_2(s), \ldots, \pi_i(s), \ldots, \pi_n(s))$ to each pure strategy combinations $s$.

Let$\beta = \times_{i \in \mathcal{N}} \beta_i : \mathcal{S} \rightarrow \mathcal{S}$ be thepure best-replycorrespondence which maps pure strategy combinations to their pure best-replycombinations. More precisely, for each player $i \in \mathcal{N}$ and pure strategy combination $s \in \mathcal{S}$,

$$\beta_i(s) = \{s_i^k \in S_i \mid \pi_i(s_i^k, s_{-i})\pi_i(s_i, s_{-i}), \ \forall s_i \in S_i\}.$$

A pure Nash equilibrium is a strategy combination $s \in S$ which is a fixedpoint of $\beta$. The set of pure Nash equilibria of $\Gamma$ is denoted by

$$\mathcal{E}(\Gamma) = \{s \in \mathcal{S} \mid s \in \beta(s)\}.$$

Throughout the paper we assume that players are *ex ante* identical andthat membership in a location is open to all players who are willing to joinit. Each player$i$ announces the location $s_i = k$ he or she wants to belong to. The playersthat announce the same location belong to the same location. **Alocation profile** is a partition of theplayer set $\mathcal{N}$. More precisely, a profile $s \in \mathcal{S}$determines a location profile i.e. apartition $\mathcal{N}(s) = (\mathcal{N}_k(s), k \in S)$ of$\mathcal{N}$ into$m$ locations where $\mathcal{N}_k(s) = \{i \in \mathcal{N} : s_i = k\}$consists of the players whochoose location$k$. Players preferences exhibit anonymity, that is a player cares onlyabout the number and not theidentity of the members in each location. This information is provided by **alocation profile** $l \in \mathcal{L}$. $l = (l_1, \ldots, l_r, \ldots l_n)$ is an $n$-dimensionalvector where $l_r$ stands for the number of locations of size-$r$.The location profile in whicheach player happens to be alone at his or her location, i.e.$l^* = (n, 0, 0, \ldots, 0)$ or equivalently $\sum_{r=1}^{n} l_r = n$, is called the**trivial profile**. The location profile in which all players live in thesame location, i.e $l^{**} = (0, 0, 0, \ldots, 1)$ or equivalently $\sum_{r=1}^{n} l_r = 1$, is called the **grand location.**A location profile is symmetric if for some $z \in \{1, 2, \ldots, n\}$,$l = (0, \ldots, l_z, \ldots, 0)$ where$l_z > 0$. We denote by $\mathcal{M} \subset \mathcal{L}$ the set ofsymmetric locationprofiles. Denote by $\mathcal{D} \subset \mathcal{L} \backslash \mathcal{M}$ the set of all location profiles such that there exist only locations of size 0,$z$ and $z + 1$. Call them quasi-symmetric locations.The set of location profiles is $\mathcal{L} = \{(l_1, \ldots, l_r, \ldots, l_n) \in I\!N_0^n \mid \sum_{r=1}^{n} l_r = m - l_0\}$ where $I\!N_0 = I\!N \cup \{0\}$.

There is a canonicalmapping $f : \mathcal{S} \to \mathcal{L}$ which assigns to each $s \in \mathcal{S}$ the location profile $\mathcal{N}(s) : f(s) = l$.

### 1.1 Externalities across locations

The model we develop involves two steps. In the first step, each player $i \in \mathcal{N}$announces a location$s_i = k \in S$. Given the $n$-tuple of announcements $s$, a location profile $\mathcal{N}(s)$ is determined and a profile $f(s)$ results. In the second step, the players engage in a strategic game where each takes advantage of the location profile that has been formed in the first step. Like Yi (1997), we resort to a reduced form of the game where the second step is replaced by continuation payoffs that depend on the outcome of the first step. We assume that player $i$'s continuation payoffs are given by a function $\pi_i : \mathcal{S} \times \mathcal{L} \to I\!R$. Player $i$ receives a payoff $\pi_i(k, l)$ if he or she belongs to the $k$th location and the prevailing location profile is $l$. Player $i$'s payoff in the entire game then is the indirect utility$u_i(s) = \pi_i(s_i, f(s))$ for $s \in \mathcal{S}$. This completes the description of the basic location game in strategic form.
Next let us specialize further and assume a particular functional form of continuation payoffs that incorporates two effects: a **public effect** which gives players alone an incentive to migrate to non-empty locations in order to avoid competition, and a **private effect** which gives players an incentive to leave a location in order to avoid the inconvenience of sharing. We will assume equal sharing of the location surplus among members. Each player whobelongs to a location $k$ of size-$z$earns:

$$\pi_i(k, l) = \pi(z, l) = z^{\alpha} \left( \sum_{r=1}^{n} l_r \right)^{\beta}, \quad \text{with} \quad \alpha, \beta < 0 \tag{1}$$

Since the game is symmetric we will write $\pi(z, l)$ instead of $\pi_i(k, l)$. Note that the payoff function (1) is decreasing for all $1 \sum_{r=1}^{n} l_r n$. If thecompetition between locations decreases, for example, because all the players of somelocations migrate to another location, outside players not involved in this process are better off. Futhermore, the payoff function is decreasing for all $1 z n$. A member of a location ofsize $z > 1$ becomes worse off if he or she leaves his or her location to join another location of equal or largersize. Note that this is not necessary true for a player who happens to be alone $(z = 1)$. The relative weight of each effect is measured by the quotient $\omega = \frac{\beta}{\alpha}$.

## 2 Stable location profiles

This section defines and examines stable location profiles for the location game with payoff function given by (1). The pre-image of a location profile $l$ is the subset of announcements $s$ which has $l$ as image. A location profile $l \in \mathcal{L}$ is stable if$f^{-1}(l) \subset \mathcal{E}(\Gamma)$. The payoff function (1) allows us to distinguish between four kinds of migration depending on the sizes, before the migration, of the location left by the migrant and of the location reached:

1. The migrant leaves a location with a population of size $z = 1$ andmoves to a location ofsize $z = 0$. Such a migration has no effect on the payoff of the migrant and thus never occurs. Such a migration will be called a migration of type 1.
2. The migrant leaves a location of size $z = 1$ and moves to a non-empty location of size $z1$. Such a migration has two simultaneous effects. First, it reduces the competition between non-empty locations, but increases the size of the location the migrant moves to. Such a migration willbe called a migration of type 2.
3. The migrant leaves a location of size $z > 1$ and moves to an empty location. Such amigration reduces the size of the location the migrant belongs to but stimulates the competition between non-empty locations. Such a migration will be called a migration of type 3.
4. The migrant leaves a location of size $z1$ and moves to a non-empty location of size $z1$. Such a migration has a positive effect on the payoff of the migrant only if the size of the location reached is smaller than the size of the location left. Such a migration will be called a migration of type 4.

1. If $\omega < 1$, then the trivial location is the only stable profile.
2. If $\omega 1$, then there exists a size of the total population suchthat the trivial location is stable and the grand location is unstable.

**Proof.** The trivial location is stable if

$$n^{\beta} 2^{\alpha} (n-1)^{\beta} \Leftrightarrow 2 \left( \frac{n}{n-1} \right)^{\omega} \tag{2}$$

which is always true for any $\omega < 1$. To see that the trivial location is the only stable profile when $\omega < 1$, consider the following argument. In order to be stable a profile, different to the trivial one, has to verify the following condition

$$z^{\alpha} \left( \sum_{r=1}^{n} l_r \right)^{\beta} \left( \sum_{r=1}^{n} l_r + 1 \right)^{\beta} \Leftrightarrow z \left( \frac{\sum_{r=1}^{n} l_r + 1}{\sum_{r=1}^{n} l_r} \right)^{\omega}$$

which is never true when $z2$.
To prove the second part of the proposition, fix some $\omega 1$. Following inequality (2), there exists a population size sufficiently large such that the trivial location is stable.
The grand location is stable if

$$n^{\alpha} 2^{\beta} \Leftrightarrow n 2^{\omega}. \tag{3}$$

Clearly, there exists a population size sufficiently large such that the inequality (3) does not hold (see the numerical examples insection 3).

•

In the sequel the superscript $c$ denotes the complement of a set. Each location profile $l \in (\mathcal{M} \cup \mathcal{D})^c$ is unstable. **Proof.** In such location profiles, there exists at least one pair of locations of size-$z$ and$z + h$ respectively with $h > 1$. In this case, a migration of type 4 is a best reply to$l$ for a player living in a location of size-$(z + h)$. Indeed, by (1) such a player has interest to migrate to a location of size-$z$. Thus $l$ is unstable.

•

Proposition 2 establishes that location profiles which are notsymmetric orquasi-symmetric are unstable. Thus, only symmetric and quasi-symmetric location profiles can be stable.The following proposition gives a necessary condition for the stability ofquasi-symmetric locationprofiles. A quasi-symmetric location profile $l \in \mathcal{D}$ with non-empty locations of size $z$ and$z + 1$ is stable only if a corresponding symmetric location profile $l' \in \mathcal{M}$ with non-empty locations of size $z + 1$ is stable. **Proof.** Regarding migrations of type 3, a quasi-symmetric location profile withnon-empty locations of size $z$ and $z + 1$ is stable if

$$z^{\alpha}\left(\sum_{r=1}^{n} l_r\right)^{\beta}\left(\sum_{r=1}^{n} l_r+1\right)^{\beta} \Leftrightarrow z\left(\frac{\sum_{r=1}^{n} l_r+1}{\sum_{r=1}^{n} l_r}\right)^{\omega}$$

and

$$(z+1)^{\alpha}\left(\sum_{r=1}^{n} l_r\right)^{\beta}\left(\sum_{r=1}^{n} l_r+1\right)^{\beta} \Leftrightarrow z+1\left(\frac{\sum_{r=1}^{n} l_r+1}{\sum_{r=1}^{n} l_r}\right)^{\omega} \tag{4}$$

Since $z < z + 1$, (4) is a condition of stability for both aquasi-symmetric location profileand a symmetric location profile with locations of size $z + 1$.

•

As a result of proposition 2, in many cases, we can restrict our analysis to symmetric location profiles. We now turn to the question of stability for symmetric profiles.At this end, we define a notion of **concentration** which permits us to compare stable location profiles under various values of $\omega$. A location profile $l \in \mathcal{L}$ is more concentrated than $l' \in \mathcal{L}$ if the number of non-empty locations is lower in $l$ than in $l'$, that is if $\sum_{r=1}^{n} l_r < \sum_{r=1}^{n} l'_r$. Let $n, m \in \mathbb{N}$ be given, with $mn$.

1. Suppose $l \in \mathcal{M} \backslash l^*$ is stable for a given $\omega$. Then it is also stable for all $\omega' > \omega$. This implies that the number of stable location profiles in $\mathcal{M}^*$ increases with $\omega$.

2. Suppose $n = 4$. The profile $(0, 2, 0, 0)$ is stable as soon as $\omega \frac{\ln 2}{\ln \frac{3}{2}}$. And the grand location is also stable as soon as $\omega = 2$. If $w < 2$, then the number of stable location profiles in $\mathcal{M} \backslash l^*$ increases with $\omega$. The new added stable location profiles are moreand more concentrated. If $w = 2$, then each location profile in $\mathcal{M} \backslash l^*$ is .
3. Suppose $n > 4$ and $\omega$ given. If a profile $l \in \mathcal{M} \backslash l^*$ such that $\sum_{r=1}^{n} l_r = \omega - 1$ is stable, then each location profile $l \in \mathcal{M} \backslash l^*$ is stable.
4. Suppose $\omega(n + 2)/2$. Then each location profile $l \in \mathcal{M} \backslash l^*$ is stable.

**Proof.** Suppose $l \in \mathcal{M} \backslash l^*$.Let a profile $l^k \in \mathcal{M} \backslash l^*$ such that $l^k = (0, \ldots, l_k, \ldots, 0)$ with $l_k > 0$.The maximum size of a location in$l^k$ is

$$k = \frac{n}{l_k}, \tag{5}$$

with $\sum_{r=1}^{n} l_r = l_k$.If $l^k$ is stable, no member of a size-$k$ location has an incentive to migrate toward an empty location. In such case

$$\left(\frac{n}{l_k}\right)^{\alpha} (l_k)^{\beta} \; (l_k + 1)^{\beta}$$

Rearranging terms, we obtain

$$n \left(\frac{l_k + 1}{l_k}\right)^{\omega} l_k \tag{6}$$

1. Note that the right hand side of the above inequalityincreases with$\omega$. Moreover, the first term in the right hand side of the inequalityis superior to 1. It follows that bytaking$\omega$ sufficiently large, we can make $1 l_k n$ as small as desired.In other words, the set of stable symmetric profiles admits more and moreelements when $\omega$increases.

2, 3 and 4. Consider the right hand side of (6) as a function of $l_k$ :

$$f_{\omega}(l_k) = \left(\frac{l_k + 1}{l_k}\right)^{\omega} l_k \quad \text{with} \quad l_k = \{1, \ldots, \frac{n}{2}\} \tag{7}$$

Consider the extension of $f_\omega$ on $\mathbb{R}_{++}$ and take the derivative with respect to $l_k$ :

$$\frac{df_{\omega}(l_k)}{dl_k} = A(l_k).(l_k + 1 - \omega) \quad \text{with}(l_k) > 0 \quad \forall l_k \in \mathbb{R}_{++}$$

We see that for each $\omega$, $f_\omega$ is a convexfunction on its domain, and attains a minimum at $l_k{}^* = \omega - 1$. The minimum argument of this function moves towards the right when $\omega$ increases. Indeed, for a given $l_k$, $f_\omega(l_k)$ increases with $\omega$.Therefore we have two distinguish three cases :

*i*) Suppose $l_k{}^* 1 \Leftrightarrow \omega 2$. Then $f_\omega$ is an increasing function on $\{1, \ldots, \frac{n}{2}\}$. If $\omega < 2$, then the number of stable location profiles increases with $\omega$ and the

new added stables profiles are more and more concentrated. Particularly, using (7), it appears that $f_\omega(2)4$ when $\omega\frac{\ln 2}{\ln \frac{3}{2}}$.If $\omega = 2$, then ${l_k}^* = 1$ and $f_2({l_k}^*) = 4$. Therefore, each location profile $l \in \mathcal{M}\backslash l^*$ is stable. This proves the second part of the proposition.
*ii*) Suppose $1 < {l_k}^* < \frac{n}{2} \Leftrightarrow 2 < \omega < \frac{n+2}{2}$. Then$f_\omega$ is decreasing before ${l_k}^*$ and increasing after. If $\omega$ is such that a location profile $l^k \in \mathcal{M}\backslash l^*$ with $l_k = {l_k}^*$ is stable (this implies that $n > 4$, see point *i*)), then each location profile belonging to $\mathcal{M}\backslash l^*$ is stable. This proves the third part of theproposition.
*iii*) Suppose $l_k = \frac{n}{2}$ and $\omega\frac{n+2}{2}$. Then $f_\omega$ is a decreasing function on $\{1, \ldots, \frac{n}{2}\}$, and

$$n < f_{\frac{n+2}{2}}\left(\frac{n}{2}\right) = \left(\frac{n+2}{n}\right)^{\frac{n+2}{2}} \frac{n}{2}$$

Thus, each location profile in $\mathcal{M}\backslash l^*$ is stable .This completes the proof.

•

Note that as a consequence of proposition 2, for some values of $n$ and $\omega$, the stable profiles may be the most and the less concentrated ones. Indeed in the case *ii*), if

$$4 < nf_\omega(l_k^*) \quad \text{with} \quad f_\omega(l_k^*) < f_{\frac{n+2}{2}}\left(\frac{n}{2}\right)$$

there exists some $n$ for which the most and the less concentrated profiles are stable while other profiles are unstable.

Proposition 2 gives a necessary condition for the stability of quasi-symmetric location profiles. This is not a sufficient condition since some quasi-symmetric location profiles corresponding to a stable symmetric location profile can be unstable. A second condition which takes into account migrations of type 2 must be added. Such a condition is relevant when at least a "loner" exists in the quasi-symmetric location profile. To see this, denote by $\mathcal{F} \subset \mathcal{D}$ the set of all location profiles such that there exist only locations of size 0, 1 and 2, and by $\mathcal{F} \subset \mathcal{F}$ the subset of such location profiles such that $l_1 > 1$. Each location profile which belongs to $\mathcal{F}$ is unstable. **Proof.** Suppose $l_1 > 1$. The payoff of a "loner" is

$$\pi(1, l) = \left(\sum_{r=1}^{n} l_r\right)^{\beta}$$

We have to distinguish two cases. First, if there exists more than one location of size1, a "loner" can move to such a location. His or her payoff is given by

$$\pi(2, l) = 2^{\alpha}\left(\sum_{r=1}^{n} l_r - 1\right)^{\beta}$$

This player has no interest to migrate if

$$\left(\frac{\sum_{r=1}^{n} l_r}{\sum_{r=1}^{n} l_r - 1}\right)^{\omega} 2 \Leftrightarrow \omega \frac{\ln(2)}{\ln\left(\frac{\sum_{r=1}^{n} l_r}{\sum_{r=1}^{n} l_r - 1}\right)} \tag{8}$$

Notice that this inequality is always true for $\omega 1$. Finally, it is easy to verify that a player living in a location of size 2 has not interest to migrate to a location of size 1.

Next, if the location profile does not contain at least two locations of size 1, thenthe only "loner" has the opportunity to migrate to a location of size 2. His or herpayoff is given by:

$$\pi(3, l) = 3^{\alpha} \left(\sum_{r=1}^{n} l_r - 1\right)^{\beta}$$

The "loner" sticks to his or her previous location if

$$\left(\frac{\sum_{r=1}^{n} l_r}{\sum_{r=1}^{n} l_r - 1}\right)^{\omega} 3 \Leftrightarrow \omega \frac{\ln(3)}{\ln\left(\frac{\sum_{r=1}^{n} l_r}{\sum_{r=1}^{n} l_r - 1}\right)} \tag{9}$$

Since

$$\frac{\ln(2)}{\ln\left(\frac{\sum_{r=1}^{n} l_r}{\sum_{r=1}^{n} l_r - 1}\right)} < \frac{\ln(3)}{\ln\left(\frac{\sum_{r=1}^{n} l_r}{\sum_{r=1}^{n} l_r - 1}\right)}$$

the "loner" has no interest to migrate when (8) is verified. Simultaneously, a migration to an empty location is not rational for a player living in location of size 2 when

$$2^{\alpha} \left(\sum_{r=1}^{n} l_r\right)^{\beta} \left(\sum_{r=1}^{n} l_r + 1\right)^{\beta},$$

that is

$$\omega \frac{\ln 2}{\ln\left(\frac{\sum_{r=1}^{n} l_r + 1}{\sum_{r=1}^{n} l_r}\right)} \tag{10}$$

Combining (8) and (10) we obtain a contradiction.

•

**Remark.** The above proposition raises the following question. Is it possible to obtain a similar result under a weaker condition, namely, that $l_1 = 1$ ? The following example shows that the answer is in the negative. Suppose $n = 5$, $l_1 = 1$ and $\omega = 5/2$. Clearly $l_2 = 2$ and the payoff to a player living in a location of size 2 is

$$\pi(2, l) = 2^{-2}3^{-5} = 972^{-1}.$$

On the other hand, the payoff of a "loner" is

$$\pi(1, l) = 3^{-5} = 243^{-1}.$$

If the loner wants to migrate his or her(expected) payoff is

$$\pi(3, l) = 3^{-2}2^{-5} = 288^{-1} < 243^{-1}.$$

Next, if a player living in a location of size 2 wants to migrate to an empty location his or her (expected) payoff is

$$\pi(1, l) = 4^{-5} = 1024^{-1} < 972^{-1}.$$

The conclusion of the above proposition fails. Thus there are admissible values of the associated quotient $\omega$ so these values guarantee that $l \in \mathcal{F}\backslash\mathcal{F}$ is stable. More precisely, it can easily be computed that a loner has no interest to migrate to alocation of size-3 when

$$w \frac{\ln 3}{\ln\left(\frac{\sum_{r=1}^{n} l_r}{\sum_{r=1}^{n} l_r - 1}\right)} \tag{11}$$

At the same time, a player living in a location of size-2 does not want to migrate to an empty location if

$$w \frac{\ln 2}{\ln\left(\frac{\sum_{r=1}^{n} l_r + 1}{\sum_{r=1}^{n} l_r}\right)} \tag{12}$$

Now, to approximate the critical interval with associated $\omega$ values such that these values assure that $l$ is stable, we define $h : [0,1) \to I\!R$ and $g : [0,1] \to I\!R$by

$$h(x) = \ln\left(\frac{1}{1-x}\right) \text{ and } g(x) = \ln(1+x)$$

By letting $x = (\sum_{r=1}^{n} l_r)^{-1}$, the functions $g$ and $h$ can be viewed as extensions of the denominators of the right hand side of (11) and (12) respectively. Taking a first-order Taylor expansion of both functions at $x_0 = 0$, that is for a large value of $n$, yields$h(x) = g(x) \simeq x$. Then, the following holds: Suppose $l \in \mathcal{F}\backslash\mathcal{F}$. For large values of $n$, if

$$\sum_{r=1}^{n} l_r \ln(2)\omega \ln(3) \sum_{r=1}^{n} l_r,$$

then $l$is stable .

## 3 Numerical Examples

In order to highlight the previous results we here include numerical examples. From the following tables it can be seen that the number of stable location profiles increases with $\omega$. The tables also confirm that the new symmetric stable location profilesadded are more and more concentrated.
For a given $z$ define $S_z = \{l \in \mathcal{M}\backslash l^* \,|\, l_z > 0, \ \forall n\}$, and $\mathcal{A}_z = \{l \in \mathcal{D} \,|\, l_z > 0_{z+1} > 0, \ \forall n\}$. In the sequel, "$s$" stands for "stable" and "$u$" for"unstable".
Case 1 : $\beta = -1$, $\alpha = -2$ and $mn$

| | Trivial | Grand | $S_2$ | $S_3$ | $S_4$ | $A_1$ | $A_2$ | $A_3$ | $A_4$ |
|---|---|---|---|---|---|---|---|---|---|
| $n=3$ | $s$ | $u$ | $\emptyset$ | $\emptyset$ | $\emptyset$ | $u$ | $\emptyset$ | $\emptyset$ | $\emptyset$ |
| $n=4$ | $s$ | $u$ | $u$ | $\emptyset$ | $\emptyset$ | $u$ | $\emptyset$ | $\emptyset$ | $\emptyset$ |
| $n=8$ | $s$ | $u$ | $u$ | $\emptyset$ | $u$ | $u$ | $u$ | $\emptyset$ | $\emptyset$ |
| $n=9$ | $s$ | $u$ | $\emptyset$ | $u$ | $\emptyset$ | $u$ | $u$ | $\emptyset$ | $u$ |

Case 2 : $\beta = -2$, $\alpha = -1$ and $mn$

| | Trivial | Grand | $S_2$ | $S_3$ | $S_4$ | $A_1$ | $A_2$ | $A_3$ | $A_4$ |
|---|---|---|---|---|---|---|---|---|---|
| $n=3$ | $u$ | $s$ | $\emptyset$ | $\emptyset$ | $\emptyset$ | $u$ | $\emptyset$ | $\emptyset$ | $\emptyset$ |
| $n=4$ | $s$ | $s$ | $s$ | $\emptyset$ | $\emptyset$ | $u$ | $\emptyset$ | $\emptyset$ | $\emptyset$ |
| $n=8$ | $s$ | $u$ | $u$ | $\emptyset$ | $u$ | $u$ | $u$ | $\emptyset$ | $\emptyset$ |
| $n=9$ | $s$ | $u$ | $\emptyset$ | $u$ | $\emptyset$ | $u$ | $u$ | $\emptyset$ | $u$ |

Case 3 : $\beta = -4$, $\alpha = -1$ and $mn$

| | Trivial | Grand | $S_2$ | $S_3$ | $S_4$ | $A_1$ | $A_2$ | $A_3$ | $A_4$ |
|---|---|---|---|---|---|---|---|---|---|
| $n=3$ | $u$ | $s$ | $\emptyset$ | $\emptyset$ | $\emptyset$ | $u$ | $\emptyset$ | $\emptyset$ | $\emptyset$ |
| $n=4$ | $u$ | $s$ | $s$ | $\emptyset$ | $\emptyset$ | $u$ | $\emptyset$ | $\emptyset$ | $\emptyset$ |
| $n=8$ | $s$ | $s$ | $s$ | $\emptyset$ | $s$ | $u$ | $s$ | $\emptyset$ | $\emptyset$ |
| $n=9$ | $s$ | $s$ | $\emptyset$ | $s$ | $\emptyset$ | $u$ | $u$ | $\emptyset$ | $s$ |

Furthermore, we can verify that, for some values of $n$ and $\omega$, the most and the less concentrated profiles are stable while other profiles are unstable. For example, if $n = 8$ and $\omega = 3.2$, then the grand location and the profile $(0, 4, 0, \ldots, 0)$ arestable while the profile $(0, 0, 0, 2, 0, \ldots, 0)$ is unstable. If $n = 10$ and $\omega = 3.8$, then the grand location and the profile $(0, 5, 0, \ldots, 0)$ are stable while the profile $(0, 0, 0, 0, 2, 0, \ldots, 0)$ is unstable.

## 4 Conclusion

In this paper we propose a two step game of coalition or city formation. In a first step, each player chooses the location in which he wants to be. This generates a location profile which is a partition of the total population into towns. The payoff function is determined in the second step by a game between the different locations. This function reflects two effects. A public effect such that payoffs decrease with the number of non-empty locations. This corresponds to the idea that there are increasing returns to scale in the production of public facilities in cities and because of this individuals are prepared to contribute more to providing them. On the other hand there is a private effect such that payoffs to the inhabitants of a particular location decrease with the size of the population at that location. This effect translates the idea that individuals can benefit from living in a location with few inhabitants and contributing little whilst profiting from the externalities generated by the highly populated locations. We compare different payoff functions with an increasing relative weight of the public effect in order to analyse the consequences for the set of stable profiles. We show that the number of stable profiles increases with the public effect but that the newly added profiles are not always more concentrated and indeed the profiles with intermediate levels of concentration are the last to become stable.

## References

1. M. H. Babiker (2001), "The abatement game: Costs, incentives, and the enforceability of a sub-global coalition", *Journal of Economic Dynamics & Control*,25, pp.1-34.
2. F. Bloch (1997), "Non-cooperative models of coalition formation in games with spillovers",in *New Directions in the Economic Theory of the Environment*,edited by Carlo Carraro and Domenico Siniscalco, pp.311-352.
3. P. Pecorino (1999), "The effect of group size on public good provision in a repeated game setting",*Journal of Public Economics*, 72.
4. R. M. Thrall, W. F. Lucas (1963), "$n$-Person Games in Partition Function Form", *Naval Research Logistics Quarterly*, 10, pp.281-298.

5. S-S. Yi, H. Shin (2000), "Endogeneous formation of research coalitions with spillovers",*International Journal of Industrial Organization*, 18, pp.229-256.

6. S-S. Yi (1997), "Stable Coalition Structures with Externalities", *Games and Economic Behavior*, 20, pp.201-237.

# Part V.

# Interactions and Knowledge

# "Simon says..." What ? Rationality of Imitation in a Simonian Perspective

**Claire Charbit and Valérie Fernandez**

Department of Economics, Management and Social Sciences of ENST, Ecole Nationale Supérieure des Télécommunications de Paris, France.

**Abstract:** In uncertain situations, decision-makers must find ways to simplify, assume and guess. One possible rule to adopt is to imitate some other agent. In this contribution we address the issue of mimetic behaviour as a possible rule for choice and action in uncertain contexts. More precisely, this paper provides an analysis of the rationality of imitative behaviours by using Simonian bounded and heuristic rationality paradigms.

We firstly look back to the concept of procedural, or rather heuristic, rationality in Simon's works in which rationality is faced as cognitive processes. This step will lead us to the conclusion that the simplifying hypothesis of the substantiality of rationality is only relevant when the choice situation is simple. So, another apparatus is needed in situations of great uncertainty and complexity, when the only possible choice is to learn.

Secondly we go deeply into the analysis of Mimetism. After having recalled some necessary conditions to make this choice possible, we show how imitation corresponds to three conceptually different statuses: information, action and means of co-ordination. Then, we present how Mimetism is the output of a deliberation following three logical moments: identification of options, appraisal and choice.

Crossing rationality and imitation analyses leads us, thirdly, to present three types of mimetic rationality, each of them defining one kind of Mimetism as a specific learning process: *positive Mimetism, Mimetism of doubt and Mimetism of uncertainty*. This taxonomy results from two main dimensions of analysis: the place of uncertainty in the process of choice, and the influence of cognitive entities on the agent's decision.

# 1. Introduction [1]

The fulcrum of the research carried out in the field of information interactions between economic agents has been the notion of network, which may be conducive to the adoption of similar behaviours. As far as we are concerned, we approach the problem from the angle of the rationality of an agent with respect to the adoption of a particular behaviour i.e. imitation. Since individuals cannot be dissociated from the networks they belong to (Kirman, 1998) (Steyer, Zimmermann, 1998), both approaches point to the analysis of mimetism as a specific form of information interaction strongly impacting on the collective choices leading to adopt such or such element (ie the economy of standardisation). Those contributions to economics and management come within the wider scope of the analysis of "mimesis" (Ossman, ed., 1998) by attempting to set up links between micro-decisions and collective results. In this article, our purpose is rather to understand why and how economic agents choose to adopt imitative behaviours. Are they rational or not ? Clearly our objective is not to provide a complete theory of mimetism but rather to propose some insights into this field, using Simon's works as a guide.

When the world is so complex that agents cannot perform necessary computations or when they cannot obtain needed information, they adopt decision-making procedures that allow them to decide quickly even in the face of these impediments. Decision makers must find ways to simplify, assume and guess. (Loasby, 1976, p.3). Rules of thumb, routines, or imitation of others are all examples related to bounded rationality. On the face of it, one has the impression of non- or sub-optimal behaviour, as defined by standard economic rationality, but as Lucas states, "In cases of uncertainty, economic reasoning will be of little value" (Lucas,1981, p. 224.).

Mimetic behaviour, in its most general definition, implies that one agent does the same thing as another agent. The notion of "doing the same thing" is deliberately general at this point, and could include charging the same price, producing the same product, using the same technology, the same supplier, the same organizational structure and so on. Each of these is a possible aspect of mimetism, and a way in which an agent could behave similarly to another agent. For us, mimetic choice is procedurally rational and this kind of rationality encompasses simplification processes (which are traditionally seen as a type of substantive rationality).

In this paper we address the issue of mimetic behaviour as a possible rule for choice and action in various contexts. We will firstly look back on the concept of procedural, or rather heuristic, rationality in Simon's works in which rationality is

1 Origins of this article draw upon the work with Robin Cowan during the spring of 1996 at ENST. The flaws that the present reflection might contain, of course are the responsibility of only the present authors.

faced as cognitive processes (Fernandez, 1998). Secondly we will present three seminal triptychs of mimetic choice. A taxonomy of mimetic behaviour follows, according to the kind of rationality which is adopted by the agent and the influence of cognitive entities on his decisions.

# 2 Rationality

When we follow the development of the concept of « rationality » in the works of H. Simon, two essential stages can be identified through the successive usage of two pairs of concepts which emphasize a progressive evolution of Simon's paradigm :

1. The first stage corresponds to H. Simon's pioneer studies in which he propounds a dual conception of rationality that opposes a concept of global rationality to a concept of limited rationality and refers to a taxonomic approach ;
2. His research work conducted since the 70s have led him to « demote » the first pair of concepts to promote a new pair which links a concept of substantive rationality to a concept of procedural rationality both referring to the limited nature of rationality considered as a cognitive process [2].

The problem posed by the latter pair of concepts undoubtedly was in embryo in H. Simon's first works. However the late occurrence in this work of the expressions « substantive rationality » and « procedural rationality », explicitly marks a change of scientific style in 1976.

## 2.1 The pair : Global rationality and bounded rationality

So it was in 1947 that Herbert Simon started his research work into the limits of rationality as he was carrying out a study of the decision making process in an administrative organization. There, he proposes a dual conception of rationality that opposes a concept of perfect or objective rationality – as described by the economic model of choices made in a situation where information is complete and perfect – to the concept of a bounded or subjective rationality which is at work in real life behaviours and which takes into consideration the limited information of the decision-maker.

2 See for instance P. Garrouste who envisages rationality as a « continuum de conceptions de la rationalité limitée » (1999, p. 138).

"A Behavioral Model of Rational Choice" published in 1955, represents a watershed in the development of the concept. After condensing and systematizing the ideas introduced in his first book, H. Simon draws up a hierarchy of models ranging from "global rationality" to "bounded rationality" (1955, pp 112-113) [3]. "Global" rationality is set up as a standard, which enables to appraise the deviation from real life behaviour.

*Models of Man, social and rational* which cam out in 1957 seems to set out the meaning of his project quite clearly (chapter IV "Rationality and Administrative Decision Making"). A world in which omniscient rationality allows for bounded rationality is superseded by a world in which bounded rationality prevails. In that type of rationality, the decision-maker devises his situation of choice. The development of this representation necessarily depends on a subject, his history and desires. The idea according to which the development of the representation of the subject's situation of choice can be described as relative and subjective, is illustrated by the example of innovative decisions within organizations (chapter VII) which includes a plea for a procedural conception of rationality (a conception launched in 1976). We can note that the idea of "procedural rationality" is also present when H. Simon develop the distinction between two types of decision process, called "programmed" and "non-programmed".

Thus, an organization, a decision-maker has a history which constitutes the cognitive capital and which allows them to have decision landmarks already expressed by themselves.

Finally, H. Simon puts forward the " complexity" of the choice situations that decision-makers have to face up to (the origin being the instability of the environment). That question is broached metaphorically (1982, pp. 289-290 : 1959, pp. 225). Indeed, H. Simon considers a bowl with an irregular shape and a system : molasses. The problem consist in predicting the behavior of molasses in the bowl i.e. the shape it will take [4].

The article entitled "Rationality" and published in 1964 in the Dictionary of social sciences gives the following definition of rationality (H. Simon,1982, pp. 405-406):

---

3 "Our whole point is that it is important to make explicit what level we are considering in such a hierarchy of models, and that for many purposes we are interested in models of "limited" rationality rather than models of relatively "global" rationality.

4 If the bowl remains steady, only one hypothesis is needed to predict its behavior : molasses "try to lover their center of gravity. It requires also a complete description of the environment, namely the shape of the bowl. Similarly, the balanced behavior of perfectly adapted organization depends on the goal pursued and its environment: it does not depend on the internal characteristics of the organization". H. Simon then proceeds to analyze the situation when the bowl is shaken hard. "To predict the behavior of an adaptive organization in a complex and fast changing environment, knowing the goal pursued is not enough. It is necessary to know the viscosity of the liquid, the speed at which the molasses fits into the shape of the bowl its goal".

"In a broad sense, rationality denotes a style of behavior (A) that is appropriate to the achievement of given goals, (B) within the limits imposed by given conditions and constraints. In particular contexts, (A) and (B) of the definition may receive more exact specification..."

That is how Herbert Simon begins his article and carries on with the idea that elements (A) and (B) can receive more precise and very diverse specifications. Finally there follows an inventory of the various definitions of the notion of "rationality".

H. Simon emphasizes the existence of two different usages of the word rationality that refer to the two different conceptions of what constitutes the rationality of a behaviour ; one in psychology, the other in economics. In the former (ancient) usage in psychology, rationality refers to processes of choice that resort to the intellective faculty.

H. Simon notices in particular that W. James (1890) calls "rationality", "the peculiar thinking process called reasoning". (p. 406) [5]. And he carries on, saying that the world "rationality" is a source of ambiguity, which explains why recent researches into cognitive psychology prefer using the term "cognitive processes" or "intellective processes". We can note that in his 1976 text where he urges to change scientific style, H.A. Simon deliberately adopted the perspective of rationality as a "cognitive process".

## 2.2 The pair : Substantive rationality and procedural rationality

"From substantive rationality to procedural rationality", written in 1973 and published in 1976 represents an important stage in the works of H. Simon.

As a matter of fact, Simon gives up the dichotomy global rationality / bounded rationality to the benefit of another theoretical pair: he characterizes his approach to decision as "procedural" as opposed to the "substantive" approach of economists. Let us recall the definitions that are deliberately "behavioral" in the wording given by H. Simon :

"Behavior is substantively rational when it is appropriate to the achievement of given goals within the limits imposed by given conditions and constraints. Notice that, by this definition, the rationality of behavior depends upon the actor in only a single respect – his goals. Given these goals, the rational behavior is determined entirely by the characteristics of the environment in which it takes place" (1976, pp. 130-131).

5 Rationality sometimes refers to processes of choice that employ the intellective faculty ; sometimes to the choice themselves. The former emphasis is typical of earlier usage in psychology, logic and ethics ; the latter emphasis predominates in economics and sociology" (Simon, 1982 ; p. 406).

"Behavior is procedurally rational when it is the outcome of appropriate deliberation. Its procedural rationality depends on the process that generated it" (1976, p. 131).

In fact, Simon associates the distinction between substantive and procedural to two kinds of rationality which find expression in two types of different situations. In situations of simple choices, a behavior can be described not only as procedurally rational but also substantively. So, in simple choice situations, we can immediately describe the situation as rational by carrying out a simplification (no mention is made of deliberation). On the other hand, in complex situations, several process of decision or behavior can be envisaged to approximately achieve the desired end ; in that case, there can only be a procedural approach to the rationality of behavior. An optimizing model of decision consisting of simplified representation of reality, gives an optimal decision only in the world of the model (1979, p. 140). It is easier to understand, then, the pragmatism that H. Simon evinced when the participated in the elaboration of optimizing programs within the framework of his work on operational research.

If references made to Simon, it is mainly to the idea of the uncertainty resulting from the complexity of the environment (1976, p. 135), an idea that was put forward as early as 1959 and illustrated by the metaphor of the bowl and molasses. H. Simon contends that the uncertainty that appears in the economic environment, does not exist in the outer environment but only in the mind of the decision-maker (1976, p. 142 ; 1982, p. 437) [6].

"Uncertainty, however exits not in the outside world, but in the eye and mind of the beholder... In such a world, their ignorance of the future prevents them from behaving a substantively rational manner. They can only adopt a rational choice procedure, including a rational procedure for forecasting or otherwise adapting of the future".

One can wonder at the relatively late occurrence in the works of H. Simon to the expression "substantive rationality" and "procedural rationality". In fact as he himself underlines it in his memoirs (1991, p. 165), the distinction between "process" and "results" already ran through his first articles (in embryo in this seminal article on bounded rationality in 1959).

H. Simon went on to point out that the passage from theories of "substantive rationality" to "procedural rationality" implies a fundamental change in "scientific style".

"The shift from theories of substantive rationality to theories of procedural requires a basic shift in scientific style, from an emphasis on deductive reasoning

6 Uncertainty does not exist but inside the mind of the individual. Choices in uncertain situations are thus outside the ambit of a substance model of rationality.

within a tight system to an emphasis on detailed empirical exploration of complex algorithms of thought" (1976, p. 147).

In *Reason in human affairs*, a book that he published in 1983, bounded rationality is shown as "intuitive" and the thesis propounded there is that it would be wrong to think that the procedural aspect of rationality boils down to though processes. The book reasserts the fact that decision making draws upon deliberate or thought processes as well as intuitive and emotional processes which direct attention. The conception of the choice situation as much as the search for alternatives are two partially not considered processes that draw upon intuition [7].

The metaphor of the "walking that makes the path" coined by Antonio Machado [8] is very often used to illustrate the "method" of the procedural rationality.

So the concept of rationality in H. Simon refers to two categories of statements. Some partake of a research into the nature of rationality. In this context, rationality is by nature "procedural". Deliberation which consists in getting acquainted with the elements of choice, collecting the necessary information and appraising the possible alternatives, encompasses therefore all the – reflective or intuitive – processes. J. Conlisk (1996, pp. 676) underlines the fact that "heuristics are rational in the sense they appeal to intuition...". Rationality is both subjective and relative. This concept of rationality as a cognitive process emphazes models of learning problem, problem solving and concept attainment. All models appear to be concerned by procedural or substantive rationality. For example while all learning is in one sense therefore procedural, there are some models of learning that appear to be more concerned with the properties of the end results of learning than with the manner by which learning takes place (Salmon, 1995, p. 242).

If the nature of rationality is one of a kind, there are two possible approaches each with its own particular field of application.

The "procedural" approach deals with a category of problems regarded as complex : "Procedural rationality is interest in situations of formulation of problems..." (Simon, 1976, p. 132)... "ill-structured situations" that place the subject in a position to formulate the problems with a view to gaining a new understanding of the situation.

Y. Giordano (1991, p. 178) underlines the fact that such situations are characterized by three properties : "uncertainty / complexity / learning".

The substantive approach to rationality restricts the rationality of man to what it produces namely a choice regarded as rational or adequate with respect to the

7 pp. 23-29, H. Simon reminds us that in psychology, an intuition refers to the act of recognising a stimulus to which a set of information is associated model of rationality.

8 Antonio Machado (1875-1939), Spanish philosopher and poet. His collection of poems « Campos de Castilla » (1907-1917) was published in two volumes including « Proverbios y cantares » from which the following lines are extracted « (walker, there is no path, you make the path when walking. The walking makes the path ».) sonnet XXIX.

given goal. Only in certain situations will it be possible to grasp the rationality of a behavior while neglecting both the processes that presided over its selection and the subject. The simplifying hypothesis of the substantiality of rationality is possible only when the choice situation is simple.

# 3 Mimetism

In this section, we will show that mimetic choice rests on three, basic triptychs:

- three necessary conditions
- three statuses of mimetic choice
- three moments of mimetic rationality

## 3.1 The three necessary conditions of mimetism

We can identify three factors necessary to make imitative behaviour possible: a relation; some degree of trust; and some identification. The first involves knowledge of one agent by the other, which can in some cases be quite circumscribed. The second is necessary because of the need for the imitator to believe that he is not being led to disaster. The third exists because of the need of the imitator to believe that the action of the imitatee is relevant to the situation of the imitator. Clearly the choice of imitation is reinforced by strong social links as they are inside a community. When someone has to make a decision, without relevant information (for instance such as ones he could have experienced n the past) he spontaneously will try to find those informations in his environment. Using a group reference facilitates individual choices and tends to substitute collective rationality to the individual one. We will deeper analyse this point in the third section.

## 3.2 The three statuses of mimetic choice

### *3.2.1 Mimetism as Information*

Mimetism can be used in the place of information, to extent that selecting the other(s) as a model is a way to get information. This is the case, for example, when an agent uses the behaviour of other agents to infer what information they had about, for example, the quality of a product. Here, mimetic behaviour arises as a response to uncertainty about the quality of information. One can see in this kind of case that mimetic behaviour is self-reinforcing. An agent sees that many have adopted a behaviour and infers that they had information that the behaviour was valid. The agent imitates. This only strengthens the effect for agents who must make the same decision in the future. Mimetic behaviour in this case transmits information (which may of course be erroneous) throughout the economy.

In a somewhat stronger sense, mimetism as a strategy, may take the place of "information-based" strategies. The idea here is two-fold. In the weaker case, agents realize that acquiring and processing basic information is more costly than imitating another agent who has performed those tasks. Here imitation takes the place of information processing, in an optimal way, through decreasing transactions costs. The imitating agent knows that to do as the other does is the optimal strategy. In the stronger case, imitation exists because information is not available. Firms imitate because, as far as they are able to tell, there is no information indicating an optimal choice among the many possible solutions.

As for the status of mimetism as a mode of information, different aspects should be emphasized. Imitating an agent, as we saw it, comes down to, first validating someone else's behaviour as relevant information. It means therefore having access to information (the other can be observed and imitated) and processing that information (the behaviour of the other is adopted for oneself). In a more procedural way, mimetic behaviour itself conveys information and learning to the one who imitates. In that sense therefore, mimetism would primarily be a means of selecting and processing information and afterwards only, a means of really obtaining it.

### *3.2.2 Mimetism as an Action*

Mimetism is a behaviour. It is a means of permitting action, especially when the actor perceives its decisional environment as a very complex one. But, even when there is no uncertainty in such cases where mimetism is clearly the best solution to adopt, this solution means: acting as the "imitated". We will not here analyse in depth all the literature on decision process, and particularly the bridge from decision to action (and the relations between actions today and future decisions). But we just want to underline that it is because mimetism is acting that it infers co-ordinated outcomes. It is because the actions of others are observable that mimetism can take place.

### *3.2.3 Mimetism as a means of co-ordination*

A well-known example of mimetic behaviour is presented in Axelrod's (1984) work on the evolution of co-operation. The "Tit for tat" strategy, employed in the prisoner' dilemma is a mimetic strategy. Tit for tat consists in playing during this period the strategy played by the opponent in the previous period, (imitating the opponent) and was shown by Axelrod to be very effective in a repeated prisoner's dilemma. Uncertainty for the agents arises from the inability to predict the actions of the other agent. The stability of the future, in this case the co-operative convention between agents, is linked to the fact that the agent playing tit for tat imitates the behaviour of the other agent. The strong degree of mimetism serves as a warrant of stability in the relationship between actions and results, and allows agents to converge on a high-payoff outcome.

Here, we can see that mimetism automatically produces one type of co-ordinated outcome, namely one in which agents perform the same actions. When an action exhibits increasing returns across agents, that is when the benefit from it increases with the number of agents doing it, agents act to capture those increasing returns, which is a sort of mimetic behaviour, and produces this sort of co-ordinated outcome. Indeed, this can be a deliberate strategy, as when agents do not know which technology is superior, or perhaps even which will dominate, but do know that it is better not to be a technological orphan (David, 1987).

Mimetism as a mode of co-ordination has extensively been investigated in literature. In fact researchers have mainly focused on the procedures of the emergence of collective choices through the mimetic decisions of agents [9]. Our contribution rather endeavours to understand the stage that comes before this evolution from micro-behaviours to micro-results. Indeed if it is obvious that mimetic choice always produces a co-ordination between agents (at least bilateral between imitator and imitated), what seems interesting to highlight is that effect can be either clearly intended by the agents or imposed by the relation existing between two agents, or completely outside the individual mimetic decision (see III).

## 3.3 Three moments of the rationality of mimetism

As a choice, mimetism is the output of a deliberation. Three logical moments constitutes it [10] :

- Identification of options : the overall information at the disposal of the agent must be conducive to formulating different options. This first moment is the moment of the conception of the choice situation. We shall see that according to the forms of mimetic choice that environment is neither obvious nor immediately understandable but has to be constructed by the decision-maker.

9 Orlean, (1998 a and b) with the "informational influence" ; Bikchandani et alii (1992) with the "informational cascades" ; Young (1996) with the emergence of conventions, etc...

10 In a sequence of "instrumental rationality" in the sense of B. Walliser (1995).

- Appraisal of options : in order to make a choice, the agent must be able to assess each possible option according to his objectives. Once again, this assessment may be questioned or even impossible.
- Choice : Eventually, in a third logical stage, the agent decides to opt for a type of behaviour which for example may be mimetic. We shall see that, for some types of mimetism, this "last" moment is in fact a way to be informed, that is, a way to know all the possible alternatives, which should in fact be part of the first logical stage [11].

These three analytical steps might, in practice, be involved and implying feedbacks. Moreover when an agent belongs to a community, playing the role of a reference basis, those three steps might be reduced just to the last one. The choice is simple, it does not need nor identification of options neither their appraisal. The choices of the other members of the community exempt the agent to thing about his own criteria of decisions.

Concluding this section, we may propose a more "precise" definition of mimetism : one agent does the same thing as another, neither by chance nor by collective decision making process. Imitative behaviour is the result of an individual choice, not of an ex-ante reciprocal agreement.

# 4 Three mimetic rationality

There are different types of mimetic behaviour [12]. Before elaborating a taxonomy, we should point out that there is a really passive way to be mimetic which could be called "coincidental mimetism". We are thinking here of a situation in which agents perform the same action but the reason is roughly that they all have the same optimization problem to solve, and thus all come to the same conclusion regarding which of the possible outcomes is preferred. All perform the same action, but without consideration of what others are doing. (Of course an even looser coincidence exists when agents are faced with different problems all of which happen to have the same solution.) We leave discussion of this situation to one side, as it seems not really like a true mimetic choice on the part of the agents.

As Simon has taught us, the choice procedures of actors always come within a heuristic reasoning. Either the agent's reasoning can be implemented, objectivized and the mimetic choice that will ensue can be described as positive; or the

11 This last process of improving knowledge in order to be able to anticipate is closed to the "knowledge rationality" of Walliser (1995).

12 See for instance the two types of Oltra and Schenk's model (1988) : "pure imitation" versus "conditional imitation", or the three types of Orlean's work (2000).

perception of the environment and the selection criteria will be fuzzy and it will be other forms of mimetism. But a such imitation remains a form of learning. In a context where everything would be known (all the possible options and the criteria to select the optimal option), it would be pointless to imitate someone else.

As far as informational environment is concerned, it could be useful to underline that mimetic behaviour cannot be adopted neither in situations of complete knowledge nor in situations of complete ignorance. In the model of Bikhchandani and alii (1992) "individuals rapidly converge on one action on the basis of some but very little information" (p. 994). In complete knowledge, the agent does not need to imitate anyone. On the other hand, mimetism requires some knowledge: at least the opportunity of someone else behaviour considered as a possible one.

## 4.1 Positive Mimetism

A significant literature now exists on fads and conformity. It addresses the phenomenon in which many agents' actions converge on the same choice, thus creating conformity. The fad aspect is important in that sometimes agents conform to what appears to be a (globally) irrational choice, but the key idea is that "everyone has good reason to conform because conforming is in each person's best interest when everyone else plans to conform." (H.P. Young, 1996 p. 105). Clearly, it is a criteria for substantive choice [13].

Young refers to conventions regarding the side of the road on which people drive, or railway gauges as examples, but this is also a very common feature of models in the technology standardization literature. In order to capture network externalities, an agent is induced to join a large network (Katz and Shapiro, 1985). This implies using the same technology as other agents, thus (apparently) imitating them. A related case, or perhaps simplified version of the same case, exists when a supplier, for example, must use the same technology, or must design his products in certain sizes, to be of use to his demander. Here the issue is not (explicitly) network externalities or increasing returns to adoption, but rather is seen more usefully as an issue of compatibility or co-ordination. In the latter case the number of agents may be small enough so that the mimetic aspect loses prominence, as negotiation becomes possible. Nonetheless, it remains the case that agents do the same thing in order to capture externalities therefrom.

In communities of practice, "The objective is... to maintain and assimilate knowledge about a practice shared by all members" (Cohendet et al., 2000, p. 6). In such cases, the choice of the agent is very substantive. As we had observed in 3.3., the member of this type of community neither needs to know what are the

13 "sanctions on deviants, positive pay-off externalities, conformity preference and communication" are listed by Bikchandani and alii (1992, p. 993 and 1009) as primary mechanisms for uniform social behaviour.

alternative options nor their value. The selection of mimetism [14] is clearly a satisfying option for the agent, because there is no question.

Dumez & Jeunemaitre (1996) observe mimetic behaviour in the cement industy, even though there are no technical exernalities. (One could argue that there exist pecuniary externalities to this behaviour). In this industry all firms tend to follow the strategies of a leading firm, often producing situations in which the market is more than saturated, and thus is not profit maximizing. Typically it is the case when one firm innovates (and becomes the leader for a short time). Other firms, regardless of the strategies they had been considering, imitate that firm. The goal of the followers is to reduce the advantages of the first mover, and to stabilize the oligopolistic structure.

Dumez and Jeunemaitre (1995) cite cases, again drawn from their study of the cement industry, in which firms use the actions of other firms as indicators of the desirability of certain strategies. The first firm to innovate is often taken as having more knowledge or information than the others, which then follow the same strategy.

In the automobile industry, J.M. Pointet has observed the mimetic strategies of car-makers with respect to the innovation introduced by the firm imitated. "L'interdépendance entre acteurs conditionne l'adoption d'un mimétisme de produit, généralement contraint, consistant à reproduire les stratégies de firmes concurrentes (réelles ou potentielles), en assimilant et intégrant dans ses propres produits des caractéristiques de produits concurrents de référence" (J.M. Pointet, 1999, p. 143). Clearly, the mimetic choice here complies with a perfectly identified rule, it nearly is a non-choice.

There are industries in which firms consider it important to be similar to their rivals. This similarity can exist on the input side, through technology or organizational structures; or it can exist on the output side, in that the products firm produce are similar. A very extreme form of following the leader exists in some industries, the fashion industry being one example. Here following the leader is more like the "raison d'être" of the firm. Department stores, for example, make their purchasing decisions based on the actions of haute couture designers. In essence, department stores exist by imitating the actions of other firms.

For all these previous examples, co-ordination is not only a result but the aim of mimetic choice. However very substantive mimetic choices exist which are not adopted to infer specific co-ordinative outcomes.

An extension of this type of mimetism might be seen in actions that are effectively means for reducing costs of search or calculation. The fact that you perform an action indicates to me that your information at the time of action told you that it was the best. I can follow you. Thus mimetic behaviour can be an instance of

14 For instance, in the frequent case of graphists choosing Apple systems.

reducing search costs, or perhaps reducing calculation costs. If our relationship is such that I believe what is good (bad) for you is good for me, then the simplest thing for me to do is to copy you (do other than what you have done). The action is driven simply by the desire to reduce transactions costs. Information about states of the world is not modified, it is simply that the cost of calculating optimal actions is avoided.

Similarly, it is a well-known fact that one of the main arguments to sell solutions offered by consulting firms is the choice that some of their reputed clients have already made [15].

A literature that is centrally, though perhaps not always explicitly, concerned with mimetism is the literature on tacit knowledge. (See Senker, 1995, for example.) The main characteristic of tacit knowledge is the difficulty of codifying it. This implies that it is difficult to learn except by imitating someone who has it. This learning is facilitated by co-operation among agents, but of course there can also be mimetic learning among agents who do not actively co-operate. A standard source of technological innovation is imitation of other firms. In this case it is typically not possible to imitate every aspect of behaviour surrounding the innovation; imitation covers only some, relatively observable, aspects. Imitating the observable aspects of an innovation is not equivalent to innovating, though. After the mimetic behaviour, the imitating firm must generate considerable knowledge, some of it through reverse engineering perhaps, some of it through learning by doing. This learning will generate the knowledge, and in particular the tacit knowledge, necessary to make the innovation effective. (For a model of this process, see for example Cohen and Levinthal, 1989).

Tacit knowledge, as we just have seen, implies imitation. In that sense it is a substantive behaviour to opt for mimetism: there is no alternative effective choice (trial and errors methods take more time) to learn.

According to the last triptych we have presented (3.3), "positive mimetism" takes place when:

- the decisional environment (the alternative options) of the agent is well-known,
- the criteria of choice (valuation of options) are also identified,
- Mimetic behaviour is the result of the selection of the "most" valuable option among the possible ones.

15 A deeper analysis must be conduced about relationship between mimetism and reputation. Boundaries between the two fields are sometimes not really clear, as it is the case for Bikchandani and alii (1992) concerning their "peer influence" analysis (p. 1014).

## 4.2 Mimetism of Doubt

Some mimetic behaviour arises from the way in which agents use the actions of other agents as a source of information (Hirschleifer, 1995 ; Orléan, 1998). If an agent's information about possible actions is incomplete he will use the actions of other agents to infer what information those agents had. Thus actions of other agents can be used to modify his views about the values of his options. Clearly, in such a situation a bandwagon can form, as the common performance of an action appears to indicate that many agents have information indicating that it is a good one. The cumulative effect is that this "public" information can overwhelm any private information, and induce an agent to follow the crowd. Here mimetic behaviour arises due to the incompleteness of agents' information. If agents were perfectly sure about the respective values of the actions, this type of mimetism would not occur. Thus it can be called mimetism of doubt.

Those kinds of phenomena often occur in financial area. For example, when the decision of one bidder attracts the others, even if it collectively conduces to negative pay-off externalities. Phenomena of reputation are often linked to this kind of rationality. We can quote peers influence in the valuation of a job applicant (which can be the result of many previous job offers), of the participation to conferences (following the reference of previous participants) and so on.

Applying our framework, mimetism of doubt means :

- again, the agent knows his decisional environment (alternative options),
- he has an idea of his criteria of choice, but this valuation can be revisited,
- under other agents influences, the agent can question his first assessment and adopts an alternative option through mimetic behaviour.

## 4.3 Mimetism of uncertainty

This last type of mimetism is more like pure mimetism. Here, an agent chooses an action because another agent (or other agents) does so. The reason for acting is the other agent's action rather than what can be inferred from it. In one type of case, this arises because there is effectively no option: the only feasible action is to imitate (even if the agent is able to judge exactly the result of the imitative action). In another case, this type of mimetism arises because of extreme lack of information. Here, the problem is that there seem to be no reasons for one action or another, and the context contains large amounts of perceived uncertainty. Imitation is a way of resolving that uncertainty [16].

16 Bikchandani et alii (1992) speak about "weak positive information" for adopters of this type of behaviour (p. 1011).

This situation can be characterized as the sort of problem that exists when trying to optimize a fitness function in a high dimensional space, since often severe uncertainty arises when agents are forced to act in many dimensions simultaneously. High dimensions are notoriously complex, especially when payoffs depend on the actions of other agents, and have the feature of creating problems that are inherently difficult to solve. One common solution among scientists is to reduce the dimensionality of the space. This can be employed also by actors in the world by making simple decisions which in effect restrain them in particular dimensions. Mimetism in this context can be interpreted as a simple decision which reduces the number of dimensions in which an agent must act, thereby reducing the difficulty of choice and deciding in an uncertain and/or complex world.

This type of mimetism was observed in an inquiry concerning the technological and organizational choice behaviour of large organizations using large data processing systems (Saloner & Steinmueller, 96 ; Charbit & Zimmermann, 99). It is clear that some information systems departments were choosing mimetic solutions when it was necessary to choose new systems. This mimetism was often the case when data processing activities were not considered as the core of the business activity of the firm. So, for this sub-problem, instead of building a completely autonomous choice, a solution was to imitate another organization, often reputed to be either facing the same kind of problems as the imitator (in terms of size, or number of transactions for example) or reputed to have one of the best information system. In some cases even the idea to renew or change the system was a mimetic one. It was "in the air" (in professional meetings for instance) that the old technology was dead and that "everybody was adopting the new one". This last source of mimetic behaviour would be an example of positive mimetism if the action was driven by the desire not to have a non-standard technology and the belief that a new standard was emerging.

Concerning strategies of evolution towards e-commerce, some companies can feel uncertain because alternative options are not clearly identified [17]. It is not possible to appraise the different options. Opting for on e-commerce specific strategy exemplified by another agent, without any knowledge about the real pay-off of this choice, is not only a way to avoid isolation but also a means of learning.

In this type of mimetic choice :

- All the possible choices available to the decision maker are not known.
- The specific criteria for the appraisal of choices are not defined.

17 Among just offering portals of information, already proposing transactions on line, ... or not enter the Internet world at all.

- Adopting someone else's choice as a model of behaviour becomes an obvious solution conveying information about possible options as well as the value of the choices.

# 5 Conclusion

In many new markets, particularly those in which technological standardization is an important issue, mimetic behaviour appears in several ways. Initially there are many possible options, and in the very beginning of the market, agents choose without notions of mimetism because early adopters are typically relatively well-informed, and have strong reasons for adopting (and indeed for the very first adopters mimetism is not an option as there is no one to imitate). Soon, however, as less sophisticated agents enter, mimetic behaviour becomes a way of deciding which of the options to take. Late in the process, as studied in the standardization literature, the variety typically has been reduced to only a small number of options. Thus mimetic behaviour becomes important as often the process leads to the survival of a single option. It is likely to be, though, that the nature of mimetism changes as the process advances. For the later adopters, who generally speaking have good information about the current states of the technologies, mimetism is of positive type. Earlier adopters, who use the behaviour of others as a source of information, or simply as a decision-making device, evince doubt or uncertainty mimetism as described below.

This example shows us that mimetic rationality can take another form with time.

We have attempted here to provide an analytical rather than a concrete typology of mimetic behaviour. Using Simon's works has allowed us to understand that the mimetic choices of agents, which follow different forms of rationality come within a heuristic perspective. In that sense, every mimetic choice is a learning process, whether it comes within a procedure that can be objectivized or whether it is "in the air" or the only means of obtaining information on the environment of the possible options.

# References

Axelrod, R. (1984): The evolution of cooperation, New York : Basic Books.

Bikchandani, S., Hirshleifer, D., Welch, I. (1992): A Theory of Fads, Fashion, Custom, and Cultural Change as Informational Cascades. Journal of Political Economy, vol. 100, n° 5, 992-1026.

Charbit, C., Zimmermann, J.B. (1999): Systèmes d'information et entreprises : convergence ou incertitude ?. Terminal, n°78.

Cohen, W.M., Levinthal, D. (1989): Innovation and Learning : the Two Faces of R&D. Economic Journal, vol 99 : 569-596.

Cohendet, P., Creplet, F., Dupouët, O. (2000): Communities of Practice and Epistemic Communities : A Renewed Approach of Organisational Learning within the Firm. Proceedings WEHIA, Marseille.

Conlisk, J. (1996): Why Bounded Rationality ?. Journal of Economic Literature, vol. XXXIV, June, 669-700.

David, P. A. (1987): "New standards fot the Economics of Standardization in the Information Age. In: P. Dasgupta and P. Stoneman (Eds): Economic Theory and Technology Policy, Cambridge University Press.

Dumez, H., Jeunemaître, A. (1995): Savoirs et décisions : réflexions sur le mimétisme stratégique. In: F. Charue-Duboc (ed), Des savoirs en Action, Contributions de la recherche en gestion, L'Harmattan, coll. Logiques de gestion, Paris.

Dumez, H., Jeunemaître, A. (1996): Information et décision stratégique en situation d'oligopole, l'exemple du secteur cimentier. Revue Economique, n°4, juillet, 995-1012.

Fernandez, V. (1998): Simons says. Forum Internet Cognition et changements organisationnels, novembre.

Garrouste, P. (1999): Apprentissage, interactions et création de connaissance. Revue d'Economie Industrielle, n° 88, 2ème trim, 137-151.

Giordano, Y. (1991): Décisions et organisations : quelles rationalités ? Economies et Sociétés, Sciences de Gestion, n°17, n° 4, 161-194.

Hebb D.O. (1949): The Organisation of Behaviour. Wiley, New York.

Hirschleifer D. (1995): The Blind Leading the Blind: Social Influence, Fads, and Informational Cascades. In: The New Economics of Human Behaviour, Cambridge Press, Cambridge.

Katz M., Shapiro C. (1985): Network Externalities, Competition and Compatibility, American Economic Review, vol. 75 : 424-440.

Kirman, A.P. (1998): Economies with Interacting Agents. In: P. Cohendet et alii (Eds), The Economics of Networks, Interaction and Behaviours. Springer, 17-51.

Le Moigne, J.L. (1995b): Sur la capacité de la raison à discerner - rationalité substantive et rationalité procédurale - d'Aristote à H.A. Simon par R. Descartes et G.B. Vico. In: L. Gérard-Varet et J.C. Passeron (Eds) Le modèle et l'enquête : les usages du principe de la rationalité dans les sciences sociales. EHESS, Paris.

Loasby B. J. (1976): Choice, Complexity and Ignorance : An Enquiry into Economic Theory and Decision-Making, Cambridge University Press.

Lucas R. (1981): Studies in Business Cycle Theory, Cambridge : MIT Press.

Oltra, V., Schenk, E. (1998): Evolution of Cooperation with Local Interactions and Imitation, in Cohendet, Llerena, Stahn and Umbhauer (eds) The Economics of Networks : Interaction and Behaviours, Springer Verlag.

Orlean, A. (2000): Comprendre les foules spéculatives : Mimétismes informationnel, autoréférentiel et normatif in Gravereau J. et Trauman J. (Eds) Crises Financières (to be published)

Orlean, A. (1998b): The Evolution of Imitation. In : Cohendet P. and alii. (Eds): The Economics of Networks, Interaction and Behaviours. Springer, 325-339.

Orlean, A. (1998a): Informational Influences and the Ambivalence of Imitation. In: Lesourne J. & Orléan, A. (Eds) Advances in Self-Organization & Evolutionary Economics, Economica, 39-56.

Ossman S., (Ed), (1998): Mimesis. Imiter, représenter, ciculer. Hermès – Cognition, communication, politique. Paris.

Pointet, J.-M. (1999): La dynamique concurrentielle différentaition/mimétisme dans l'industrie automobile européenne. In: J. Krafft (ed) Le processus de concurrence, Economica, Paris.

Saloner, G. Steinmueller, W.E. (1996): Demand for Computer Products and Services by Large European Organizations. Final Report of a cooperative study, Merit, Maastricht.

Senker, J. (1995): Tacit Knowledge and Models of Innovation, Industrial and Corporate Change, vol 4(2).

Salmon, M. (1995): Bounded Rationality and Learning ; Procedural Learning. In: Learning and Rationality in Economics, Kirman A. and Salmon M. (Eds), 236-275.

Simon, H.A. (1947): Administrative Behavior - A Study of Decision Making Process in Administrative Organization, Macmillan Company ; édition française Administration et processus de décision 1983, Economica, New York.

Simon, H.A. (1955): A Behavioral Model of Rational Choice. Quaterly Journal of Economics, n° 69, 99-118 (repris dans (1982), vol. 2).

Simon, H.A. (1957): Models of Man Social and Rational, Wiley, New York.

Simon, H.A. (1964): Rationality, in J. Gould et W.L. Kolbb (eds.), A dictionary of social sciences, Glencoe, Free Press, pp. 573-574 ; repris dans Simon (1982) vol. 2, 405-407.

Simon, H.A. (1976): From Substantive to Procedural Rationality. In: LATSIS S.J. Method and Appraisal in Economics, Cambridge University Press, London, 129-148 (repris dans (1983), vol. 2, 424-443). Traduction française De la rationalité substantive à la rationalité procédurale, Pistes, 3, 1992, 25-43.

Simon, H.A. (1982): Models of Bounded Rationality, vol. 2 Behavioral Economics and Business Organization, MIT Press, Cambridge.

Simon, H.A. (1983): Reason in Human Affairs, Basil Blackwell, Oxford.

Simon, H.A. (1991): Models of my life, Basic Books, New York.

Steyer, A., Zimmermann, J.B. (1998): On the Frontier : Structural Effects in a Diffusion Model based on Influence Matrixes. In: P. Cohendet et alii (Eds), The Economics of Networks, Interaction and Behaviours. Springer, 223-240.

Walliser, B. (1995): Rationalité instrumentale et rationalité cognitive. In: L. Gérard-Varet et J.C. Passeron (sous la dir.) Le modèle et l'enquête : les usages du principe de la rationalité dans les sciences sociales, EHESS, 309-322, Paris.

Young, H.P. (1996): The Economics of Conventions, Journal of Economic Perspectives, vol.10, n°2, Spring, 105-122

Young, H.P. (1998): Individual Learning and Social Rationality, European Economic Review, 42, 651-663.

# Interacting Individuals and Organizations : a Case Study on Cooperations Between Firms and Research Laboratories

**Michel Grossetti* and Marie-Pierre Bès****

*Cers-CNRS UMR 5117, Université Toulouse Le Mirail, Allée A. Machado, 31 058 Toulouse, 05 61 50 36 69, michel.grossetti@univ-tlse2.fr

** Cers-CNRS UMR 5117, Université Toulouse Le Mirail, Allée A. Machado, 31 058 Toulouse, 05 61 50 36 70, bes@univ-tlse2.fr

## Introduction[1]

The issue of which kind of entities are really interacting is specifically central and difficult in the study of cooperations between research laboratories and firms, because these partnerships involve various kinds of agents.

In France, a contract of cooperation between a research laboratory and a firm is signed both by the firm and by a big research organization like a university or the CNRS (the French National Center for Scientific Research). The contract usually mentions the name of the research laboratory, as is defined by the university or the CNRS, and also the name of the research leader. Regarding the research side, there are three different levels of action : the liable organization ; the research laboratory as an internal organization within the university or the CNRS ; and at last a single researcher. It is also known (Amiot, 1996) that there is a fourth level, which is the small research team (a dozen of researchers) working with the scientific leader. As for the firm, there are also several levels, especially for a group : the group ; a local establishment actually cooperating with the laboratory ; a specific team within this establishment ; and at last, the industrial leader of the cooperation.

Which is the appropriate scale of analysis for the study of this kind of interactions between heterogeneous agents ? Who are the agents really interacting (individuals, small teams, organizations) ? What are the processes through which decision making shifts from a level to another ?

---

[1] Thanks to Jean-Michel Carrère who accepted to correct the english version.

We will address these questions on the basis of the results of an empirical study[2] of cooperations between firms and CNRS laboratories in the field of engineering, conducted in order to understand the genesis of cooperations. To do so, we have reconstructed 130 stories – from the beginning to the end – of cooperations (with a contract at one time, but not necessary during all of the whole story), involving 27 scientific leaders, 19 laboratories and 81 industrial partners (firms or interprofessionnal organizations). These stories were reconstructed by using interviews with the participants, both in the laboratory and in the firm. A minimum of two interviews (and a maximum of five) were used for each story. The stories have various durations (from six months to forty years) and are set in various fields (computing, electronics, fluid mechanics, chemical engineering, etc.) and various institutional contexts (existence of a technical milieu with associations or government organizations). The questions were mainly focused on the genesis of any cooperation and also on the processes of routinization and institutionalization resulting in a shift of the level of action and relationship (shifting from individual level to the organizational one for example). We will discuss the various kinds of meeting using the notions of embeddedness (Polanyi, Granovetter, White), and decoupling (White), and more generally the theory of Harrison White's "Identity and Control" (1992).

We will discuss the following questions on the basis of our empirical study :

What units of interaction are relevant in that kind of cooperation ? We make the general hypothesis that the relevant action units can change in time from the individual level to the organizational level or conversely in becoming personal after a start within an inter-organizational frame. Our second hypothesis on this issue is that the relationship between scientific organizations and firms form an exchange sphere partially embedded in other activity spheres : the higher education system, the labour market and less importantly the non professional activities..

These questions and hypotheses are especially involved in the analysis of spatial proximity effects, that is the fact that they are more likely if the partners are located in the same urban area[3]. The two main explanations for spatial proximity effects are, on the one hand, the constraint of face to face interaction for tacit knowledge exchange (Zucker, Darby and Armstrong, 1994, repeated by Callon et Foray, 1997), and on the other hand, the effect of personal networks in the relations between organizations (Powell et Brantley, 1994 ;

[2] This paper presents the first results of a study made with the support of the CNRS program "Les enjeux économiques l'innovation".

[3] Jaffe, 1989 ; Audretsch et Feldman, 1996 ; Zucker, Darby et Armstrong, 1994 ; Estades, Joly et Mangematin, 1995 ; Grossetti, 1995, etc.

Saxenian, 1994 ; Grossetti, 1995). Here, the choice between analysis units (organizations, individuals, connection between the two) is a crucial point. We make the hypothesis that the proximity effects are explained by the embeddedness of the relationship between organizations within interpersonal, professional or non professional networks.

# 1. Our theoretical frame : Embeddedness and decoupling

It seems to us that the theoretical framework proposed by White (1992), although we adopt only some of his ideas[4], is helpful in clarifying some aspects of the problem of choice on the appropriate action unit in a collaboration context. First, it doesn't postulate as a start the superiority of one particular action unit (individual, firm or state) but considers the problem of constructing identities in interaction confronted with the actions of other identities. In some cases, the germane action units will therefore be the individuals and in other cases, the collective entities. An identity has a story and can grow or dissolve itself in the interaction dynamics. Second, the embeddedness concept developped by White is more complicated than the one that was made famous by Granovetter. For White, the embeddedness is a process with a reverse, the decoupling, that is the interactions abstraction in some particular institutions and in some regulatory ways, which he calls styles. So, the embeddedness and decoupling processes are very similar to the framing/overflowing notions, as defined by M. Callon (1999) after Goffman ; with the framing process corresponding to the decoupling and overflowing to the embeddedness. As Callon suggests, it's possible to link the issue of framing/overflowing or decoupling/embeddedness with the notions of market internalization and externalities, as developed by economists. A synthesis of these correspondences can be shown in the following table :

| Conceptual frame | Autonomy of on exchange sphere | dependence upon other exchange spheres or forms |
|---|---|---|
| White / Granovetter | decoupling | embeddedness |
| Callon | framing | overflowing |
| standard economist Theory | internalization | externalities |

[4] Thanks to Alain Degenne for his help in the comprehension of theorical framework elaborated by Harrison White trough a working paper very clear not yet published.

The analysis framework we adopt is similar to White's and considers two action levels, the organizations level and the individuals level, and three kinds of embeddedness/decoupling process : the first type corresponds to the relationship embeddedness between organizations within inter-individuals networks (example : a relationship between individuals, giving rise to an exchange between their respective organizations). The second type is the embeddedness of inter-individual relationship in collective contexts, that is the fact that personal relationships more often develop within collective frames before becoming autonomous from those frames (for example, two PH students from the same scientific team, have entered an individual relationship in this collective frame, and keep in touch although one has joined a commercial firm and the other has stayed in the laboratory).

The third kind, which results from the first two on a wider level in exchange spheres, is the embeddedness between one sort of exchange (the relationship between public research and firms) within another sort of exchanges (the work market for example).

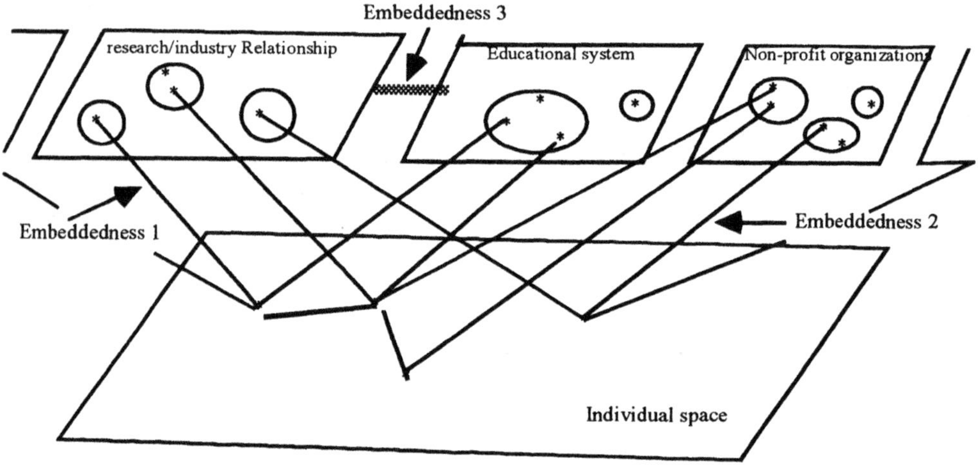


**two spaces, three kinds of embeddedness**

The issue of the autonomy of this particular exchange sphere is an important point of discussion. Some authors defend the thesis of the autonomy of cooperations between firms and research laboratories, due to the specific nature of knowledge in R&D activities. Others defend the embeddedness thesis, in which this sphere partially depends on other spheres (labour market, educational system) or (horresco referens for some) on non-professional social activities. Raising the issue of which of the action units are considered allows to separate between two problems which are often mixed up : the problem of the embeddedness between activities spheres and, the problem of the embeddedness between organization relationships within interindividuals exchange.

## 2. Some Collaborations Stories

We have chosen to study the collaborations between some CNRS research laboratories within the "Engineering Sciences" department, where public research also have to conduct a mission in firms support. We analysed more than a hundred collaborations, which at some time or other were based on a contract.

The method consists first in collecting information from the CNRS data base "protocole"[5] which lists all the contracts signed between the CNRS units and outside partners (about 14 000 contracts signed with firms between 1987 and 1998). We selected some researchers who were scientifically responsible for some of these contracts and we asked them to describe several collaboration experiences in order to collect data on the genesis and on the evolution of the collaboration. In fact, we not only insisted on the relationship genesis between organizations but we also tried to go back in time to the prior individual. Each story collected from one researcher was completed by interviews made with industrial partners or other participants (other researcher, doctorate funded in the frame as part of this collaboration, help given by employees from external organizations, colleagues, and so on).

The basic unit , the history of which we have been trying to reconstruct is not the contract, as it appears in the CNRS information data, but the collaboration, which may go through a succession of contracts as well as non-contractual periods or different institutional forms (a common research laboratory for example). To define the beginning and the end of a story, we used a criteria of relational continuity, that is the fact that the contacts between the teams, in the research laboratory and in the firm, keep on going, even if the people change or if the content evolves. The important fact is the opportunity to observe some intermediary operation between the individuals, involved in both organizations.

We conducted 27 initial interviews with researchers, which generated 130 stories, which themselves were completed with 119 interviews (73 with industrial partners, 46 with other researchers or participants). The researchers initially interviewed belong to teams located in Toulouse (11), Bordeaux (5), Montpellier (3), Clermont-Ferrand (2) and Grenoble (6);

---

[5] Thanks to Joseph Baixeras and to the CNRS relationship firms service for allowing us to access to some data information.

most of the SPI (sciences for Engineering) are represented : electrotechnics (1) electronics (7), automatics and robotics (3), computing (5), chemical engineering (4), fluid mechanics (7).

The 130 relationships involve 81 partners, in which 38 major companies (73 relationships), 35 small firms (38 relationships) and 8 are public organizations of technological research (CNES, CNET, and so on) which we have chosne to include because they played the part of true industrial partners in the stories collected. 36% of these relationships are located in the same region (the research group and the firm are located in the same region), 38% associate a provincial laboratory and a firm located in the Paris region and 26% a research laboratory and a partner from another region or country.

## 3. Which explanation for spatial proximity ?

The two hypotheses that we presented above are very different. In the first one, the proximity effects are active all along the cooperations, while in the second one, they are active only during the genesis of the cooperations. In the first one, the relation can involve different individuals for each organization, in so far as they can have face to face interactions, while in the second one, a relation can disappear if the involved individuals are substituted by other people. In the first one, the relevant unit of action remains the firm (a firm can choose to move to get nearer to a partner in order to facilitate tacit knowledge exchange) while in the second one, there are two kinds of units of action, the organizations and the individuals. In the first one, the relation is decoupled or framed, autonomous from the flow of personal interactions outside the involved organizations. In the second one, on the contrary, the relation can be embedded in this flow, at least at the beginning. In the first one, we can remain within the framework of innovation economy while in the second one, we need to use a sociological approach.

After more than a year or study, we totally dropped the first hypothesis and we have several new reasons to advocate the second one.

### 3.1. The constraint on tacit knowledge exchange doesn't explain proximity effects

Let's recall that the tacit knowledge explanation for proximity effects, which is advocated by several authors, postulates that people from firms must have the opportunity to go to the laboratories in order to get researchers' tacit knowledge, and that is what makes local relations more likely.

We have seen no cases when engineers or firm members had regularly gone to their cooperating laboratory more than the 3 or 4 usual yearly coordination meetings, a rate that practically doesn't vary and that makes no difference between local cooperations and the other ones. When a cooperation needs considerable exchange, which is not always the case, there generally is a PHD student who works at various periods either in the laboratory or the firm. When the partners are far from each other, the technical problems can be solved by specific methods like making a double model of the studied system in the laboratory and in the firm (in electrical engineering for example).

There is another reason to refuse tacit knowledge explanation : in our data we have no correlation between local cooperations and the contents of these cooperations as we coded it[6]. It doesn't mean that there is no tacit dimension in these cooperations but it doesn't explain the proximity effect[7]s.

## 3.2. The ways that cooperations begin

In a study of laboratories from agronomics national institute or research (INRA) using questionnaires, (Estades, Joly et Mangematin, 1996) conclude that there are three logics at the start of cooperations between laboratories and firms : a "proximity" logic, when the local milieu and personal networks are central ; a "market" logic when a firm have defined a precise problem and look for the best partner by formal ways (journals, etc.) ; the "club" logic, when there is another organization that puts the partners together. We will see that local relations don't necessarily come from personal networks and that personal networks can result in non local relations. We will also see that it is not necessarily the firm that takes the initiative of looking for a partner. It can be the laboratory. Last, the third organization can be there without originating the relation. We have made up three categories of cooperation genesis that don't take into account the fact that cooperations are local or not : the network logic ; the institutional logic and the market logic.

### 3.2.1. The "network" logic

As we expected, a great part of the cooperations (48 cases, 44%) have their origin in a chain of personal ties between the two leaders who will later sign the contract. In each case, it is possible to isolate one key personal tie that explains the making of the cooperation. It can be classified in two sub-categories of network logics, the first one that

---

[6] We used 7 categories of contents : measures, characterization ; modeling or improvement of a process or system ; adapting a system to new components, adaptation a system to a new domain of application ; technological watching.

[7] We think about knowledge on the partner behaviour, on the scientific disciplines concerned by the cooperation, on habits to cooperate, and son on.

involves professional relations (former co-workers, former students on the same courses, professor and former student), and the second one, that involves private relations (kin ties, or friend ties). The two logics define two different kinds of embeddedness : in the first case, cooperations between research laboratories and firms are embedded in the teaching activity of universities or engineering schools, while in the second case, it is a much more general embeddedness. The following table shows that in our stories it seems to be the first kind of embeddedness that dominates, particularly the relations that take their origins in the teaching activity and the labor market. However, non professional ties are not rare, and this shows that even research, which is such a specialized activity, can be embedded to some extent in non professional relations.

**Categories of personal ties used in network logics of connexion**

| Key ties categories | Key ties sub-categories | Number | Proportion of network connections | Proportion of connections total (among 110 coded stories ) |
|---|---|---|---|---|
| Non professional ties | Kinship, childhood (5)<br>Non professional organizations, friends (4) | 9 | 19% | 8% |
| Teaching relations | Former students (9)<br>Teacher/ former student(10) | 19 | 40% | 17% |
| Professional relations | Former co-workers in research (12)<br>Former co-workers in firms(8) | 20 | 41% | 19% |
| Total | | 48 | 100% | 44% |

We must insist on the fact that this embeddedness is limited to the genesis of the cooperations. A cooperation which takes its origin in personal networks can move to a more formal way, involving new people and developing in a way that keeps no traces of its originates. In that case, the relation is decoupled from its initial context.

The making of cooperations between organizations on the basis of personal ties is a kind of embeddedness that operates a move in the scale of action since what is happening at the organization level results from of logics from another level of action (individuals and their relations).

### 3.2.2. "Institutional" logics

In some of the stories (20, 18%), the cooperation between a laboratory and a firm results from the participation of the two leaders in a meeting organized by an third organization, from national or local administration. For example, a firm executive and a researcher that can influence their organizations and don't know each other are invited to participate in an expert committee. To participate in such a group (generally from 10 to 30 people) that have to give collective advice and to make collective decisions implies interactions that can result in bilateral cooperations projects.

**Institutional connections**

| Categories of institutions | Sub-categories | Number | Proportion of institutional connections | Proportion of all connections(among 110 coded stories) |
|---|---|---|---|---|
| National level | National administration(CNRS committees, etc.) (11)<br>State technological industry (EDF, etc.) (3) | 14 | 70% | 13 % |
| Regional level | Regional administration of transfer organizations(5)Local establishment of national administration (ANVAR) (1) | 6 | 30% | 5% |
| Total | | 20 | 100% | 18% |

Let's remark that here are only selected the cases when institutional organizations originate the cooperations and not all of the cases when institutions where involved at one time or another to finance or to do something else in the cooperation. For example, in the seventies, “the Direction générale de la recherche scientifique et technique” (DGRST) fund research projects only if they involved an association between a laboratory and a firm. This policy resulted in a number of cooperations.

### 3.2.3. “Market” logic

The third kind of connection (42, 38%) involves neither an external organization nor a chain of personal ties. It can result from the initiative of a researcher or a firm executive that is looking for a partner by using formal means (reading scientific journals, consulting data bases, etc.). When he finds it, he will contact it to propose a meeting or a demonstration. The cooperation can also result from a organized meeting (congress, colloquium). The difference with the institutional connection is the number of people that attend the meeting. While in the institutional connection, the groups are small and people have to work together, in a congress, there are much more people and the future partners have to choose to contact each other. We also count as market logic, the kind of story when the connection results from a student getting a training period in a firm by using formal means.

**"Market" connections**

| Categories of market connections | Number | Proportion of market connections | Proportion of all connections (among 110 coded stories) |
|---|---|---|---|
| Publications, reputation | 22 | 52% | 20% |
| Congress, meetings | 13 | 31% | 12% |
| Training periods market, intermediaries | 7 | 17% | 6% |
| Total | 42 | 100% | 38% |

## 3.3. Proximity effects can be explained by embeddedness or cooperations in personal networks, especially in professional ties

We are now able to table connection logics with the local or non local character of the cooperations.

**Connection logics and partners regions**

| Connection logicsPartners regions | Networks | Institutions | Market | Total |
|---|---|---|---|---|
| Local cooperations(the firm is in the same region than the laboratory) | 24(60%) | 8(20%) | 8(20%) | 40(36%) |
| The firm is in Paris region | 17(41%) | 7(17%) | 17(42%) | 41(37%) |
| The firm is in another region or another country | 7(24%) | 5(17%) | 17(59%) | 29(27%) |
| Total | 48(44%) | 20(18% | 42(38%) | 110100% |

(Khi2 = 11,89, p=0,018)

These results suggest clearly that proximity effects can be explained by the embeddedness in local personal networks. However, we must insist that if the correlation is very significant, one must not make a confusion between local cooperations and networks logic of connection, since this logic applies for a great part to non local cooperations. The relative importance of network logic of connection with Parisian firms can be explained by the importance of the Parisian labor market for engineers and PHD from provincial laboratories, school or universities.

Let's note that connection logics are not correlated with the laboratories regions, neither with the kind of firm (small or big), neither with the technical contents, neither with the length of the cooperations (see Bès & Grossetti, 2000b).

We consider that these data well sustain the first and second hypothesis, i.e. the explanation of proximity effects by spatial structure of personal networks and the relative embeddedness of cooperations between laboratories and firms in the teaching activity and in the labor market. We still have to examine the third hypothesis about the variation of action units.

## 4. The research team level and some decoupling processes

The analysis of the stories that we have collected shows that the most relevant group level is the research team having two to fifteen researchers, organized around a scientific leader, which was already seen by Michel Amiot (1996) or Terry Shinn (1980). On the firm's side, the level is almost the same, a specialized team in a big establishment, the manegement team and some engineers in a small enterprise. One can observe at this level regular exchanges of information and handing over between people in research and development operations. We didn't always observe true teams or working groups in the laboratories and in the enterprises. Sometimes it is only one researcher or engineer that cooperates, sometimes with the help of a student.

If we take into account the two levels of the individual and the team, we obtain four configurations that can exist at one time in a cooperation : 1) one researcher and one member of a firm ; 2) one researcher and an industrial team ; 3) a research team and one executive from a firm ; 4) a research team and an industrial team. In the last three cases, there can be, and generally is, an intermediary person.

We can analytically distinguish four kinds of processes, always linked in real cases, that result in decoupling the relation from the flow of interindividuals interactions. These processes can result in sharing contacts by a team and the setting up a formal relationship.

The first decoupling process, the “collectivization”, is the result of the teams internal organization and of the organizational characteristics of the cooperation project itself. This can lead to a kind of task division, some people becoming specialists in interface activities (contract seeking, participation in scientific associations, etc.). Traditionally, these tasks were assumed by the scientific leader, along with the scientific management. This process, that we can call "collectivization" has its reciprocal, the personalization, when a researcher or an engineer, who has participated in a cooperation in a team, takes it over so that if he would moved to another laboratory, the cooperation would end. Almost half of the 19 teams that we studied have a real collective organization. The other can be described as totally individualized or only partly collective.

The second decoupling process, the “formalization” one, well documented by Cassier (1997), is the legal framework for the contract that defines resource sharing and what everyone is supposed to do. The contract contributes framing individual action and collective action (from the team, the laboratory, the CNRS, the firm, the industrial group). A great part of the contract negotiation is dedicated to the making up internal agreements (with the legal department or the finance department for example) on the basis of the first agreement between the two cooperating teams.

The third process, that we can call "materialization", is the action of all the technical devices, materials and equipment that allow common work : modeling, making of specific software, instrumentation, specific data, materials, intermediary reports. These material intermediates allow the outmatching of permanent interaction between participants and are necessary to task division. It is an important dimension of scientific and technical work, well documented in the works of Callon (1989), Latour (1994) and other researchers following this line of analysis.

These three processes, that we have found in our stories, are well analyzed by the sociology of organizations or the innovation studies. We would now like to insist on a fourth process that we found is absolutely central in the cooperation between laboratories and enterprises and which is much less often mentioned in the studies of these cooperations. This process, that we call "personification", is the involvement in the cooperations of specific people that embody for a time the relationship between a laboratory and a firm : students doing training periods in the firm, PHD students working on common topics with a grant from the firm, researchers being transferred to a firm or on the contrary, firm members placed for a time in the laboratory. These people belong for a time to both the organizations. One could say that they belong to the cooperation organization and to the cooperation project. In more than 9 cases on 10, the cooperation is based on the work of these intermediate people.

Like material intermediates, human intermediates contribute to decouple the relations from the involved organizations. At the same time, in so far as students tend to be recruited by the cooperating firms and as people on transfer maintain relationships with their former co-workers, this fourth process also results in an embeddedness in personal networks and can contribute to taking down the studied decouplings made by the three other processes.

## Conclusion

These results can reinforce the theoretical framework that we have presented at the beginning of this paper. The interest of this framework is to define several processes of embeddedness processes involving different units of action. The proximity effects explanation involves the embeddedness of organizations relations in personal networks, the embeddedness of personal networks in different spheres of activity and, as a result of these two processes, the embeddedness of one particular sphere of activity in other spheres

(here particularly the embeddedness of cooperations between firms and laboratories in the teaching activities and the labor market)

The analysis of our stories shows the relevance of two levels of action for the understanding of cooperations between laboratories and firms : the individuals and the small teams, with frequent shift of initiative from one to another. In that particular case, it is difficult to make models with only one level of action without studying the specific effects of the other level.

## References

Amiot, M. (1996): Le laboratoire entre équipes et réseaux. poids des techniques et conflits de légitimités. Archives Européennes de Sociologie 37:2, 271-319

Audretsch, D.B., Feldman, M.P. (1996): R&D Spillovers and fhe Geography of Innovation and production. The American Economic Review 86:3, Juin, 630-640

Bès, M.-P., Grossetti, M. (2000): Collaboration trajectories between CNRS research laboratories In Engineering Sciences and Firms : genesis and evolution, European Association For Evolutionary Political Economy Conference, November 2-5, Berlin

Callon, M. (1998): Introduction : the Embeddedness of economic Markets in Economics. In : Callon, M. (Ed.): The Laws Of The Markets. Blackwell Publishers/The Sociological Review. Oxford, U.K, 1-57

Callon, M. (1999): Une contribution de la sociologie à l'analyse des externalités. essai sur la notion de cadrage/débordement., In: Foray, D., Mairesse, J. (Eds.): Innovations et Performance. Editions EHESS, Paris, 399-432

Cassier, M. (1997): Compromis institutionnels et hybridations entre recherche publique et recherche privée. Revue d'économie Industrielle 79, 191-212

Estades, J., Joly, P-B., Mangematin, V. (1996): Dynamique des relations industrielles dans les laboratoires d'un grand organisme public de recherche : coordination, apprentissage, réputation et confiance. Sociologie du Travail 3, 391-408

Feldman, M. (1994): The Geography of Innovation. Kluwer, Dordrecht

Granovetter, M. (1985): Economic action and social structure : the problem of embeddedness. American Journal of Sociology, 91, 481-510

Granovetter, M., Swedberg, R. (Eds) (1991): The Sociology of Economic Life. Westview Press, Boulder, San Francisco, Oxford

Grossetti, M. (1995): Science, Industrie et Territoire. Presses Universitaires du Mirail, Coll. Socio-Logiques, Toulouse

Grossetti, M. (1998): La proximité en Sociologie : une réflexion à partir des systèmes locaux d'innovation. In: Bellet, M., Kirat T., Largeron C. (Eds.): Approches mutiformes de la proximité. Coll. Interdisciplinarité et nouveaux outils, Hermès, Paris, 83-101

Grossetti, M. (2001): Les effets de proximité spatiale dans les relations entre organisations : une question d'encastrements, Espaces Et Sociétés 101-102 (forthcoming)

Latour, B. (1994): « Une Sociologie sans objet ? Remarques sur l'interobjectivité ». Sociologie du Travail 4, 587-607

Powell, W.W., Brantley, P. (1992): Competitive cooperation in Biotechnology : Learning through Networks ?. In: Nohria, N., Eccles, R.G. (Eds): Networks and Organizations : Structure, Form and Action. Boston, Harvard Business School Press, 366-394

Powell, W.W. (1994): Interorganizational Relations. In: International Encyclopedia of Business and Management. Routledge, Londres.

Saxenian, A. (1994): Regional Advantage. Harvard University Press.

Shinn, T. (1980): Division du savoir et spécificité organisationnelle. Revue française de Sociologie, 21:1, 3-35.

White, H.C. (1992): Identity and Control. A Structural Theory of Action. Princeton University Press, Princeton.

Zucker, L.G., Darby M.R., Armstrong, J. (1994): Intellectual Capital and the Firm : the Technology of Geographically Localized Knowledge Spillovers. *NBER Working Paper Series,* Working Paper 4946, NBER, Cambridge, Mass.

# Organisational Innovation, Communities of Practice and Epistemic Communities: the Case of Linux [1]

**Patrick Cohendet, Frederic Creplet, Olivier Dupouët**

BETA, UMR CNRS 7522, Université Louis Pasteur, 61 Avenue de la Forêt Noire 67085 Strasbourg - tel: 33.3.90.41.41.74/ fax: 33.3.90.41.40.50

cohendet@cournot.u-strasbg.fr ; creplet@cournot.u-strasbg.fr ; dupouet@cournot.u-strasbg.fr

## Introduction

A growing number of works in the literature consider the process of production and circulation of knowledge within the firm as is the key determinant of the capability of the organisation to innovate (Nonaka and Takeuchi, 1995; Leonard-Barton, 1995; von Krogh, Roos and Kleine, 1998, etc...). It is widely agreed that the "cognitive architecture" of knowledge within the firm (the way knowledge is produced, stored, exchanged, transmitted, retrieved) strongly influences the process of organisational learning, and in turn the innovative process. As Nonaka and Takeuchi (1995) emphasised, the process of creation of knowledge within firms relies on two main dimensions; the first one is the "epistemological dimension": the critical assumption is that human knowledge is created and expanded through social interaction between tacit knowledge and explicit knowledge ("knowledge conversion"); the second one is the "ontological dimension" which is concerned with the levels of knowledge creating entities (individual, group, organisational and inter-organisational). In line with the assumptions initially made by Argyris and Schön (1978), the organisation supports creative individuals or provides contexts for them to create knowledge. "*Organizational knowledge creation, therefore, should be understood as a process that organizationally amplifies the knowledge created by individuals and crystallizes it as a part of the knowledge network of the organization. This process takes place within an "expanding community of*

[1] This work benefited from the financial support of the TIPIK European programme on codification of knowledge. The authors thank all the researchers taking part in this programme for their remarks and suggestions and are obviously the only liable for mistakes or omissions.

*interaction", which crosses intra and inter-organizational levels and boundaries*" (Nonaka and Takeuchi, 1995).

This contribution will explore the ontological dimension of the process of creation of organisational knowledge, by focusing on a specific aspect of it: the role played by "communities". Our view is the following: If we fully acknowledge that the organisational creation of knowledge results from a permanent interaction between the individual and organisational levels, we consider that these interactions are shaped and determined at an intermediate level, the level of communities. We argue that the firm is composed of a myriad of overlapping "communities" (functional work groups, project teams, networks, communities of practice, epistemic communities, etc..), each of which presents a dominant mode of learning and collective behaviour. The organisational learning is thus viewed as a complex process of interaction between heterogeneous communities rather than as a direct interplay between an organisational knowledge structure and heterogeneous individual agents. As Brown and Duguid (1991) quoted, "*within an organisation perceived as a collective of communities, not simply of individuals, in which enacting experiments are legitimate, separate community perspectives can be amplified by inter-changes among communities. Out of this friction of competing ideas can come the sort of improvisational sparks necessary for igniting organisational innovation. Thus large organisations, reflectively structured, are perhaps well positioned to be highly innovative and to deal with discontinuities. If their internal communities have a reasonable degree of autonomy and independence from the dominant worldview, large organisations might actually accelerate innovation*".

Having underlined the role played by heterogeneous communities in organisational learning, we will emphasise the growing importance of two specific communities in the process of creation of knowledge within firms: communities of practice and epistemic communities. These communities are the ones that trigger the process of knowledge creation, but they differ in the ways they are organised to produce knowledge. On the one hand, epistemic communities are structured in order to deliberately produce new knowledge. In doing so, it will be shown that they are characterised by the building of a procedural authority. On the other hand, communities of practice develop new knowledge through actions "*and co-ordinate activities of individuals and groups in doing their real work as it is informed a particular organisational or group context*" (Cook and Brown, 1999). These communities are not structured to produce new knowledge deliberately. For them, the creation of organisational knowledge should be considered as a by-product.

Of course, there are many other types of communities that participate in the production and circulation of knowledge within the firm. For instance project teams have always been considered as a key organisational device to bring together different forms of heterogeneous functional knowledge held by agents in order to develop new knowledge. However in an emerging knowledge based industrial context, our fundamental assumption is that the nature and the quality of the process of creation of knowledge essentially depend on the mutual interactions between communities of practice and epistemic communities, and on the ways they interact with other types of communities.

As a practical example, this contribution will analyse the case of Linux, the operating system for computers. To a large extent, the development of Linux could be interpreted as a process of creation of knowledge that relied on the functioning and interplay of communities. More precisely, we suggest to interpret the unfolding of events as a shift from one type of particular social setting (community of practice) to another one (epistemic community). Since we contend that learning processes are intertwined with social structures, this shift entails a symmetric shift in the learning modes at stake. We then address the issue of a possible generalisation of the lessons learned from this story.

## 2. A typology of communities within the firm: the definition of the concepts of epistemic communities and communities of practice

The need to think in terms of communities when the creation of organisational knowledge is at stake comes from a pitfall in the classical Argyris and Schön's vision of organisational learning. In the face of the heterogeneity of the individuals in the firm and of their associated cognitive maps (Huff, 1990) it seems difficult to use Argyris and Schön's model to precisely account for the interactions existing among them. Indeed, this model overlooks the problems associated with social and cultural idiosyncrasies that act as barriers to knowledge exchanges. This approach intends to consider the organisation at a global level whereas the firm is made of numerous different knowledge spaces too heterogeneous to be dealt with in a uniform way (Nonaka and Konno, 1998; Brown and Duguid, 1991; Blackler and McDonald, 2000). As a consequence, the possibility of a shift from individual learning to organisational learning is sometimes difficult to grasp. Such a bridge between these two types of learning needs to be clarified by the definition of the mechanisms related to it. This presupposes that the bridge we mention cannot occur in a regular or systematic manner. The mechanisms at work in the linking of organisational and individual levels are mainly made of communication, transparency and integration as well as cross-fertilisation, interactions of several types, sharing of experiences, problem solving, etc.

The reason why Argyris and Schön's vision is somewhat incomplete is that it lacks an intermediate level between the individual learning and the organisational learning. We consider this level to be the level of communities, where in a specific group context, the elementary process of collective generation of knowledge may occur. According to us a firm, and more generally any kind of social organisation, is made of several types of overlapping communities. Each community is characterised by specific traits that shape the ways this community learn and participate in the development and expansion of organisational knowledge: the objective they pursue, the agents composing them, their dominant cognitive activity, their recruitment rule, their way of producing knowledge, their dominant learning mode and what holds them together. The ways, for instance, routines are experienced, explored, replicated within the organisation depends on the nature of the community which "operates" the routine.

When considering the communities, it appears that some of them are rather knowledge creation oriented and some other are action oriented; some are defined and controlled by specific hierarchical mechanisms, others are more autonomous. To clarify the distinctions between communities, we present in a first part, the best identified forms of communities existing within a firm, namely functional groups, teams and networks. In a second part, two newer concepts are presented, epistemic communities and communities of practice, and their differences with more traditional collective entities are displayed. These two concepts are then used to give an account of the Linux story from a cognitive standpoint.

## 2.1 "Classical" Communities in firms

*Functional groups* are traditional communities characterised by homogeneous agents sharing a disciplinary specialisation (such as marketing, finance, accounting, etc…). These communities are the basis of the division of work. They play the key role in the functional structures of the firm as described by Chandler (1977). They are also present both in divisional structures and matrix structures of the firm. In this type of community, the production of knowledge within the firm is unintended. The original knowledge is defined in the codebooks of the respective disciplines, and agents communicate with one another with codes and local jargons developed in their own discipline. The dominant learning mode is learning by doing, and the recruitment procedure is based on the recognition of the mastering of the discipline (diploma) by the hierarchy of the firm.

*Teams* are communities made of heterogeneous agents having particular disciplinary knowledge that they have to integrate with the functional knowledge of others. These agents interact through collective routines they set in order to function together as a unit, without any negotiation. They have to fulfil a given task and generally are not deliberately constituted to produce new knowledge. The members are recruited by the team leader and are held together by a common goal and job requirements. The dominant mode of learning there is learning by interacting (Lundvall, 1992; Von Hippel, 1988). One important particular case is the project team (Giard and Midler, 1993; Midler, 1996). This latter is set to fulfil one-shot goals and the decisions made during the process are irreversible. In addition, such a team faces great uncertainty at each stage toward the objective. These particular teams meet the characteristics stated above, but, in addition, they have to reach their goals within a limited period of time.

*Network* is the last classical form of structure binding individuals together. The objective of a network is to allow a mutually negotiated specialisation. It is made of heterogeneous agents whose cognitive activity is to exchange knowledge. They interact through informal and formal meetings and the recruitment rule is mutual trust. People are held together in networks by the need to gain complementary knowledge. The main mode of learning is learning by exchanging and the knowledge is produced either on purpose or not, depending on the kind of network considered.

## 2.2 Epistemic Communities and Communities of Practice

Epistemic communities and communities of practice are the most relevant types of groups for the purpose of this paper, since they are the place where the most dramatic knowledge creation is likely to occur. The key point, within the scope of this paper is that epistemic communities are truly oriented toward new knowledge creation, whereas communities of practice are oriented toward the achievement of an activity. In this latter case, knowledge creation is an unintended spill-over.

### Epistemic Communities

*Epistemic communities* can be defined as small groups of "*agents working on a commonly acknowledged subset of knowledge issues and who at the very least accept a commonly understood procedural authority as essential to the success of their knowledge activities*" (Cowan *et al.*, 1998). Epistemic communities can thus be defined as a group of agents sharing a common goal of knowledge creation and a common framework allowing the shared understanding of this trend. The goal of epistemic communities is thus simultaneously outside and above the community's members. The concept of "epistemic communities" was developed in particular in the realm of international relations (Haas (1992); Adler and Haas (1992))[2]. Using this concept to address the issue of codification of knowledge, Cowan *et al.* (1998) suggest that any codification activity implies the existence of codes that are understandable by the communicating actors. Following this approach, an epistemic community may then be concretely defined as a framework within which codification can occur.

What defines a community is thus the existence of a procedural authority that can be explicit or not. However, it must be different from the kind of authority held by a "guru" to ensure a certain autonomy of the members. Moreover, the procedural authority conveys the idea of progress toward the cognitive goal set by the community. The belonging of members will thus be evaluated with respect to this procedural authority. It should be noted that this procedural authority can *a priori* emerge from the interactions among members. In that case, the organisational closure is either realised, or imposed from the outside and then not realised. In the former case, the epistemic community is self-organised and then close in this respect to a community of practice. This remark is important since it shows evidence of the possibility for one form of community to evolve into the other.

Epistemic communities are structured around a goal to be reached and a procedural authority endowed by themselves (or with which they were endowed) to fulfil that goal. Notions of autonomy and identity are thus weaker than in the case of communities of practice (see below), thus favouring the group's creativity (Kao, 1998; Leonard-Barton, 1995). Thus, the community increases its ability to seize future opportunities. This form of organisation spawns knowledge creation by favouring the synergy of individual varieties. We find here the principle of "required variety" stated by Ashby (1956). Individuals accumulate knowledge according to their own experiences. The quality of this knowledge depends on two factors. The first is the variety of individual experiences in interaction. The second factor is the "knowledge of the experience". This is consistent with the idea of a rational ability of experience feedback within which the validation is made according to the procedural authority: what is evaluated is the contribution of the agent to the cognitive goal with regard to the criteria set by the procedural authority.

---

[2] Related concepts are also to be found in sociology of science. In this domain, one may mention Barber (1952) who asserts that scientists tend to create self-regulated communities and Knorr-Cetina (1981) who developed the concept of scientific communities. One may also quote Beyssade (1998) who studies linguistics and, in this field, uses the notion of epistemic community and stresses the importance of a common language as a cement for such communities.

Because of the heterogeneity of the agents, the objective of knowledge creation for the sake of knowledge, the first task of epistemic communities is to create a "*codebook*". Hence, knowledge circulating within epistemic communities is explicit (but not codified since it remains mainly internal to the community (Baumard, 1999).Because of the lack of deeply shared values, it appears that the knowledge creation mode is much like a form of externalisation (conversion of tacit into explicit knowledge (Nonaka and Takeuchi, 1995).

Validation of the cognitive activity of an agent is made with respect to the procedural authority. What is evaluated is the contribution to the endeavour toward the goal to be reached, according to the criteria set within the procedural authority. Within an epistemic community, agents are bound together by their commitment to enhance a particular set of knowledge. The recruitment rule is thus defined with regard to the contribution an agent makes to fulfil this goal (this goal is likely to be partly given and partly emergent (Blackler and McDonald, 2000)[3]).

**Communities of Practice**

The concept of *communities of practice* was introduced by Lave and Wenger (1990) who, by focusing on individuals' practices, identified groups of persons engaged in the same practice, communicating regularly with one another about their activities. Members of a community of practice essentially seek to develop their competencies in the practice considered. Communities of practice can then be seen as a means to enhance individual competencies, they are oriented toward their members (Lave and Wenger, 1990; Brown and Duguid, 1991). This goal is reached through the construction, the exchange and the sharing of a common repertoire of resources (Wenger, 1998).

Wenger (1998) and Brown and Duguid (1991; 1998) state that self- organisation is an essential characteristic of communities of practice. According to Lesourne (1991), self-organisation is the ability of a system to acquire new properties by organising itself or by modifying by itself its own organisation (Lesourne, 1991). Self-organisation confers to the system an adaptive ability to evolve without any constraint of authority nor any determinism. The system is then autonomous and sets a boundary with respect to the other functions of the firm. It creates a sort of "organisational closure" in the terminology of the theory of self-organisation. This

---

[3] Epistemic communities emerge in uncertain context calling for the creation of a new paradigm (which is not the case for communities of practice) (Haas, 1992; Whiteneck, 1996). We are then close to the community of young researchers overcoming the old paradigm in Kuhn's theory (1962).

idea is important since it underlines the cross functional nature of communities of practice within the firm.

More precisely, autonomy and identity of communities, the key characteristics of self-organisation allow the collective acquisition and processing of stimuli from the environment (Wenger, 1998; Dibiaggio, 1998). Identity and autonomy are essential for the agent to define him-/herself with respect to his/her environment and for the members of the community to behave collectively.

The self-consciousness is also visible in the mutual commitment of the community. It is built around activities commonly understood and continually renegotiated by its members. A community's member feeds it with his/her experience and, in turn, relies on the knowledge capitalised by the community to carry out his/her activity. These processes take the shape of "war stories" (Brown and Duguid, 1998) that members tell when they gather. They thus develop a jargon understandable by the members only. It is thus a mutual commitment that binds agents in a social entity, ensure cohesion of the community and recruitment of new members.

Lave and Wenger (1991) interpret the practice of these communities as the vector of learning, that is in turn the building of an individual entity. Hence, the evaluation of an individual is made by the community of practice as a system and is focused both on the values adopted by the individual and on the progress made in his/her practice, the two being co-constitutive.

Within communities of practice, the privileged knowledge is thus essentially the know-how (Brown and Duguid, 1991), which is tacit and socially localised. The nature of knowledge is due to the objective and the structure of the communities of practice. As a result, the community tends to send no messages toward the outer world. Messages are almost exclusively exchanged among the members of such a community.

## 2.3 A typology of communities within the firm

We sum up what has been said above in the table 1 below inspired by Snyder and Wenger's (2000) one.

**Table 1 : A typology of communities within the firm**

| | Objective | Agents | Cognitive Activity | Recruitment rule | Knowledge Production and Dominant Learning Mode | What holds the Community together |
|---|---|---|---|---|---|---|
| Functional Group | Ensure a given function | Homogeneous | Disciplinary Specialisation | Hierarchical | Unintended Learning by doing | Education Task definition |
| Team | Achieve a given task | Heterogeneous | Integration of functional knowledge | Team's leader | Unintended Learning by interacting | Job requirements Common goals |
| Network | Mutually negotiated specialisation | Heterogeneous | Knowledge exchange | Mutual trust | Intended or unintended depending on the network Learning by exchanging | Need to access complementary knowledge |
| Community of Practice | Increase the skills in a given practice | Homogeneous | Accumulate knowledge about a given practice Circulation of best practices | Members who select themselves | Unintended Learning in working | Common passion for the practice |
| Epistemic Community | Produce knowledge | Heterogeneous | Construction of knowledge or language or messages Circulation of codified knowledge | By peers | Intended Searching | Respect of a procedural authority |

# 3. Interaction between Heterogeneous Communities

As mentioned in the introduction, one of the major hypotheses of this article is that the understanding of the process of organisational learning cannot be disconnected from the cognitive architecture of the firm, in particular the way knowledge is produced and distributed within the firm. When considering the cognitive architecture of the firm, we argue that the evolution of communication and information technologies and the tendency towards a knowledge-based economy lead to distinguish two historical modes of cognitive architecture within firms, that is two successive modes where organisational learning has to be interpreted differently:

The first mode is a traditional vision of separation between the communities in charge of a deliberate production of knowledge and the communities in charge of using knowledge for the regular production activities. In this first mode the core of the formation of organisational learning resides in the interaction between the two "hierarchical communities", the functional work groups and the multidisciplinary teams.

The second mode relies on the hypothesis that there are less and less separation between the communities in charge of producing new knowledge and the communities in charge of using and transferring knowledge. In this second mode we will argue that the core of the formation of organisational learning resides in the interaction between the two "non-hierarchical" communities, communities of practice and epistemic communities.

## 3.1 Mode 1: The Focus on the Interactions between Functional Groups and Multidisciplinary Teams

Most of the classical works on organisational learning (Argyris and Schön, 1978; Midler 1996) refer to a specific context, that we define by "mode 1". In this context, the focus is on the interaction between the two "hierarchical communities" (functional groups and multidisciplinary teams). The main problem at stake is the coupling between on the one hand communities specialised in a given field of knowledge, and on the other hand communities in charge of integrating different bodies of knowledge to accomplish specific tasks. For instance, intense discussions in the literature on matrix forms of organisation, and on the respective Japanese and western models of organisation referred to such a situation. « *The British approach is based on the principle of individual task and functional specialisation, and the product development cycle is managed on a sequential basis. The job boundary of the British engineers are much more narrowly defined within their specialist arenas and their role in cross-functional co-ordination is limited. The Japanese approach is based on the principle of undifferentiated job demarcation and decentralised horizontal co-ordination. The product development process is overlapping. Japanese engineers are expected to engage in direct cross-functional liaisons and in scanning market information. Within this model, the co-ordination function becomes partly embedded in the role of engineers. In the Japanese firms, the ability to co-ordinate, to process and to share information is regarded as an important part of an engineer's skills and role competence.* » (Lam, 1999). In such a perspective, the problem of the diffusion of knowledge is not only the integration of different bodies of specialised knowledge by the multi-disciplinary teams, it is also the problem of "restitution" of the knowledge acquired by members of multi-disciplinary teams when they come back to the functional groups to which they belong.

In mode 1 the role allocated to the other communities is in general a marginal one. Communities in charge of production of new knowledge (epistemic communities) are disconnected from the activities in charge of using knowledge. In the traditional vision of the firm the process of transformation of knowledge is viewed as evolving from separate departments in charge of producing new (deliberate) knowledge or handling and distributing information to the other departments that assimilate and use this new knowledge to improve their current activities. These latter departments could to some extent produce some new knowledge from their routine activities, but this was a non-deliberate form of production of knowledge that emerged as a by-product of production through learning by using or learning by doing.

The communities of practice naturally exist in the mode 1. Agents who share the same common interest for a given practice could be found within teams or functional groups for instance. However the circulation of best practices within a given community of practice remains essentially local. The absence of powerful links of communication between agents limit the circulation, exchange and interpretation of best practices to some confined areas of the firm.

## 3.2 Mode 2: The Focus on the Interaction Between epistemic communities and communities of practice

We argue that the production system is progressively tending towards a mode 2, where the differentiation between deliberate and non-deliberate forms of knowledge production is becoming strongly blurred. In a knowledge-based context the essence of the coherence of the firm precisely relies on the ways the communities in charge of the production of knowledge and the communities in charge of using knowledge are deliberately interacting and organising simultaneously the production and circulation of knowledge

As Lundvall (2000) pointed out, the emergence of new forms of learning such as « experimental learning » makes the differentiation between « on-line » and « off-line » learning activities less and less relevant. This implies a complete reconsideration of the separation between epistemic communities and communities of practice. Experimental learning takes place « on-line », that is to say during the process of producing a good, and consists in deliberately experimenting during the production process.[4] The ability to proceed to on-line learning affects all the types

[4] "By doing so, one creates new options and variety. This form of learning is based on a strategy whereby experimentation allows for collecting data, on the basis of which the best strategy for future activities is chosen. For example, a professor can undertake pedagogical experiments; the craftman can seek new solutions to a problem even during the fabrication

of communities that where not focused on the deliberate production of knowledge. However this ability, coupled with the ability to use intense forms of communication, will contribute to enhancing the specific role of communities of practice, as triggers of organisational changes within the firm.

The management of the collectively distributed knowledge within the organisation that brings together epistemic communities and communities of practice is thus one of the cornerstones of the coherence of the firm in a knowledge-based context. The development of different modes of interaction between the two types of communities (that rely in particular on specific processes of codification of knowledge) becomes critical. They determine the problem of knowledge fit or integration across boundaries, by stressing the powers of longer term "languaging" devices. Thus, in the area of reconciling external and internal knowledge, they stress, for example, the role of informal networks (e.g. membership of associations, clubs) and "boundary spanning" individuals (e.g. brokers and intermediaries, employee exchanges between firms). Along similar lines, Nooteboom (1999) emphasises the role of third party "go-betweens" as vital brokers of innovation who help to sediment trust, maintain unique secrets, resolve conflicts, reveal mutual advantages and introduce innovation without destabilising established competencies within each firm. Sustained over time, these devices are able to help firms maintain a balance between development and replication of knowledge.

The language of replication, selection, imitation and variety helps in making a case for learning trajectories and in explaining the fine balance between path-dependency and renewal in organisational evolution. However, it does little in terms of revealing the fine grained processes through which learning occurs, in general and along particular trajectories, as a grounded and distributed process (Gibbons *et al.*, 1994). Equally, it has a tendency to under-emphasise learning as a process that involves trial and many errors, chance discoveries, mistakes, and open-ended experimentation. Ironically, it fails to take seriously the argument that firms are "experimental learning machines" (Eliasson, 1994) which operate in uncertain circumstances because of both their bounded knowledge and the "*creative destruction processes of the experimentally organised economy*" (p. 184). Thus the

---

process. The possibility of moving this type of learning in many activities represents an important transition in the historical emergence of the knowledge-based economy. In effect, as long as an activity remains fundamentally based on learning processes that are routine adaptation procedures and leave no room for programming experiments during economic activity, there remains a strong dichotomy between those who deliberately produce knowledge and those who use and exploit it. When an activity moves to higher forms of learning, and where the individual can programme experiments and obtain results, the production of knowledge becomes much more collectively distributed... With the emergence of experimental learning, the feedback and reciprocal links that tie « on-line » learning processes and in house R&D together- and whereby a potential creative activity effectively contributes to the production of knowledge- become crucial" (Lundvall, 2000).

daily hazard is to "*act prematurely on a very incomplete information base*" (p. 184). This is not to say of course that all is left to chance as we shall see below, but it does emphasise that learning, or any other condition of survival for that matter, is neither guaranteed nor that easily "arranged" (Metcalfe, 1998).

In many regards, the perils of failure at every turn are avoided through daily practices of inter-play between procedural and recursive knowledge among individuals and groups within firms. This daily practice is the source of learning, such that knowledge can be said to be generated through practice, through social interaction, through action. It is time to visit anthropological understandings of learning processes in firms. In their seminal article on communities of practice, Brown and Duguid (1991) argue that learning and innovation only too often are situated practices in the everyday humdrum of interaction with one's peers and with the environment. They explain: "*Alternative world views, then, do not lie in the laboratory or strategic planning office alone, condemning everyone else in the organisation to a unitary culture. Alternatives are inevitably distributed throughout all the different communities that make up the organisation. For it is the organisation's communities, at all levels, who are in contact with the environment and involved in interpretative sense making, congruence finding, and adapting. It is from any site of such interactions that new insights can be co-produced. If an organisational core overlooks or curtails the enacting in its midst by ignoring or disrupting its communities-of-practice, it threatens its survival in two ways. It will not only threaten to destroy the very working and learning practices by which it, knowingly or unknowingly, survives. It will also cut itself off from a major source of potential innovation that inevitably arises in the course of the working and learning" (p. 76).*

## 4. An Illustrative Example: the Linux Case

The story of Linux has been extensively quoted and cited as an instance of a new way to produce goods and has been considered as the herald of a "new economy" (Dan-Nguyen and Pénard, 1999). We will here give an interpretation of this story using the two concepts developed above, namely communities of practice and epistemic communities. This particular interpretation allows us to highlight a shift in the way organisational learning occurs thanks to the new information and communication technology.

## 4.1 Linux: A Community of Practice (Partly) Turned into an Epistemic Community

### 4.1.1 The Initial State:

As mentioned by Kogut and Turcanu (1992), Linux was developed by a sub-part of the community of hackers. According to the definitions stated above, the community of hackers qualifies as a community of practice. The analysis of the community of hackers below is based on the FAQ (Frequently Asked Questions) "How to Become a Hacker" written by Eric Raymond (1999). In our opinion, this text provides a rich and precious sociological material upon which we can rely to characterise this particular community. To sustain our viewpoint, we will then review the material provided by Raymond and compare it with the definitions we set. The objective of hackers is to solve problems, refine their competencies and exercise their intelligence. Besides, one of the essential rules is that "*nobody should have to solve a problem twice*". Hackers must then share and make their productions publicly available. Thus, members of the hackers' community seek to develop their personal competencies. In order to ease this task, they gather in a community within which resources (e.g. pieces of codes) are publicly shared.

According to Raymond, "*if you are part of this culture [i.e. of expert programmers and network wizards], if you have contributed to it and if other people belonging to it know you and call you a hacker, then you are a hacker*". This is a co-option phenomenon. Members are chosen according to their commitment to and participation in the identity of the community. One can then assert that this community is self-organised since no external determinism influences the shape of its structure. By the same token, it is the commitment to this culture that binds members together. This commitment is nonetheless subordinated to the necessary acquisition of programming skills. To know HTML, to master Unix, etc. are parts of the minimum prerequisite to be ushered into that community. We find here the two notions of shared values and required individual skills that we highlighted in the definition of communities of practice.

The hackers' culture is one of gift giving and reputation building. Accordingly members who give the more are rewarded with the greater reputation. Nonetheless, the quality of members' contributions is evaluated by those having equal or greater technical skills. What is judged is both individuals' technical abilities and behaviours. Within this community, messages circulate intensively among members. By contrast, nothing is emitted toward the outside world. Raymond even asserts that "*to be socially excluded help remaining concentrated on the really important things such as hacking and thinking*".

To sum up, the community of hackers meet all the characteristics of a community of practice such as we defined this concept above in terms of objective, cement of the community, dominant cognitive activity, recruitment rules and their dominant learning mode. However, the introduction of the Linux project within this community has changed its nature and ways of functioning in various ways as we will see below.

### 4.1.2 Linux and the Community of Hackers

Originally, Linux was a mere project of a student, Linus Torvalds. It is an operating system based on Minix, a small operating system for PC inspired from Unix. Linus Torvalds wrote the kernel and published the source code on the Internet to receive feedback in order to enhance his programme. The community of the developers of Linux thus did not start from scratch: Linus Torvalds had already written the core of the programme. Moreover, he had written it in a language well mastered by all the potential developers (the C language). The objective was thus not purely cognitive in the first place but operational. People who agreed to participate were not seeking to enhance the body of knowledge in computing science but to improve their skills and to gain reputation. In this respect, they were like a community of practice. Nonetheless, it turned out that the aim was to create something more general than competencies, something placed both outside and above the individuals. Indeed, Linux is an intended and conscious innovation and thus implies new knowledge and new ways to articulate knowledge. In that sense the Linux community was already much like an epistemic community.

Anybody having the required technical skills and the desire to participate could bring his/her contribution. The Linux community thus emerged and organised itself spontaneously. In this sense, the Linux community is self-organised and looks like a community of practice. However, in order to cope with the flux of contributions, a committee in charge of evaluating the incoming contributions was created. This comity was then apart and above the bunch of developers. There was thus the creation of a procedural authority to judge what constituted a progress toward the objective. It should be noted that the objective evolved from the enhancement of a student project to the building of a competitor for Windows NT.

In the case of Linux, the codebook was already written and was part of the project, since it was the programming language and the algorithmic necessary to achieve the project. So, in that sense, the internal messages (exchanged among members in order to fulfil the task) and the external messages (a final product which was a free software) were conflated. Moreover, the developers were also the first users. Hence, at least during the early period senders and receivers were the same. The diffusion of Linux was not the original objective, but the growing success of the software and the use of the Internet as a means of communication diffused Linux beyond the circle of its first users.

The Linux community matches certain criteria of epistemic communities, but in a non-planned manner. The Linux community is born out of the community of hackers which is a community of practice. It had, however, a tendency to become epistemic, though it did not fully qualify as an epistemic community: it gained a hybrid shape borrowing from both epistemic communities and communities of practice. This shift was certainly made possible at such a scale by the use of the Internet. The use of this far-reaching means of communication, together with a code understood by all the relevant actors (the C language) made the quick gathering and commitment of the developers possible. The interest of the project (many hackers, for various reasons, are interested in having a scalable operating system running on PC (Jullien, 1999) also contributed to the commitment of programmers.

## 4.2 Entailed Change in Learning Modes

Within communities of practice, as stated above, knowledge is created in an unintended way. Members of such communities are primarily concerned with the enhancement of their own personal skills. Because practice is the central concern, they only seek to enhance their activities. In Cook and Brown‘s (1999) words, the stress is put on the knowledge coming to play during activity. The learning mode at work here is thus fully of the type "learning in working". Most of the knowledge produced cannot be separated from the activity, the few pieces of knowledge separable from activity are by-products.

Moreover, this process is only directed at the individual level. Indeed, to a large extent, it is the individual who chooses which activity to pursue and hence what kind of knowledge to produce and how to produce it. The community is a support to learning but not in a prescriptive way. At the social level, there is thus no knowledge creation in a voluntary way.

Things became quiet different once the Linux project was introduced inside the community of hackers. As mentioned above, the introduction of a collective project on the one hand, and of a procedural authority on the other, sparked the emergence of a different type of knowledge, and, accordingly of new learning modes. Indeed, the knowledge created is no longer individual-oriented. Instead, the knowledge is now voluntarily put outside the individual and made both explicit and available to all members of the community. Unlike what occurred within communities of practice, the trend here is to create and accumulate a stock of knowledge about the operating system. Besides, the procedural authority guides the process of knowledge creation and accumulation at the community level, rewarding the "good" contributions and discarding the "bad" ones.

To reach the goal stated by the procedural authority, new learning modes arise. Although learning in practice still exists in the Linux community, two new learning types now co-exist along with it, namely learning by interacting and experimental learning.

Learning by interacting undoubtedly appeared due to the extensive use of the Internet within the development of Linux. It would not have been possible to work out the project without this means of communication. Communities of hackers, understood as communities of practice, did exist before the coming of the Internet. They were however limited by space and time constraints (Kogut and Turcanu, 1992). Hackers then interacted with one another, but these relationships could only exist within local geographical areas. The possibility to release these local constraints allows for new behaviours. It is now possible to create links between far away communities having the same concerns, values, cultures, etc. and to initiate fruitful exchanges (Steinmueller, 2000). This phenomenon occurred during Linux development. Hackers from all over the world have then been able to virtually gather together, exchange ideas and build something together on the same common ground. Knowledge creation occurred in these interactions where pieces of knowledge were woven together and where new building blocks were added incrementally. However, it should be noted that communications were centralised, since contributions were posted to a committee that evaluated them and then diffused them back to the community.

The second new type of learning is experimental learning as presented by Lundvall (2000) and described above. As mentioned earlier, at least during the first period of Linux's life, the developers were also the users. Hence, the project benefited from a workbench upon which to test new ideas and propose improvements.

Besides, the users were skilled and able to point to potential problems and/or to fix them. Members of the community was then able to set efficient search processes. Trial-and-error processes could be implemented in a useful way, thanks to both the almost instantaneous means of communication allowing for frequent feedback and the competence of the users able to return useful comments. Here again, it is unlikely that this learning mode could have been implemented without the use of the Internet. The Internet provided the speed and the large broadcasting necessary to carry out experimental learning efficiently.

## 4.3 What is specific in the Linux case

Having described the story of Linux, the next question is to what extent the model explained above can be applied in different contexts. That is, can the creation of new products by virtual communities be generalised. First, one should be aware that the development of Linux has really specific characteristics that cannot easily be reproduced. However, we contend, in the conclusion, that some of the features of the Linux model of development can be replicated under certain particular conditions.

The development of the software was made possible through the Internet because the language used for its production was already understood by the participants. The Internet only allows the diffusion of one type of knowledge, namely codified knowledge. According to Cowan and Foray (1997), three elements are necessary to codification: a commonly accepted grammar, a shared vocabulary and a repository of sentences made from the two previous elements. Clearly, a programming language such as C, algorithmic skills and the kernel written by Linus Torvalds fulfil these three conditions.

But the fact that codified knowledge exists with regard to this particular project is not sufficient. In addition, there must exist a group of people able to understand it and to use it in a co-ordinated manner. Blackler *et al.* (2000) assert that to be engaged in a collective activity is enough to guarantee co-ordination and mutual understanding. However, in the case of Linux, this was clearly not enough. As Steinmueller (2000) states, in order to understand the code, there must exist a tacit knowledge of how to read and write the code. Moreover, in order to use the code collectively must mean sharing a culture, a collective understanding of where one are heading to and how to get there. This is what Weick (1995) calls beliefs-driven sense-making. That is, it is the collectively shared beliefs that allow for a collective effort toward a goal. It would have been impossible to make several thousands of persons learn a language through the Internet. Hence, in that case, the aspects the Linux community inherited from the hackers community of practice are essential to its success. It seems that, in order to work out such a project, members have to stem from the same community of practice. In addition we could note that hackers greatly contributed to the development of the Internet. Hence, in a way, the Internet is their natural environment.

In addition to the problem of acquisition of a common knowledge and a shared language by thousands of people, several other particularities of the case should be pointed out. First, it seems impossible to start a project from scratch that way. Indeed, Linus Torvalds provided the first kernel from which programmers were able to start. Something has to be proposed to the community to allow it to grab the project. It is likely that it is impossible to spark a bottom-up process to decide what kind of product to build.

Moreover, as Raymond (1999) points out, the proposal of the project must appear feasible and interesting to the persons to whom it is proposed. If people do not see any interest in the proposal, they will not work on it. It must be either a product that can be useful for them in the future or a project that allows them to enhance their skills and display their know-how in order to gain reputation.

A related question is, if this mode of development was to be transposed to the market world, will people accept to freely develop a product that would next generate private profit for a firm. Raymond (1999) seems to think that the key values of the community of hackers (gift giving and reputation seeking) would overcome this problem. Namely, he thinks that money is not what make hackers work and that, as long as they get rewards in terms of reputation, they will accept to develop software.

Netscape actually tried to implement the open source model to the development of a generation of its browser. The result was only partially a success. Raymond explains this by the facts that Netscape still wanted to protect the codes (with a specific license) and that the project was ill-defined and had many flaws in the first place. But maybe another explanation is that hackers would be reluctant to work for free for a firm that does not match their culture.

However, Netscape has now released the source code of *Communicator*. The idea beneath is to impose a *de facto* standard upon which various applications could be built and commercialised. According to this example, as Zimmerman (1999) puts it, "*the free software model, or more precisely, the open source standard, blaze the trail of a conciliation between private individual interest and collective efficiency. Its efficiency and relevance is not exclusive to the building of a standardisation understood as the diffusion of standards viewed as public goods, as opposed to the one resulting from dominant positions and from the exploitation of monopoly power.*"

Lastly, along with Raymond (1999) we should point out the importance of the co-ordinator attitude and charisma. Linus Torvalds has the great ability to acknowledge others' good ideas. Beside, he had frequent exchanges with the other programmers and engaged in a really interactive relationship with them. These features of Linus Torvalds' personality seem to have played a major role in the success of Linux.

The fact that the project was handled by a community of practice sharing knowledge, language and value and comfortable in the Internet environment is an important factor of success. Equally important is the fact that the product was a software, thus easily shared and handled through the Internet. Besides, the fact that the definition of the objective matched the culture of the developers also helped in carrying out the process successfully. All these are serious constraints to any attempt to transpose the model in other contexts, especially in non-software industries. However, we contend that some features can fruitfully be imitated by other branches of activity.

# 5. Conclusion: Can the Linux case be generalised?

It seems rather difficult to spawn out of nothing a virtual organisation based on the lessons learned from Linux. More modestly, however, in an already existing organisation, it is possible to encourage some features that made the success of Linux. In an already existing organisation, their is already to some extent a shared culture and loci where common knowledge exist. Therefore, as far as communities of practice can be identified in a firm, it is possible to encourage behaviours close to the one observed in Linux in such communities. In the case of a dispersed community, the Internet can play a significant role, since a common body of knowledge upon which building new knowledge already exists by definition.

A good internal communication is thus likely to be a key element in the success of the implementation of a Linux like pattern of development. In addition, it is crucial for managers wanting to implement methods borrowed from the study of the Linux case to take care of the communities of practice existing within their firms. However, communities of practice are in essence self-organised, it is thus not possible to create them or to direct them in a usual way. The best that can be done is to set a favourable environment to their creation and development.

Moreover, the top management of a firm has the ability to state the global direction in which they wish to engage the organisation. Hence, it is possible to set a global objective that members of the firm will follow. This brings up the idea of the importance of the role of the entrepreneurial vision in the knowledge creation process (Nonaka *et al.*, 2000). This can be a substitute for the role played by Linus Torvalds' charisma. This is reinforced by the stress put by Weick (1995) on the importance of care (or "heed" as he put it) in organisations.

Another fundamental issue is to determine to what kind of innovation the Linux model can be useful. Chesbrough and Teece (1996) set a fundamental dichotomy in the nature of innovations. They distinguish on the one hand the systemic innovations. Systemic innovations are innovations that imply the total redefinition of the product. On the other hand there are the autonomous innovations. These are innovations that can be pursued in one part of the product without entailing modifications for the system as a whole.

In order to pursue one kind of innovation or the other, firms should adopt a particular setting. Brusoni *et al.* (2000) characterise firms along two criteria: responsiveness and distinctiveness. Responsiveness corresponds to the ability for a firm to control and centralise its various components. It has distinctiveness if it is able to possess several decentralised centres of competence. During the cycle of development of a product, organisation requires either responsiveness or distinctiveness. Distinctiveness is useful in the development of autonomous innovation whereas responsiveness is worth using for systemic innovation. These two criteria are of course linked to the notion of routines of exploration *versus* routines of exploitation.

As we saw, the "bazaar" style (as Raymond puts it) is well designed for exploration in a given (though loose) framework. It is thus particularly fit for exploration of already existing but ill-designed products. Hence, this model could be particularly useful in the case of autonomous innovation. Indeed, a firm able to give some autonomy to communities in order to explore some parts of a product and enhance it should gain some competitive advantage.

However, in order to be able to incorporate these decentralised innovations in the global product, firms should have some responsiveness abilities. This responsiveness is not only organisational. It is not enough to have a strongly centralised organisation to integrate the various pieces of knowledge developed locally. It is, in addition, essential to set standardised cognitive interface between the different blocks of knowledge. It should be noted that, through the use of object oriented languages, the software industry is able to do so. Indeed, one of the features of the object-oriented languages is that they allow programmers to develop the part of the programme they have in charge without bothering about its integration into the bigger software.

## References


Adler E, Haas P (1992) Conclusion: Epistemic Communities, World Order, and the Creation of a Reflective Research Programme. In : International Organisation, vol 46, n°1.

Argyris C, Schön D (1978) Organisational Learning: a Theory of Action Perspective. Addison-Wesley, Reading.

Ashby WR (1956) An Introduction to Cybernetics. Chapman and Hall, London.

Barber B (1952) Science and the Social Order. Glencoe, The Free Press.

Baumard P (1999) Tacit Knowledge in Organizations. Sage, London.

Beyssade C (1998) Sens et savoir: Des communautés épistémiques dans le discours. Presses Universitaires de Rennes.

Blackler F, McDonald S (2000) Power, Mastery and Organizational Learning. Journal of Management Studies, Forthcoming.

Brown JS, Duguid P (1991) Organizational Learning and Communities of Practice: Toward a Unified View of Working, Learning and Innovation. In: Organization Science, Vol. 2, n°1, pp 40-57.

Brown JS, Duguid P (1998) Organizing Knowledge. In: California Management Review Vol. 40, n°3, pp 90-111.

Brusoni S, Pavitt K, Prencipe A (2000) Knowledge specialisation and the boundaries of the firm : Why do firms know more than they do ?. Paper presented at the conference Knowledge Management : Concepts and Controversies, University of Warwick.

Chandler AD (1977) The Visible Hand: the Managerial Revolution in American Business. Harvard University Press.

Chesbrough W, Teece D (1996) When is Virtual Virtuous ? Organizing for Innovation. In : Harvard Business Review, January-February.

Cohendet P, Llerena P (1999) La conception de la firme comme processeur de connaissance. In: Revue d'Economie Industrielle, n°88, 2° tri., pp 211-237.

Cook SDN, Brown JS (1999) Bridging epistemologies: The Generative Dance between Organizational Knowledge and Organizational Knowing. Organization Science, vol 10, n°4, pp 381-400.

Cowan R, Foray D (1997) The Economics of Codification and the Diffusion of Knowledge. In: Industrial and Corporate Change, vol 6 (3), pp 594-622

Cowan R, David P, Foray D (1998) The Explicit Economics of Knowledge Codification and Tacitness. Mimeo TIPIK project.

Curien N (1992) Economie et Management des Entreprises Réseaux. In: ENSPTT, Economica.

Dang-Nguyen J, Pénard JT (1999) Don et coopération dans internet: une nouvelle économie?. In: Terminal n° 80/81, Les logiciels libres: de l'utopie au marché.

Dibiaggio L (1998) Information, Connaissance et Organisation. Thèse de doctorat, université de Nice-Sophia Antipolis.

Dibiaggio L (1999) Apprentissage, Coordination et Organisation de l'Industrie. Une Perspective Cognitive. In: Revue d'Economie Industrielle, n°88, pp 111-136.

Eliasson G (1994) The Theory of the Firm and the Theory of Economic Growth. In Magnusson L (eds) Evolutionary and Neo-Schumpeterian approaches to Economics, Kluwer, Dordrecht.

Foray D, Lundvall BA (1997) The Learning Economy. Implications on Innovation, Growth and Employment. In : Guilhon B *et al.*, Economie de la Connaissance et Organisations, L'Harmattan, Paris, pp 16-38.

Giard V, Midler C (1993) Pilotages de Projet et Entreprises – Diversités et Convergences. In: ECOSIP, Economica.

Gibbons M, Limoges C, Nowotny H, Scwartzman S, Scott P, Trow M (1994) The New Production of Knowledge: The Dynamics of Science and Research in Contemporary Societies. Sage Publications, London.

Haas P (1992) Introduction: Epistemic Communities and International Policy Coordination. In: International Organization, vol 46, n°1, pp 1-37.

Huff A (1990) Mapping Strategic Thought. John Wiley and Sons.

Jullien N (1999) Linux: la convergence du monde Unix et du monde PC. In: Terminal n° 80/81, Les logiciels libre: de l'utopie au marché.

Kao J (1998) Organiser la créativité. Village Mondial.

Knorr-Cetina K (1981) The Manufacture of Knowledge, An Essay on the Constructivist and Contextual Nature of Science. Pergamon Press.

Kogut B, Turcanu A (1992) Distributed Knowledge in Software Development: The Emergence of E-Innovation. Working Draft Presented at theCondor seminar, Ecole Polytechnique, Paris..

Kuhn T (1962) The Structure of Scientific Revolutions. University of Chicago Press, Chicago.

Lam A (1999) Tacit Knowledge, Organisational Learning and Societal institutions. Working paper, University of Kent, Canterbury.

Lave J, Wenger EC (1990) Situated Learning: Legitimate Peripheral Participation. Cambridge University Press New York, NY.

Leonard-Barton D (1995) Wellsprings of Knowledge: Building and Sustaining the Sources of Innovation. Harvard Business School Press, Cambridge MA.

Lesourne J (1991) Economie de l'ordre et du désordre. Economica, Paris.

Lundvall BA (1992) (ed) National Systems of Innovation: Towards a Theory of Innovation and Interactive Learning. Pinter Publishers.

Lundvall BA (2000) Unlocking Knowledge. Contribution to the OECD report on Knowledge and Society, Paris.

Metcalfe S (1998) Innovation as a policy problem: new perspectives and old ones on the division of labour in the innovation process. In: *Mimeo*, Centre for Research on Innovation, University of Manchester.

Midler C (1996) L'auto qui n'existait pas. InterEditions.

Nonaka I (1994) A Dynamic Theory of Organizational Knowledge Creation. In: Organisation science, vol 5, n°1, pp 4-37.

Nonaka I, Konno N (1998) The Concept of Ba : Building for Knowledge Creation. In: California Management Review, vol 40, n°3, Spring.

Nonaka I, Takeuchi H (1995) The Knowledge-Creating Company: How the Japanese Companies Create the Dynamic of Innovation. Oxford University Press New York, NY.

Nonaka I, Toyama R, Konno N (2000) SECI, Ba and Leadership: a Unified Model of Dynamic Knowledge Creation. In: Long Range Planning 33, pp 5-34.

Nooteboom B (1999) Inter-Firm Alliances. Routledge, London.

Raymond E (1999) The Cathedral and the Bazaar. O'Reilly.

Snyder WM, Wenger EC (2000) Communities of Practice : the Organizational Frontier. In: Harvard Business Review, January February 2000.

Steinmueller WE (2000) Will new information and communication technologies improve the codification of knowledge. In: Industrial and Corporate Change, vol 9, issue 2, pp 361-376.

Von Hippel E (1988), The Sources of Innovation. Oxford University Press, New York.von Krogh, Roos and Kleine (1998)

Weick K (1995) Sensemaking in Organizations. Sage.

Wenger E (1998) Communities of Practice; Learning as a Social System. In: Systems Thinker, June.

Whiteneck D (1996) The Industrial Revolution and Birth of the Anti-Mercantilist Idea: Epistemic Communities and Global Leadership. In: Journal of World-Systems Research, vol 2, n°1.

Zimmermann JB (1999) Logiciel et propriété intellectuelle: du copyright au copyleft. . In: Terminal n° 80/81, Les logiciels libres: de l'utopie au marché.

# Knowledge Creation, Knowledge Diffusion and Network Structure

Robin Cowan[1,3] and Nicolas Jonard[2,3]

[1] MERIT, University of Maastricht,
P.O. Box 616, 6200 MD Maastricht, The Netherlands
[2] CNRS, BETA, Université Louis Pasteur,
61 Ave. de la Forêt Noire, 67000 Strasbourg, France
[3] International Institute of Infonomics (IIoI),
P.O. Box 32, 6333 PS Heerlen, The Netherlands

**Abstract.** This paper models knowledge creation and diffusion as processes involving many agents located on a network. Knowledge diffusion takes place when an agent broadcasts his knowledge to the agents to whom he is directly connected. Knowledge creation arises when agents receive new knowledge which is combined with their existing knowledge stocks. Thus both creation and diffusion are network-dependent activities. This paper examines the relationship between network architecture and aggregate knowledge levels. We find that knowledge growth is fastest in a "small world", that is, when the underlying network structure is relatively cliquish (dense at a local level) yet has short paths. This corresponds to a locally-connected graph which includes a few long-distance connections or shortcuts.

## 1 Introduction

One reason to reject the representative agent model is the observation that economic agents are not identical. This simple fact creates serious problems for the underpinnings of the standard model.[1] Agents are heterogeneous in many ways, but an important one stems from the fact that any agent in a large population interacts directly with only a very small number of other agents. Thus a potentially important source of agent heterogeneity stems from the "neighbourhood" in which an agent operates. To represent this feature of economic interactions, we can model the population of agents as being located on a network. This gives a natural structure which captures the fact that an agent's direct connections are few relative to the total population, and thereby distinguishes one agent from another.

In this paper we examine the relationship between the architecture of the network of agents and its aggregate performance. The issue here is rooted in the economics of innovation and growth. It is commonplace since the growth accounting work of Solow in the 1950s that technological change is central

[1] On this, see Kirman (1992).

to economic growth. There are two aspects to technical change: knowledge creation and knowledge diffusion. Early concepts of knowledge treated it as a public good, non-rivalrous, and non-exclusive; while expensive to create, it is very cheap to duplicate, and thus trivial to disseminate or diffuse. (See for example Arrow 1962; Nelson 1959.) Recently, though, this view is considered incomplete. While it may nicely capture certain aspects of codified knowledge, there is much knowledge that does not have these properties. In particular, work on tacit knowledge (Cowan and Foray 1997; von Hippel 1998) and absorptive capacity (Cohen and Levintahl 1989), has emphasized that knowledge diffusion is not a trivial activity. Indeed, empirical studies of knowledge flows using patent data have shown that the ease of knowledge flow is negatively related to the distance over which it travels, implying that knowledge is not freely available to the entire population once it has been created. (See the works of Jaffe and his collaborators, e.g. Jaffe et al. 1993.)

The model in this paper captures both of these notions. Knowledge, when created, is not globally available. It is transmitted through face-to-face interactions. Further, because agents are located on a network, the requirement of face-to-face interaction creates a natural notion of distance — the number of interactions needed to pass the knowledge from originator to final recipient. The central issue in this paper is the relationship between network structure or architecture and the ability of the system to create and diffuse knowledge rapidly. We model knowledge creation and diffusion taking place within a population of agents located on a network, and examine the growth of aggregate knowledge levels. Diffusion of knowledge clearly increases aggregate knowledge levels simply through agents acquiring some existing knowledge. But recent work on innovation as the recombination of existing ideas suggests another benefit from diffusion. As an agent receives knowledge or information he is able to integrate it with his existing stock, and create new knowledge. Knowledge diffuses by an agent broadcasting (or perhaps more accurately narrow-casting) his knowledge to those he is directly connected with. Rebroadcasting diffuses the knowledge throughout the economy. In this framework one agent's mistakes or discoveries will benefit those with whom he interacts, and innovations take place as a result of this broadcasting as recipients re-combine the new knowledge with their existing knowledge. But knowledge diffused this way can only be beneficial to those agents who are at least partly capable of understanding and integrating it. Thus there is a threshold value for dissimilarity in agents' knowledge levels below which no transmission is possible — if $i$ and $j$ are too dissimilar they cannot learn from each other.

For this economy we measure aggregate performance as the mean knowledge level over all agents. The parameter we use to characterize network architecture is the degree of spatial regularity in the inter-agent connections through which knowledge flows. At one extreme of the space of networks there is a regular structure in which every agent is connected to his $n$ nearest neighbours, whereas at the other extreme we have an irregular network in which each agent is connected to, on average, $n$ agents located at random in the network. We are dealing with networks that are identically local in the sense that their density is low and constant, but they have different structural features which, we will see, influence collective properties in a non-trivial way. Our interest is in the space of networks between these two extremes, and we find that one region stands out: "small world" networks generate the fastest knowledge growth. As defined formally by Watts and Strogatz (1998), the small world combines a strong degree of local cohesiveness with a small fraction of long distance links permitting knowledge to be circulated rapidly among distant parts of the network.

# 2 The Model

In our economy many agents are located on a graph, each agent having direct connections with a small number of other agents. Each agent has a knowledge endowment represented by a real-valued vector. At random times an agent broadcasts his knowledge.[2] Knowledge is broadcast to all those agents with whom the sender has a direct connection. If the knowledge level of the potential recipient is not too dissimilar from that of the broadcaster, knowledge is received and assimilated. Formally, if $i$ broadcasts to $j$, then in any knowledge category in which $i$ exceeds $j$, $j$'s knowledge increases. In some cases, agents have imperfect absorptive capacity, and only assimilate part of what they receive. In others, they are "super-absorptive": they absorb everything they are sent and are able to use it to create new knowledge, increasing their knowledge levels further.

## 2.1 Knowledge interaction

Each agent is characterized by a knowledge vector $v_i$ which evolves over time as the agent receives information broadcast by other agents. Formally, let $v_{i,k}^t$ denote agent $i$'s knowledge endowment in category $k \in \{1, \ldots, K\}$ at time $t$. Agent $i$ broadcasts to every $j \in \Gamma(i)$ (equivalently $i \in \Gamma(j)$ as the graph is non-directed), if dissimilarity between $i$ and $j$ is low enough. By dissimilarity, we mean the relative distance between $i$ and $j$ in terms of knowledge, which we write as

$$\Delta(i,j) = \max\left\{r, \frac{1}{r}\right\} - 1, \tag{1}$$

where $r = |v_i| / |v_j|$ and $|\cdot|$ is the standard Euclidean norm of a vector. For each agent $j \in \Gamma(i)$, provided $\Delta(i,j) < \theta \in (0, \infty)$, $i$ makes his knowl-

[2] The nature of knowledge, and its relation to data and information is complex. we avoid these complexities except to the extent that we recognize the importance of and difficulty in transmitting tacit knowledge.

edge available to $j$.[3] For every knowledge category, $k$, when $i$ broadcasts, $j$'s knowledge increases according to

$$v_{j,k}^{t+1} = v_{j,k}^{t} + \alpha \cdot \max\left\{0, v_{i,k}^{t} - v_{j,k}^{t}\right\}, \tag{2}$$

for all $k$ without any consequent loss of knowledge to agent $i$. The parameter $\alpha$ captures an important aspect of knowledge diffusion and transfer. When $\alpha < 1$ knowledge is only partly assimilable. This notion has been examined as an issue of absorptive capacity by Cohen and Levinthal (1989), or Cowan and Foray (1997). Hence broadcasting results in partial acquisition of knowledge by the recipient, as well as a partial diminution of the distance between broadcaster and recipient. In a regime of "collective invention" (Allen 1983), however, knowledge is characterized as super-additive, i.e. $\alpha > 1$. Here, unobserved (by the analyst) complementarities in the knowledge stocks of $i$ and $j$ imply that when $j$ receives $i$'s knowledge he is able to improve upon it, innovating by combining his knowledge with the knowledge newly acquired.[4]

## 2.2 The interaction structure

Consider $N$ agents existing on an undirected connected graph $G(I, \Gamma)$, where $I = \{1, \ldots, N\}$ is the set of vertices (agents) and $\Gamma = \{\Gamma(i), i \in I\}$ the list of connections (the vertices to which each vertex is connected). Formally $\Gamma(i) = \{j \in I \setminus \{i\} \mid d(i,j) = 1\}$, where $d(i,j)$ is the length of the shortest path (geodesic) from vertex $i$ to vertex $j$. Only agents separated by one edge can interact, and when $i$ broadcasts, only those agents in $\Gamma(i)$ are potential recipients.

Our interest in network architecture differs from that seen in the majority of the literature in that we do not vary the density of the network. The family of graphs we consider here contains a constant number $n \cdot N/2$ of edges . Our concern is with the degree of regularity in the structure. The following heuristic (or 're-wiring' procedure) is employed: Create the regular lattice structure. With probability $p$ re-wire each edge of the graph. That is, sequentially examine each edge of the graph; with probability $p$ disconnect one of its vertices, and connect it to a vertex chosen uniformly at random. In the algorithm we ensure both that vertices are not self-connected by this

[3] As an illustration, if $|v_j| = 1$ and $\theta = 0.4$ then agent $j$ can learn from agent $i$ provided $|v_i| \in [0.714, 1.4]$.

[4] Note that $\alpha$ could be intepretated as parameterizing the tacitness of knowledge. If $\alpha < 1$ even in the absence of a dissimilarity constraint the failure to absorb all available knowledge can arise because codified, broadcast knowledge needs to be interpreted and this interpretation intimately involves tacit knowledge, which the receiving firm is unlikely to have. When $\alpha > 1$, the dissimilarity constraint performs the same function. Agents with similar codified knowledge are likely to have similar, if not the same, tacit knowledge. See Cowan et al. (2000) for a discussion of codification and tacitness.

procedure, and that there are no duplications, i.e. no two vertices are connected more than once.[5] For large graphs, this procedure ensures that the connectivity is preserved and that the average number of edges per vertex is constant at $n$. By this algorithm we tune the degree of randomness in the graph with a single parameter $p \in [0, 1]$, hence the label $G(I, n, p)$ for graphs in this family.

The structural properties of $G(I, n, p)$-graphs can be captured by the concepts of average path length and average cliquishness. To illustrate, in friendship networks, the path length is the number of friendships in the shortest chain connecting two agents, whereas cliquishness reflects the extent to which friends of one agent are also friends of each other. Recall $d(i, j)$ is the length of the shortest path between $i$ and $j$. The average path length $L(p)$ is

$$L(p) = \frac{1}{N} \sum_{i \in I} \sum_{j \neq i} \frac{d(i,j)}{N-1} \tag{3}$$

and average cliquishness $C(p)$ is given by

$$C(p) = \frac{1}{N} \sum_{i \in I} \sum_{j,l \in \Gamma(i)} \frac{X(j,l)}{\#\Gamma(i)\,(\#\Gamma(i) - 1)/2}, \tag{4}$$

where $X(j,l) = 1$ if $j \in \Gamma(l)$ and $X(j,l) = 0$ otherwise. The evolution of path length and clique size with $p$ is depicted on Figure 1, for a graph of $N = 500$ vertices, each vertex having on average $n = 10$ connections. (The graph depicts average values over 50 replications.) For the sake of clarity, we plot the normalized values $L(p)/L(0)$ and $C(p)/C(0)$.

The upper curve (thick black) in Figure 1 is the normalized average cliquishness index $C(p)/C(0)$ for $p \in [0, 1]$. It remains almost constant when $p$ is reasonably small and falls slowly for large values of $p$. By contrast, average path length (thin black) as measured by $L(p)/L(0)$ falls quickly for very small $p$ values and flattens out near 0.01. As emphasized by Watts and Strogatz, there is a non-negligible interval for $p$ over which $L(p) \simeq L(1)$ yet $C(p) \gg C(1)$. This interval, in which high cliquishness and low path length coexist, constitutes the small world region.

## 3 Numerical experiments

We are interested generally in the relationship between the structure of the network across which knowledge diffuses and the distribution power of the innovation system. It is natural therefore to examine the evolution of knowledge levels in this economy. We can do this by simulating the economy and relating long run knowledge levels to the value of $p$ in our re-wiring algorithm.

[5] This is the re-wiring procedure employed by Watts and Strogatz (1998), in their seminal work on small worlds.

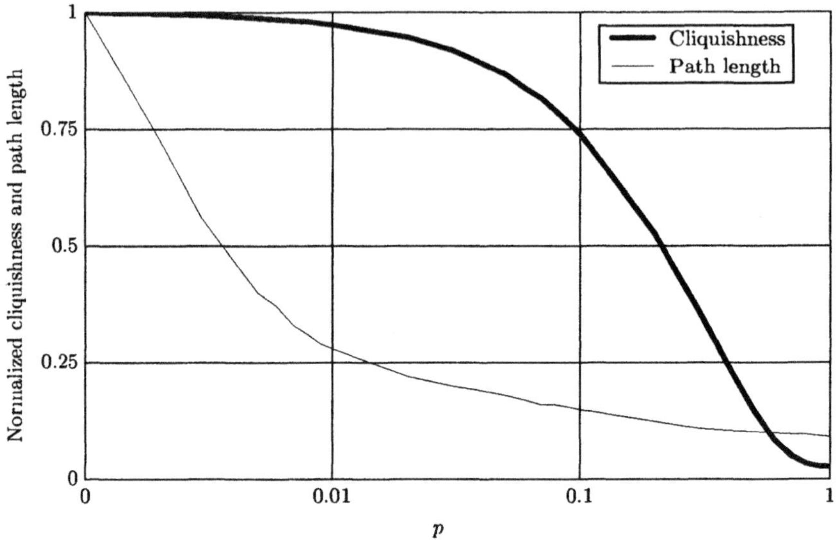


**Fig. 1.** Average cliquishness and average path length as functions of the degree of randomness in network structure, $p$

Of particular interest is whether the model displays small world properties. Define agent $i$'s average knowledge level at time $t$ as $\mu_i^t = \frac{1}{K}\sum_k v_{i,k}^t$. The average level of knowledge in the economy at time $t$ is

$$\mu^t = \frac{1}{N}\sum_{i\in I}\mu_i^t \tag{5}$$

and the variance in knowledge allocation is

$$\frac{1}{N}\sum_{i\in I}(\mu_i^t)^2 - (\mu^t)^2. \tag{6}$$

We consider a third measure, having to do with the degree of spatial order in a system of locally interconnected components. To check whether our broadcast economy generates spatially auto-correlated knowledge allocations, we compute the Moran coefficient defined as follows (McGrew and Monroe 1993). Dropping time indexes, the coefficient is written

$$S = \frac{1}{\sigma^2}\sum_{i\in I}\sum_{j\neq i} w_{i,j}(\mu_i - \mu)(\mu_j - \mu), \tag{7}$$

where

$$w_{i,j} = \frac{1/d\,(i,j)}{\sum_{i\in I}\sum_{j\neq i} 1/d\,(i,j)}. \tag{8}$$

This statistic is written in terms of the type of cross-products found in the standard correlation, but instead of two variables, the pairs of adjoining vertices are used. Values of the Moran coefficient $S$ tend to be in $[-1, +1]$, though are not restricted to this range. Values near $+1$ indicate similar values tend to cluster; values near $-1$ indicate dissimilar values tend to cluster; and values near $-1/(\#I - 1)$ (going to 0 as $\#I$ gets large) indicate values tend to be randomly scattered.

We consider an economy with $N = 500$ agents, each having, on average $n = 10$ direct connections to other agents. Within a single simulation run the network structure is fixed, so each possible economy is represented by a graph from the family $G(I, 10, p)$ with $I = \{1, \ldots, 500\}$.[6] Each agent is endowed with a 5-category knowledge vector, with each element randomly initialized from a uniform distribution $U[0, 1]$. In simulation time, in each period one randomly selected agent broadcasts his knowledge. The knowledge is received by all those who are directly connected to him and are sufficiently similar. Formally, the broadcast of agent $i$ is received by the set of agents $\{j \in \Gamma(i) \mid \Delta(i,j) < 0.2\}$. We consider a large number $\tau$ of periods, to ensure that each agent takes part in a reasonably large number of knowledge transmissions, both as sender and receiver. With $\tau = 30,000$, the average agent broadcasts 60 times. We run 50 replications for each value of $\alpha$ and $p$.

We are interested in the effects of two parameters. The first is $p$, which determines the network structure. This we vary from 0 to 1 in units convenient to display on a log scale. The second is $\alpha$, which determines the extent to which knowledge can be assimilated. We vary $\alpha$ from 0.96 to 1.2 with an increment of 0.01. Values of $\alpha < 1$ indicate a regime in which tacit knowledge is important, and knowledge received is only absorbed with some effort. Thus the effects of receiving a broadcast are a partial absorption by the recipient and thus a diminution of the dissimilarity of sender and receiver. When $\alpha > 1$ we are in a regime in which knowledge is super-additive. The recipient of knowledge is able to leap-frog the sender. This corresponds to Allen's (1983) description of collective invention, in which knowledge is relatively simple to absorb and easy to improve upon.

## 4 Results

The first set of results we examine has to do with the efficiency and equity of knowledge creation and diffusion. We then turn to the spatial behaviour of the system.

---

[6] The issue of the endogeneity of the network structure is not addressed here. This is typical in this literature. Notable exceptions, in which endogenous network strutures are the focus of the analysis, are Bala and Goyal (1998); Jackson and Wolinsky (1996); Plouraboue et al. (1998).

### 4.1 Efficiency and equity in knowledge diffusion

We examine the efficiency of the network structure in terms of long run average knowledge levels. The statistic we use is the average knowledge level in the economy $\mu^T$. The parameter space can be partitioned into two distinct regions corresponding to $\alpha < 1$ and $\alpha > 1$. In the first, new knowledge is not created, and changes in aggregate knowledge level arise purely through diffusion. In the second there is both creation and diffusion. We consider them separately.

**Diffusion** Consider first the case in which absorptive capacity is less than perfect — agents are able to assimilate only part of the knowledge with which they are presented. This corresponds to the region of the parameter space where $\alpha < 1$. Here, a purely diffusive mechanism is at work — knowledge is diffused, but not generated. In this regime, network structure has no apparent effect on long run knowledge levels. All values of $p$ produce the same long run state of the economy. The effect of changing average path length between agents is to change the speed of convergence — shorter path lengths imply faster convergence.

Contrasting results concerning this region of the parameter space are presented in Cowan and Jonard (1999), who show that a small-world architecture dominates other forms of organization. This result is obtained in a model in which knowledge is diffused not by broadcast but by barter trades among agents. Broadcasting, as opposed to barter, eliminates the need for a double coincidence of wants for knowledge diffusion. Eliminating this requirement eliminates the advantage of cliquishness, which lies in the fact that in a cliquish world a failure of the double coincidence of wants does not have dramatic consequences for there are many other possible (short) paths along which a piece of knowledge can travel.

**Creation and Diffusion** In the second region of the parameter space, where $\alpha > 1$, we get a joint process of knowledge creation and diffusion. Each agent incorporates the knowledge he receives into his existing stock and becomes more knowledgeable, hence achieving higher efficiency than before transmission.

Figure 2 depicts the relationship between the overall knowledge level and the re-wiring probability $p$ and absorptive/innovative capacity $\alpha$. We plot normalized rather than absolute values, to preserve legibility.[7] For $\alpha$ lower than 1.05, no clear distinction exists between ordered and random graphs. With finer graining, we see evidence of a smooth change starting at $\alpha = 1$. By contrast, when $\alpha$ exceeds approximately 1.05, clear patterns emerge even with coarse graining of the parameter space, and a sharp efficiency peak

[7] What we plot is actually $\mu^T$ divided, for each $\alpha$-value, by the maximal value it takes for $p \in [0, 1]$, i.e. $\mu^T(\alpha, p) / \max_{p'} \mu^T(\alpha, p')$.

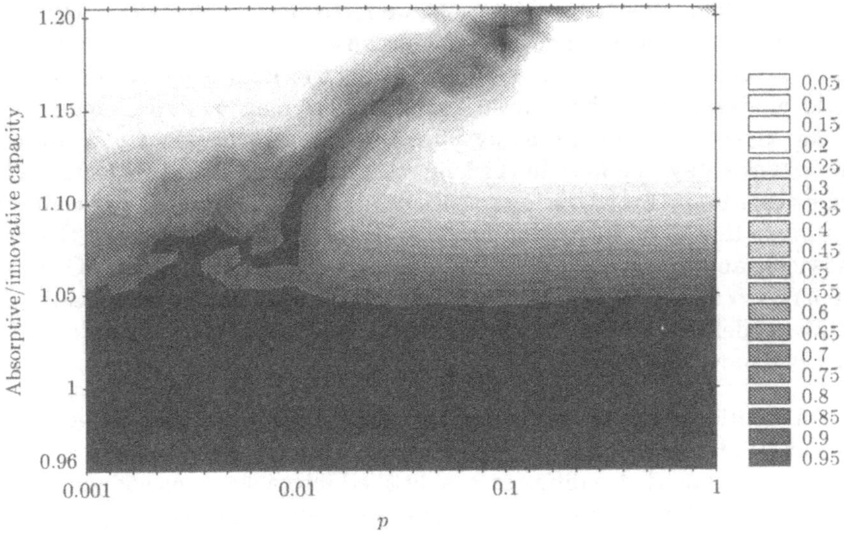


**Fig. 2.** Knowledge levels in the $(p, \alpha)$-space

obtains in the small world. At the same time, the knowledge-maximizing $p$-value (degree of randomness) increases monotonically with $\alpha$, though it remains confined to the small-world region, i.e. the area between $p = 0.01$ and $p = 0.1$. The maximizing $p$-value always lies in this region, but it is worth mentioning that most of the mass of the distribution itself is also located in this region. Outside the small-world region, for every value of $\alpha$, long run average knowledge levels are much lower than the maximum. Hence a small-world architecture, i.e., one in which there is at the same time a significant amount of local order and a few random connections, achieves the best overall performance in our broadcast economy.

A remark is in order here. In our experiments we ignored values of $\alpha$ greater than $1 + \theta$. Recall that an agent who receives information not only incorporates all of it but improves upon it (by $\alpha - 1$). This implies that after receiving a transmission, the knowledge level of the recipient is higher than that of the sender. If $\alpha > 1+\theta$ the new knowledge level of the recipient would increase so much that the sender now fails to pass the dissimilarity threshold vis-à-vis the former recipient. That is, if I send you knowledge, you improve so much that I cannot receive from you. This seems unreasonable, so we have excluded this possibility by restricting $\alpha$ to be less than $1 + \theta$.

**The interplay of path length and cliquishness** In order to explain the mechanism underlying this result, we focus on the effects of cliquishness and path length on system performance. As an initial step in that explanation, consider the role of dissimilarity. In the knowledge re-combination world analyzed here, the innovative jump a recipient makes increases with his lag vis-à-vis the sender as long as their distance in terms of knowledge endowments stays below $\theta$. In the long run, this means that if we focus on two agents repeatedly broadcasting to each other, their knowledge level converges to a value that is an increasing function of both the initial level of the more knowledgeable agent and the initial gap between the two agents. Of course the larger $\alpha$ is, the higher the limit they jointly achieve also is. So large gaps in this model lead to high long-run levels, unless initial gaps are so large that they preclude transmission.

As mentioned earlier, there are two dimensions when knowledge growth is considered, namely creation and diffusion. Though in the model knowledge creation and knowledge diffusion are distilled into a single episode, it is intuitive that cliquishness mainly influences knowledge creation and path length drives knowledge diffusion. We shall now examine the importance of these two forces in knowledge growth. Figure 3 summarizes three typical outcomes of the process of knowledge broadcast which we comment in turn.

The regular world is one in which an agent's potential recipients are recipients of each other. So when a piece of knowledge is released, a self-reinforcing local mechanism is at work, and produces fast localized knowledge growth within a set of neighbouring agents. This is the knowledge generation part of the dynamics, which mainly takes place at the intra-clique level. It entails unequal knowledge growth across clusters of agents. Strong inequalities arise between groups that advance quickly and groups which lag behind. The first (upper) graph in Figure 3 represents long-run knowledge states in the regular world.

At the other extreme we have the random world in which in only a few transmissions knowledge is passed through the whole population due to short travel distances. In this situation there is a strong tendency for knowledge levels to homogenize, and we have seen that this is bad for knowledge creation because homogeneity is bad for recombination. Knowledge diffusion is efficient (hence the homogenization) but disruptive in that it squeezes the dispersion of people in terms of knowledge, thereby leaving little chance for knowledge creation. The random world is the lowest part of Figure 3, and homogeneity is patent.

Between these extremes there is a trade off between knowledge creation and knowledge diffusion, and the small world turns out to be the knowledge-growth maximizing architecture. It shares most of the advantages of the two extremes. It connects distant and possibly heterogenous parts of the graph, creating the possibility for large innovative jumps. Shortcuts bring together parts of the graph in which, at the same time, local reinforcement

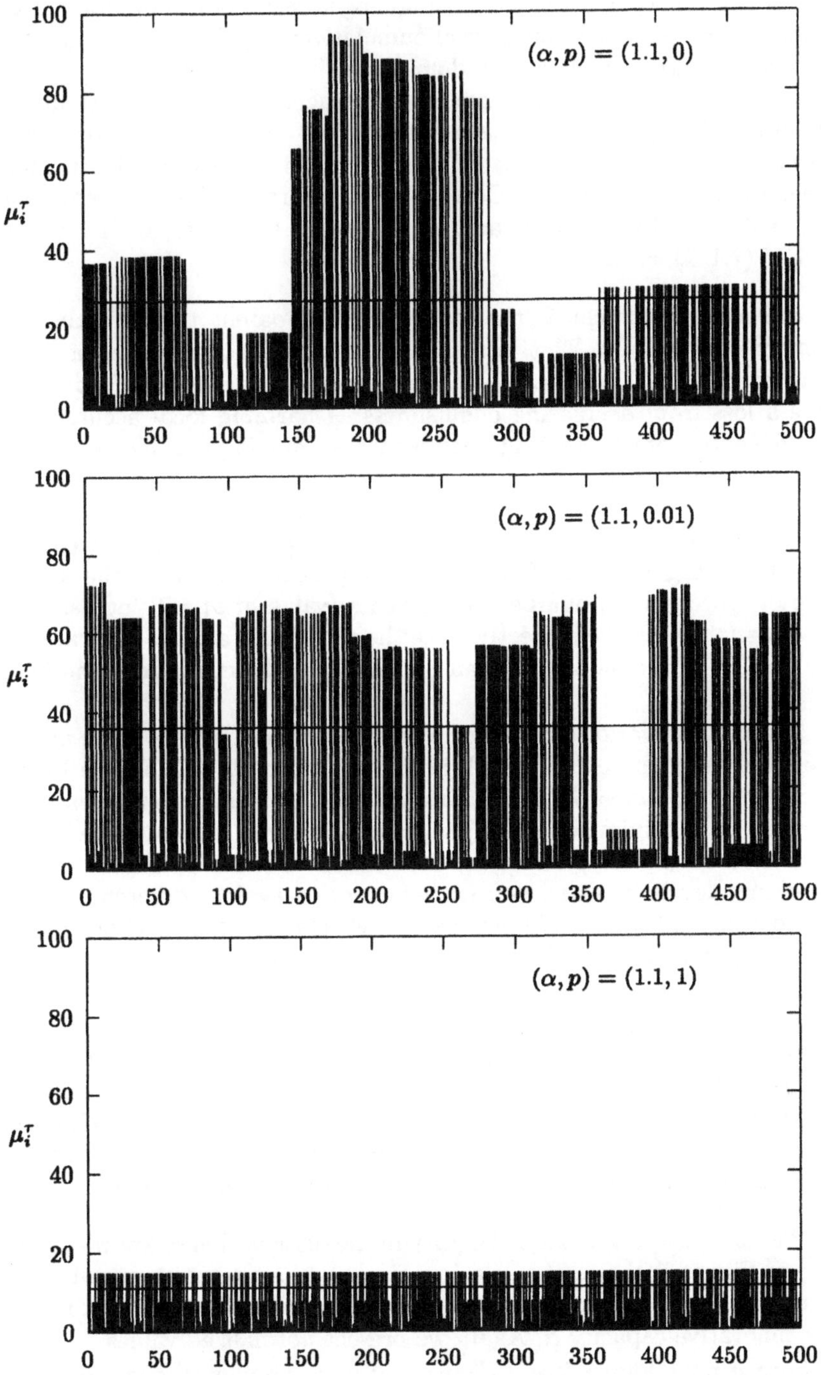


**Fig. 3.** The average individual long-run knowledge levels in three typical simulations runs for $\alpha = 1.1$ and values of $p$ corresponding to a regular world, a small world and a random graph

is at work. There is a tendency to homogenize overall knowledge levels but without weakening too much local cumulativeness, thus we end up with levels that are larger than in a random graph, but relatively homogeneous even at the global level. Homogenization is stronger as soon as shortcuts are introduced, though as long as there are not too many the local reinforcement stemming from large degrees of cliquishness is preserved. This is shown in the middle part of Figure 3. Corresponding average values are represented in Figure 3 by horizontal lines, and we have $\mu^{\tau}(1.1, 0) = 26$, $\mu^{\tau}(1.1, 0.01) = 36$ and $\mu^{\tau}(1.1, 1) = 11$.

A second comment is that there is an apparent trend in the optimal number of shortcuts. This arises in the marginal contribution of a random connection as $\alpha$ is varied. When a shortcut is added, there is at the same time a loss from decreasing cliquishness — harming local accumulation — and benefits from decreasing path length — making feasible the possibility of connecting highly knowledgeable parts of the graph. As $\alpha$ increases these two effects grow in importance, but what matters is actually their relative importance. If the marginal gain from decreasing path length increases faster than the marginal cost of decreasing cliquishness, then (if the solution is interior) the optimal number of shortcuts (value of $p$) will increase with $\alpha$. This cannot be checked directly since the two effects are not separable within the context of the model. We can perform an indirect check in the following way, however.

In a regular world, path lengths are long, cliquishness is at its maximum. Thus the main driver of growth in knowledge levels will be the local reinforcement that arises from cliquishness. By contrast, in the random word cliquishness is minimized, as is path length, so growth in average knowledge levels will be driven by the fact that diffusion is rapid. Plotting long run knowledge levels in these two extreme worlds against different levels of $\alpha$ will give an indication of how the two effects (increasing cliquishness and decreasing path length) change in magnitude with $\alpha$. This we do in Figure 4.

From this figure, it is clear that the benefits of cliquishness increase slowly but steadily with $\alpha$, as evidenced by the thin black curve; the benefits of short paths rises rapidly when $\alpha$ becomes relatively large (thick black curve growing at a more than exponential rate). This suggests that effects of decreasing path length does indeed increase faster than the costs of decreasing cliquishness. Figure 4 shows that at some point, i.e., for large enough $\alpha$-values, the benefits from shorter paths exceed the losses from reduced cliquishness.

We turn now to the issue of equity in the distribution of knowledge across agents. Figure 5 shows the relationship between the economy-wide knowledge variance as a function of the re-wiring probability $p$ and the absorptive/innovative capacity $\alpha$. Again we present normalized values.

Surprisingly, Figure 2 and Figure 5 are very similar (though dispersion is lower in Figure 5), with highest inter-agent differences observed for a degree

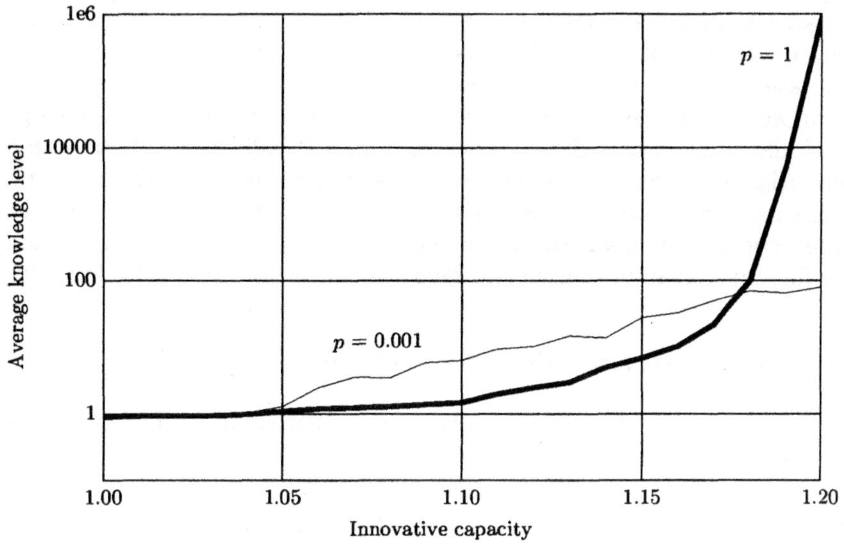


**Fig. 4.** The population average long-run knowledge level in two polar cases ($p = 0.001$ and $p = 1$), as a function of $\alpha$

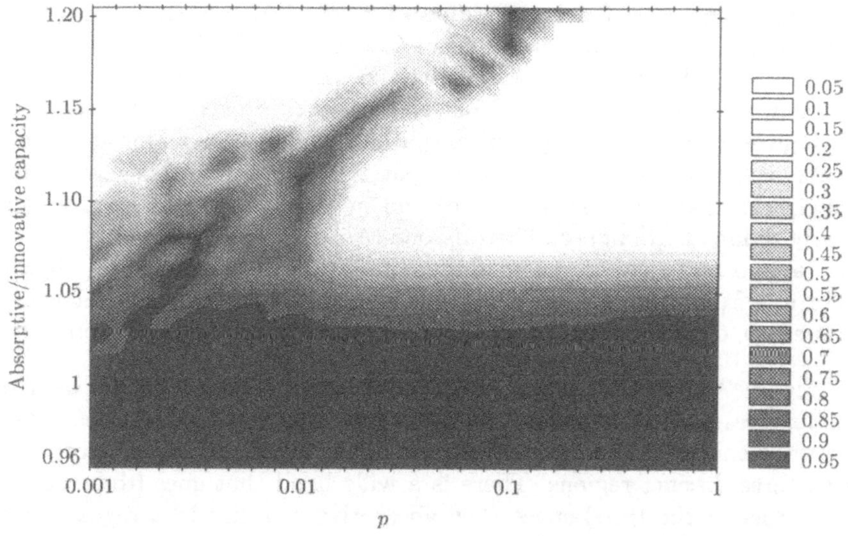


**Fig. 5.** Long run heterogeneity in knowledge allocation in the $(p, \alpha)$-space

of randomness monotonically increasing with $\alpha$ while remaining confined to the area between $p = 0.01$ and $p = 0.1$.[8] We therefore come to the somewhat uncomfortable conclusion that the small world region at the same time generates the best overall performance in terms of how much knowledge is produced by the system, and the worst overall performance when homogeneity of allocation is considered desirable. Figure 3 indicates that the source of this variance is that some agents get left behind. Roughly the same number of agents get left behind in each world, and stay close to their initial knowledge levels, but in the small world, the agents who do advance advance rapidly and far, thus creating a large gap between themselves and those left behind.

## 4.2 The spatial allocation of knowledge

Spatial correlation of knowledge levels can be considered either in geographical space or in the space of the network itself. In general, a positive spatial correlation exists if agents 'near each other' have comparable knowledge vectors. By contrast, negative correlations obtain when knowledgeable agents are the neighbours of laggards and no clustering of knowledge exists. Correlation in the geographic space means taking as the distance between nodes $i$ and $j$ the simple absolute difference $|i - j|$ (with the adequate modulo). In that sense, node $i$ is very close to nodes $i \pm 1$, slightly less close to nodes $i \pm 2$ and so on. A priori, since knowledge generation and diffusion takes place over the network, there is no reason in general to expect spatial correlation in geographic space. For small values of $p$ however, geographic space has a very similar topology to network space (the spaces differ in roughly $p$ percent of the edges). Thus, if there are non-trivial correlations for small $p$ values in the network space, there should be echoes of them in geographic space.

Figure 6 shows the geographical spatial correlation as a function of $p$ and $\alpha$. As expected, there is virtually no spatial correlation in this space. All the values are small (in absolute value) and in general not statistically significantly different from zero. There is one region that stands out in contrast. For large $\alpha$ and small $p$, many of the correlations, while small, do differ significantly from zero.[9] On the time scale we consider, though, this relationship between $p$, $\alpha$ and spatial correlation contains considerable randomness.

Figure 7 gives the Moran coefficient $S$ as a function of the degree of randomness $p$ and the absorptive/innovative capacity $\alpha$, when the network metric is considered. A very different picture obtains in the network space, with three distinct regions. There is a wide band that goes from the lower left corner of the $(p, \alpha)$-space (low absorptive capacity in a regular world) to the upper right corner (significant super-additivity in a random world)

[8] The coefficient of variation of knowledge levels follows the same pattern, indicating that the pattern in the variance is not driven simply by re-scaling.

[9] This was checked using a standard two-tailed $t$-test.

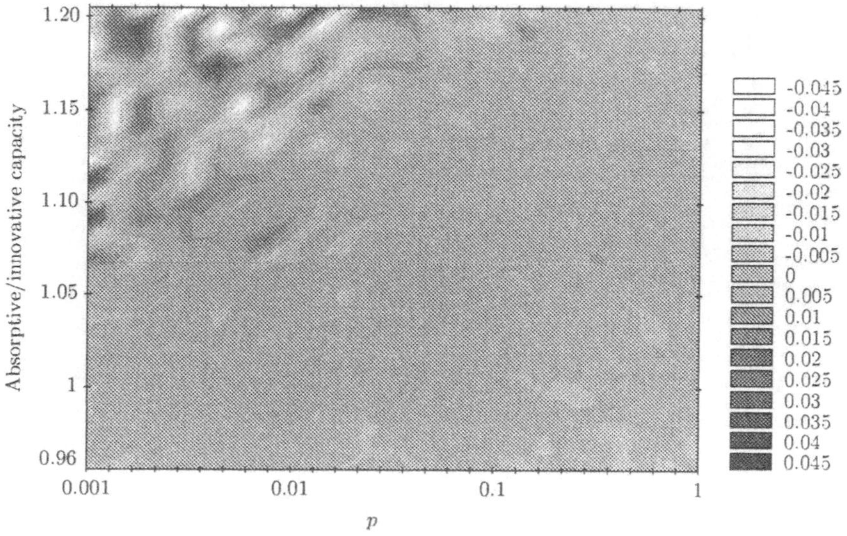


**Fig. 6.** Spatial correlation in the geographical space

where knowledge is more or less randomly allocated among agents. There is a region in which the world is random and absorptive capacities are imperfect, in which a (small) negative correlation is observed. Finally, positive spatial correlation obtains when the network of agents' relationships is cliquish and innovation rates are high, as we would expect.

The correlation in network space for small $p$ values was indeed echoed in geographic space. The echo is very weak however. Even a small discrepancy between the space in which knowledge moves and the space in which correlations are measured creates a very different impression on the existence and strength of clustering.

It is clear from the results presented in Figures 6 and 7 and the discussion, that there is no small world effect on spatial correlation or knowledge clustering. Cliquishness is a manifest source of knowledge clustering and introducing shortcuts to possibly distant parts of the graph only tends to impede clustering.

## 5 Conclusion

The use and creation of knowledge is central to economic growth and development. The fact that knowledge diffusion takes time forces analysts to acknowledge that agents are heterogeneous in economically important ways. Modelling this heterogeneity, and understanding its consequences for eco-

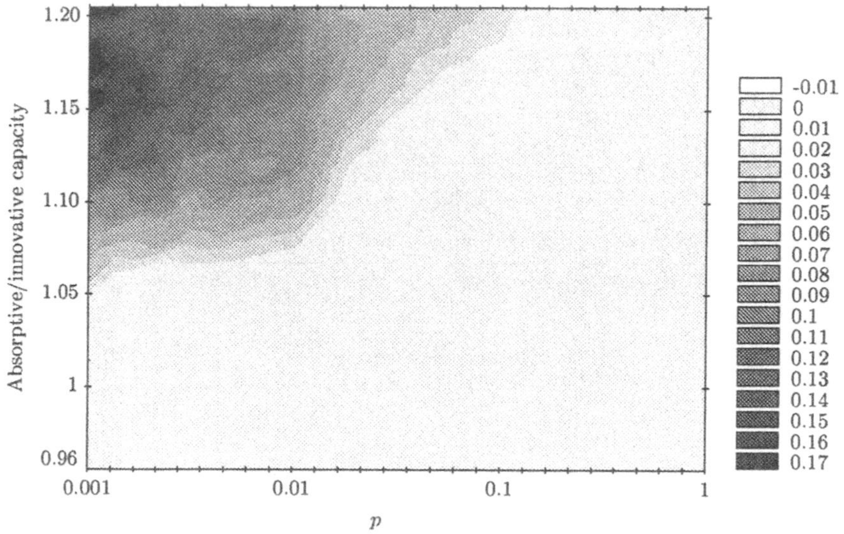


**Fig. 7.** Spatial correlation in the network space

nomic performance is thus a central task in the attempt to understand any dynamic economy. The model in this paper has explicitly included in the diffusion process the idea that knowledge is not merely information, and thus is diffused through specific contacts between individual agents. Repeated interactions are valuable in facilitating diffusion, and a group of agents can, by collectively sharing knowledge advance together. Groups of agents sharing knowledge more or less freely is something observed historically, and Allen (1983) has coined the phrase "collective invention" to describe it. Common to these episodes is that this organization of innovation generates rapid technical advance. One of the important features of collective invention is the sharing of information among a broad, though typically localized, group of agents. It is this feature that we have emphasized here, showing how the results of this sharing is affected by the network structure over which communication takes place. Unless technical opportunities are extremely large, and thus innovation is very straightforward, small-world network structures produce fastest knowledge growth rates. Cliquishness is in general a good thing, but the ability to bring knowledge into the clique from outside has a vital role. As technological opportunities grow, though, there can be too much cliquishness — because innovation involves capital investments that depreciate over time, it can happen that with a rapid innovation rate and strong super-additivity in knowledge, an agent can be left behind by members of his clique, as they use knowledge he has generated but which has created for him a temporary lock-in due to his capital investments. In this sort of situation a very wide variety in knowledge sources is important for an agent to keep in the race.

## References


1. Allen, R. (1983) Collective Invention, Journal of Economic Behavior and Organization, 4, 1–24
2. Arrow, K. (1962) The Economic Implications of Learning-by-Doing, Review of Economic Studies, 29, 155–173
3. Bala, V., and S. Goyal (1998) Learning from Neighbours, Review of Economic Studies, 65, 595–621
4. Cohen, W., and D. Levinthal (1989) Innovation and Learning: The Two Faces of Research and Development, The Economic Journal, 99, 569–596
5. Cowan, R., P. David, and D. Foray (2000) The Explicit Economics of Knowledge Codification, paper prepared for the European Contract "TIPIK: Technology and Infrastructure Policy in the Knowledge-Based Economy", programme TSER, DG XII (1998/2000)
6. Cowan, R., and D. Foray (1997) Quandaries in the Economics of Dual Technologies and Spillovers from Military to Civilian Research and Development, Research Policy, 24, 851–868
7. Cowan, R., and N. Jonard (1999) Network Structure and the Diffusion of Knowledge, Discussion Paper 99028, MERIT/Maastricht University, Maastricht, The Netherlands
8. Jackson, M. O., and A. Wolinsky (1996) A Strategic Model of Social and Economic Networks, Journal of Economic Theory, 71, 44–74
9. Jaffe, A., M. Tratjenberg, and R. Henderson (1993) Geographic Localization of Knowledge Spillovers as Evidenced by Patent Citations, Quarterly Journal of Economics, 63, 577–598
10. Kirman, A. P. (1992) Whom and What Does the Representative Individual Represent?, Journal of Economic Perspectives, 6, 117–136
11. McGrew, C., and J. Monroe (1993) An Introduction to Statistical Problem-Solving in Geography. W. C. Brown Publishers.
12. Nelson, R. R. (1959) The Simple Economics of Basic Scientific Research, Journal of Political Economy, 67, 297–306
13. Plouraboue, F., A. Steyer, and J.-B. Zimmermann (1998) Learning Induced Criticality in Consumers Adoption Pattern: A Neural Network Approach, Economics of Innovation and New Technology, 6, 73–90
14. Von Hippel, E. (1998) The Sources of Innovation. Oxford University Press.
15. Watts, D., and S. Strogatz (1998) Collective Dynamics of Small-World Networks, Nature, 393, 400–403

Vol. 405: S. Komlósi. T. Rapcsák. S. Schaible (Eds.), Generalized Convexity. Proceedings, 1992. VIII, 404 pages. 1994.

Vol. 406: N. M. Hung, N. V. Quyen, Dynamic Timing Decisions Under Uncertainty. X, 194 pages. 1994.

Vol. 407: M. Ooms, Empirical Vector Autoregressive Modeling. XIII. 380 pages. 1994.

Vol. 408: K. Haase, Lotsizing and Scheduling for Production Planning. VIII. 118 pages. 1994.

Vol. 409: A. Sprecher. Resource-Constrained Project Scheduling. XII. 142 pages. 1994.

Vol. 410: R. Winkelmann, Count Data Models. XI, 213 pages. 1994.

Vol. 411: S. Dauzère-Péres, J.-B. Lasserre, An Integrated Approach in Production Planning and Scheduling. XVI, 137 pages. 1994.

Vol. 412: B. Kuon, Two-Person Bargaining Experiments with Incomplete Information. IX, 293 pages. 1994.

Vol. 413: R. Fiorito (Ed.), Inventory, Business Cycles and Monetary Transmission. VI, 287 pages. 1994.

Vol. 414: Y. Crama, A. Oerlemans, F. Spieksma, Production Planning in Automated Manufacturing. X, 210 pages. 1994.

Vol. 415: P. C. Nicola, Imperfect General Equilibrium. XI, 167 pages. 1994.

Vol. 416: H. S. J. Cesar, Control and Game Models of the Greenhouse Effect. XI, 225 pages. 1994.

Vol. 417: B. Ran, D. E. Boyce, Dynamic Urban Transportation Network Models. XV, 391 pages. 1994.

Vol. 418: P. Bogetoft, Non-Cooperative Planning Theory. XI. 309 pages. 1994.

Vol. 419: T. Maruyama, W. Takahashi (Eds.), Nonlinear and Convex Analysis in Economic Theory. VIII, 306 pages. 1995.

Vol. 420: M. Peeters, Time-To-Build. Interrelated Investment and Labour Demand Modelling. With Applications to Six OECD Countries. IX, 204 pages. 1995.

Vol. 421: C. Dang, Triangulations and Simplicial Methods. IX. 196 pages. 1995.

Vol. 422: D. S. Bridges, G. B. Mehta, Representations of Preference Orderings. X, 165 pages. 1995.

Vol. 423: K. Marti, P. Kall (Eds.), Stochastic Programming. Numerical Techniques and Engineering Applications. VIII, 351 pages. 1995.

Vol. 424: G. A. Heuer, U. Leopold-Wildburger, Silverman's Game. X. 283 pages. 1995.

Vol. 425: J. Kohlas, P.-A. Monney, A Mathematical Theory of Hints. XIII. 419 pages, 1995.

Vol. 426: B. Finkenstädt, Nonlinear Dynamics in Economics. IX, 156 pages. 1995.

Vol. 427: F. W. van Tongeren, Microsimulation Modelling of the Corporate Firm. XVII, 275 pages. 1995.

Vol. 428: A. A. Powell, Ch. W. Murphy, Inside a Modern Macroeconometric Model. XVIII, 424 pages. 1995.

Vol. 429: R. Durier, C. Michelot, Recent Developments in Optimization. VIII, 356 pages. 1995.

Vol. 430: J. R. Daduna, I. Branco, J. M. Pinto Paixão (Eds.), Computer-Aided Transit Scheduling. XIV, 374 pages. 1995.

Vol. 431: A. Aulin, Causal and Stochastic Elements in Business Cycles. XI, 116 pages. 1996.

Vol. 432: M. Tamiz (Ed.). Multi-Objective Programming and Goal Programming. VI, 359 pages. 1996.

Vol. 433: J. Menon, Exchange Rates and Prices. XIV, 313 pages. 1996.

Vol. 434: M. W. J. Blok, Dynamic Models of the Firm. VII, 193 pages. 1996.

Vol. 435: L. Chen, Interest Rate Dynamics, Derivatives Pricing, and Risk Management. XII, 149 pages. 1996.

Vol. 436: M. Klemisch-Ahlert, Bargaining in Economic and Ethical Environments. IX, 155 pages. 1996.

Vol. 437: C. Jordan, Batching and Scheduling. IX, 178 pages. 1996.

Vol. 438: A. Villar, General Equilibrium with Increasing Returns. XIII, 164 pages. 1996.

Vol. 439: M. Zenner, Learning to Become Rational. VII, 201 pages. 1996.

Vol. 440: W. Ryll, Litigation and Settlement in a Game with Incomplete Information. VIII, 174 pages. 1996.

Vol. 441: H. Dawid, Adaptive Learning by Genetic Algorithms. IX, 166 pages.1996.

Vol. 442: L. Corchón, Theories of Imperfectly Competitive Markets. XIII, 163 pages. 1996.

Vol. 443: G. Lang, On Overlapping Generations Models with Productive Capital. X, 98 pages. 1996.

Vol. 444: S. Jørgensen, G. Zaccour (Eds.), Dynamic Competitive Analysis in Marketing. X, 285 pages. 1996.

Vol. 445: A. H. Christer, S. Osaki, L. C. Thomas (Eds.), Stochastic Modelling in Innovative Manufactoring. X, 361 pages. 1997.

Vol. 446: G. Dhaene, Encompassing. X, 160 pages. 1997.

Vol. 447: A. Artale, Rings in Auctions. X, 172 pages. 1997.

Vol. 448: G. Fandel, T. Gal (Eds.), Multiple Criteria Decision Making. XII, 678 pages. 1997.

Vol. 449: F. Fang, M. Sanglier (Eds.), Complexity and Self-Organization in Social and Economic Systems. IX, 317 pages, 1997.

Vol. 450: P. M. Pardalos, D. W. Hearn, W. W. Hager, (Eds.), Network Optimization. VIII, 485 pages, 1997.

Vol. 451: M. Salge, Rational Bubbles. Theoretical Basis, Economic Relevance, and Empirical Evidence with a Special Emphasis on the German Stock Market.IX, 265 pages. 1997.

Vol. 452: P. Gritzmann, R. Horst, E. Sachs, R. Tichatschke (Eds.), Recent Advances in Optimization. VIII, 379 pages. 1997.

Vol. 453: A. S. Tangian, J. Gruber (Eds.), Constructing Scalar-Valued Objective Functions. VIII, 298 pages. 1997.

Vol. 454: H.-M. Krolzig, Markov-Switching Vector Autoregressions. XIV, 358 pages. 1997.

Vol. 455: R. Caballero, F. Ruiz, R. E. Steuer (Eds.), Advances in Multiple Objective and Goal Programming. VIII, 391 pages. 1997.

Vol. 456: R. Conte, R. Hegselmann, P. Terna (Eds.), Simulating Social Phenomena. VIII, 536 pages. 1997.

Vol. 457: C. Hsu, Volume and the Nonlinear Dynamics of Stock Returns. VIII, 133 pages. 1998.

Vol. 458: K. Marti, P. Kall (Eds.), Stochastic Programming Methods and Technical Applications. X, 437 pages. 1998.

Vol. 459: H. K. Ryu, D. J. Slottje, Measuring Trends in U.S. Income Inequality. XI, 195 pages. 1998.

Vol. 460: B. Fleischmann, J. A. E. E. van Nunen, M. G. Speranza, P. Stähly, Advances in Distribution Logistic. XI, 535 pages. 1998.

Vol. 461: U. Schmidt, Axiomatic Utility Theory under Risk. XV, 201 pages. 1998.

Vol. 462: L. von Auer, Dynamic Preferences, Choice Mechanisms, and Welfare. XII, 226 pages. 1998.

Vol. 463: G. Abraham-Frois (Ed.), Non-Linear Dynamics and Endogenous Cycles. VI, 204 pages. 1998.

Vol. 464: A. Aulin, The Impact of Science on Economic Growth and its Cycles. IX, 204 pages. 1998.

Vol. 465: T. J. Stewart, R. C. van den Honert (Eds.), Trends in Multicriteria Decision Making. X, 448 pages. 1998.

Vol. 466: A. Sadrieh, The Alternating Double Auction Market. VII, 350 pages. 1998.

Vol. 467: H. Hennig-Schmidt, Bargaining in a Video Experiment. Determinants of Boundedly Rational Behavior. XII, 221 pages. 1999.

Vol. 468: A. Ziegler, A Game Theory Analysis of Options. XIV, 145 pages. 1999.

Vol. 469: M. P. Vogel, Environmental Kuznets Curves. XIII, 197 pages. 1999.

Vol. 470: M. Ammann, Pricing Derivative Credit Risk. XII, 228 pages. 1999.

Vol. 471: N. H. M. Wilson (Ed.), Computer-Aided Transit Scheduling. XI, 444 pages. 1999.

Vol. 472: J.-R. Tyran, Money Illusion and Strategic Complementarity as Causes of Monetary Non-Neutrality. X, 228 pages. 1999.

Vol. 473: S. Helber, Performance Analysis of Flow Lines with Non-Linear Flow of Material. IX, 280 pages. 1999.

Vol. 474: U. Schwalbe, The Core of Economies with Asymmetric Information. IX, 141 pages. 1999.

Vol. 475: L. Kaas, Dynamic Macroeconomics with Imperfect Competition. XI, 155 pages. 1999.

Vol. 476: R. Demel, Fiscal Policy, Public Debt and the Term Structure of Interest Rates. X, 279 pages. 1999.

Vol. 477: M. Théra, R. Tichatschke (Eds.), Ill-posed Variational Problems and Regularization Techniques. VIII, 274 pages. 1999.

Vol. 478: S. Hartmann, Project Scheduling under Limited Resources. XII, 221 pages. 1999.

Vol. 479: L. v. Thadden, Money, Inflation, and Capital Formation. IX, 192 pages. 1999.

Vol. 480: M. Grazia Speranza, P. Stähly (Eds.), New Trends in Distribution Logistics. X, 336 pages. 1999.

Vol. 481: V. H. Nguyen, J. J. Strodiot, P. Tossings (Eds.). Optimation. IX, 498 pages. 2000.

Vol. 482: W. B. Zhang, A Theory of International Trade. XI, 192 pages. 2000.

Vol. 483: M. Königstein, Equity, Efficiency and Evolutionary Stability in Bargaining Games with Joint Production. XII, 197 pages. 2000.

Vol. 484: D. D. Gatti, M. Gallegati, A. Kirman, Interaction and Market Structure. VI, 298 pages. 2000.

Vol. 485: A. Garnaev, Search Games and Other Applications of Game Theory. VIII, 145 pages. 2000.

Vol. 486: M. Neugart, Nonlinear Labor Market Dynamics. X, 175 pages. 2000.

Vol. 487: Y. Y. Haimes, R. E. Steuer (Eds.), Research and Practice in Multiple Criteria Decision Making. XVII, 553 pages. 2000.

Vol. 488: B. Schmolck, Ommitted Variable Tests and Dynamic Specification. X, 144 pages. 2000.

Vol. 489: T. Steger, Transitional Dynamics and Economic Growth in Developing Countries. VIII, 151 pages. 2000.

Vol. 490: S. Minner, Strategic Safety Stocks in Supply Chains. XI, 214 pages. 2000.

Vol. 491: M. Ehrgott, Multicriteria Optimization. VIII, 242 pages. 2000.

Vol. 492: T. Phan Huy, Constraint Propagation in Flexible Manufacturing. IX, 258 pages. 2000.

Vol. 493: J. Zhu, Modular Pricing of Options. X, 170 pages. 2000.

Vol. 494: D. Franzen, Design of Master Agreements for OTC Derivatives. VIII, 175 pages. 2001.

Vol. 495: I Konnov, Combined Relaxation Methods for Variational Inequalities. XI, 181 pages. 2001.

Vol. 496: P. Weiß, Unemployment in Open Economies. XII, 226 pages. 2001.

Vol. 497: J. Inkmann, Conditional Moment Estimation of Nonlinear Equation Systems. VIII, 214 pages. 2001.

Vol. 498: M. Reutter, A Macroeconomic Model of West German Unemployment. X, 125 pages. 2001.

Vol. 499: A. Casajus, Focal Points in Framed Games. XI, 131 pages. 2001.

Vol. 500: F. Nardini, Technical Progress and Economic Growth. XVII, 191 pages. 2001.

Vol. 501: M. Fleischmann, Quantitative Models for Reverse Logistics. XI, 181 pages. 2001.

Vol. 502: N. Hadjisavvas, J. E. Martínez-Legaz, J.-P. Penot (Eds.), Generalized Convexity and Generalized Monotonicity. IX, 410 pages. 2001.

Vol. 503: A. Kirman, J.-B. Zimmermann (Eds.), Economics with Heterogeneous Interacting Agents. VII, 343 pages. 2001.

GPSR Compliance

*The European Union's (EU) General Product Safety Regulation (GPSR) is a set of rules that requires consumer products to be safe and our obligations to ensure this.*

*If you have any concerns about our products, you can contact us on ProductSafety@springernature.com*

In case Publisher is established outside the EU, the EU authorized representative is:

Springer Nature Customer Service Center GmbH
Europaplatz 3
69115 Heidelberg, Germany

**Batch number: 10844669**

Printed by Printforce, the Netherlands